THE FUNGI OF
NORTH EAST WALES

**Other titles by Bruce Ing
from the same publisher**

*Biodiversity in the North West:
The Mildews of Cheshire,
Lancashire and Cumbria* (2020)

*Biodiversity in the North West:
The Slime Moulds of Lancashire
and Cumbria* (2020)

*Biodiversity in the North West:
The Slime Moulds of Cheshire* (2011)

*The Exciting World of
the Slime Moulds* (2008)

THE FUNGI OF NORTH EAST WALES

A Mycota for Vice-Counties
50 (Denbighshire) and 51 (Flintshire)

Bruce Ing

University of Chester Press

First published 2020
by University of Chester Press
University of Chester
Parkgate Road
Chester CH1 4BJ

Printed and bound in the UK by the
LIS Print Unit
University of Chester

Cover designed by the
LIS Graphics Team
University of Chester

© University of Chester, 2020
except Figure 5 © Roger Brown

The moral right of the author
of this work has been asserted

All Rights Reserved
No part of this publication may be reproduced, stored in a retrieval system or transmitted in any form or by any means without the prior permission of the copyright owner, other than as permitted by current UK copyright legislation or under the terms and conditions of a recognised copyright licensing scheme

A catalogue record for this book is available
from the British Library

ISBN 978-1-910481-12-7

CONTENTS

Acknowledgements — vi

List of illustrations — vii

List of abbreviations — viii

PART 1

Introduction
1.1 The area covered by this book — 1
1.2 Physical environment — 1
1.3 Vegetation and fungal habitats — 5
1.4 Mycologically important sites — 7
1.5 Fungal recording in North East Wales — 17
1.6 Sources of records — 19
1.7 List of individual collectors — 20
1.8 Classification and identification of fungi — 20
1.9 Conservation of fungi — 23
1.10 Layout of entries in the species accounts — 24
1.11 Bibliography — 25

PART 2

Species Accounts
The Fungi of North East Wales
Myxomycota — 34
Plasmodiophoromycota — 54
Peronosporomycota — 55
Chytridiomycota — 61
Glomeromycota — 62
Zygomycota — 63
Ascomycota – Pezizomycotina — 66
Ascomycota – Saccharomycotina — 146
Ascomycota – Taphrinomycotina — 146
Basidiomycota – Agaricomycotina — 149
Basidiomycota – Pucciniomycotina — 274
Basidiomycota – Ustilaginomycotina — 288

Illustrations — 292

ACKNOWLEDGEMENTS

This book would not have been as comprehensive without the generosity of all the collectors (see 1.7), especially Charles Aron, Roger Brown, Joe Phillips, Graham Rose and Dave Sykes, whose records include many 'firsts' for the region. They have shared their data and their enthusiasm with me and it is a pleasure to record my indebtedness to them. It is also a pleasure to thank the following specialists who have given their opinion on numerous specimens and helped considerably with the identification of uncommon or unusual species over the thirty-five years of this study:

> Professor Henry Beker, Brussels (*Hebeloma*); Ernest Emmett, Kingussie (*Mycena*); Alick Henrici, RBG, Kew (agarics, corticioid fungi); Alan Hills, Oxford (boletes); Geoff Kibby, London (boletes, *Cortinarius, Russula*); Dr Paul Kirk, RBG, Kew (Hyphomycetes, Mucorales); Dr Thomas Laessoe, Copenhagen (agarics, ascomycetes); Professor David Pegler, lately of RBG, Kew (polypores); Dr Peter Roberts, lately of RBG, Kew (clavarioid, corticioid and jelly fungi); the late Maurice Rotheroe, Lampeter (*Coprinus s.l., Russula*); Dr Derek Schafer, Buckinghamshire (*Coprinus s.l.*); Dr Brian Spooner, lately of RBG, Kew (ascomycetes); the late Dr Jan Vesterholt, Denmark (*Cortinarius, Hebeloma, Lactarius*); Professor Roy Watling, lately of RBG, Edinburgh (agarics); Professor Anthony Whalley, lately of Liverpool John Moores University (Xylariaceae).

I am grateful to Dr Joan Daniels for allowing me to include the records from Fenn's, Bettisfield and Whixall Mosses National Nature Reserve. Numerous participants at local forays have contributed records and are gratefully acknowledged as are the Rangers at all the Country Parks in the region for their continuous support.

The librarians and staff of the libraries at the Linnaean Society and the Natural History Museum, London, have given tremendous help, often at short notice, with obscure references, as have the staff at the Flintshire Record Office and the Flintshire County Library.

I am most grateful for the assistance of Sarah Griffiths and her colleagues at the University of Chester for their skill and patience.

My son, Gareth Lewis-Ing, has provided valuable, and highly necessary, help and advice on all matters to do with computer technology. My wife, Eleanor, has been a tower of strength in supporting the work for more than fifty years. It would be easy to take her support for granted – I do not! Without her encouragement this book would never have been finished!

LIST OF ILLUSTRATIONS

Figure 1:	The boundaries of vice-counties 50 and 51	292
Figure 2:	The local authority areas	293
Figure 3:	The 10 km squares of the National Grid	294
Figure 4:	*Erysiphe guarinonii*, a rare powdery mildew on laburnum, from Mold	295
Figure 5:	*Sarcoscypha austriaca* and its rare variety *lutea*, from Abergele (Roger Brown)	295
Figure 6:	*Mitrophora semilibera*, an uncommon morel, on coal waste, Ewloe	296
Figure 7:	*Morchella crassipes*, an uncommon morel, from Cilygroeslwyd Wood	296
Figure 8:	*Taphrina pruni*, causing pocket plums on blackthorn, from Mold	297
Figure 9:	*Amanita phalloides*, the Death Cap, from Erddig	297
Figure 10:	*Cyathus stercoreus*, a rare bird's-nest fungus, from Point of Ayr	298
Figure 11:	*Coprinus picaceus*, the rare Magpie Ink Cap, from Alyn Valley	298
Figure 12:	*Hericium erinaceum*, the rare Monkey-head, from Erddig	299
Figure 13:	*Psathyrella ammophila*, the Dune Brittlestem, from Point of Ayr	299
Figure 14:	*Hyalopsora polypodii*, the rare rust on Brittle Bladder fern, from Loggerheads	300
Figure 15:	*Puccinia smyrnii*, the rust on Alexanders, from Point of Ayr	300
Figure 16:	*Exobasidium karstenii* galling Bog Rosemary, from Fenn's Moss	301

LIST OF ABBREVIATIONS

AONB	Area of Outstanding Natural Beauty
AV	Alyn Valley woodlands
BO	Bodelwyddan Castle Estate
BP	Bishop's Wood, Meliden
BW	Big Wood, Hendre
CC	Coed Coch, near Betws-yn-Rhos
CE	Celyn Woods, Northop
CF	Clocaenog Forest
CN	Coed y Felin Nature Reserve
CR	Critically endangered (known only from one or two sites in the UK)
DD	Data deficient and requiring more research to assess their threat levels
DCC	Denbighshire County Council
DU	Ddol Uchaf Nature Reserve, Ysceifiog
DEFRA	Department of Environment, Food and Rural Affairs
EN	Endangered (currently known from five or fewer sites in the UK, having been more widespread in the past)
ER	Erddig Estate, near Wrexham
EX	Probably extinct (not recorded anywhere for fifty years)
FCC	Flintshire County Council
FE	Forest Enterprise
GW	Gladstone Estate woodlands, Hawarden
HC	Halkyn Common
HM	Waun-y-Llyn Country Park, Hope Mountain
LC	Common and widespread species giving least concern
LH	Llyn Helyg woodlands
LNR	Local Nature Reserve
LO	Loggerheads Country Park
MA	Maeshafn
MF	Moel Famau
ML	Glan-yr-Afon, Marli, Abergele
MO	Mold area
NEWW	North East Wales Wildlife
NF	Nant-y-Ffrith
NM	Nercwys Mountain
NT	Known from few sites but not in serious decline, near threatened
NT	National Trust
NWWT	North Wales Wildlife Trust
P	Private ownership (not listed)
PA	Point of Ayr
PP	Nant Mill Country Park, Coedpoeth
RDL	Red Data List
RE	Regionally extinct (not recorded in Wales for fifty years)
RSPB	Royal Society for the Protection of Birds
SNP	Snowdonia National Park
SSSI	Site of Special Scientific Interest
v.c.	Vice-county or vice-counties

List of Abbreviations

VU	Currently known from five or fewer sites, either having been more widespread in the past, or known from few sites, but without evidence of decline; vulnerable
WCBC	Wrexham County Borough Council
WT	Woodland Trust
WW	Wepre Woods Country Park

PART 1

INTRODUCTION

1.1 The area covered by this book

North East Wales is an arbitrary area comprising mostly the counties of Denbighshire and Flintshire, the County Borough of Wrexham and the eastern part of the County Borough of Conwy. These areas have been modified or changed quite considerably in recent decades, often for political rather than administrative expediency. When the present fungus survey began there were two counties, Denbighshire and Flintshire, which more or less coincide with the area under study. These were combined, with some minor changes around Corwen, into the county of Clwyd in 1975. The present day administrative areas do not coincide with these older counties, whose boundaries have been reshaped drastically. Naturalists have long discarded administrative counties as being too ephemeral and have used for more than a century the vice-county (v.c.) system proposed by Watson (1847–59). Thus Denbighshire, as it was in Watson's time, is v.c. 50 and Flintshire (less the detached areas in Maelor, which went to v.c. 50) is v.c. 51. This book covers the whole of the (unchanging) v.c. 50 and 51. There is, however, a small anomaly in the southern part of the Clocaenog Forest, where the area around Bod Petrual is actually in v.c. 48 (Merioneth) although it is in present-day Denbighshire and was also in Clwyd. As this mycologically important spot was not covered by Aron (2005) in his excellent book on the fungi of North West Wales it is included here. A large piece of old Denbighshire, in the Tanat valley, is now incorporated into Powys, and includes the north side of the Pistyll Rhaeadr valley; it remains part of v.c. 50 and is part of our region. Maps 1–3 show the relationship between v.c., the different local authority areas and the National Grid squares.

1.2 Physical environment

Geology

Travelling westward across North Wales from the border with Cheshire to Snowdonia is like moving backwards in geological time as each successive rock type on the journey is older than the last. The western rim of our area is mostly formed by the Conwy Valley but this also marks the boundary between the rocks of Ordovician and Silurian age. The only significant areas of Ordovician rocks within our region are the mudstones of the Berwyn range, which have a few bands of volcanic ash inlaid.

The Silurian rocks dominate the west and south of the region, mostly as shales, grits and greywackes (muddy sandstones) of the Wenlock series, especially important on the Denbigh moors, and the shales, flags and grits of the Ludlow series, especially in the Clwydian Range but also in parts of the Denbigh Moors. In the Llangollen area the shales have been quarried for roofing slates. The shales have relatively few fossils but graptolites are important indicators. Devonian rocks are absent from North East Wales.

The Carboniferous rocks occur as three major groups, limestone, Millstone Grit and the Coal Measures. Carboniferous limestone outcrops on the peninsula culminating in the Great Orme (just outside our area) form the prominent hill of Bryn Euryn near Rhos-on-Sea. The western band of the limestone runs from Colwyn Bay to Abergele then south-east to Denbigh, where it outcrops on the western side of the Vale of Clwyd, being well developed south of Ruthin. The eastern band is seen as the distinct scarp between Dyserth and Prestatyn which then runs south-east, via the western outskirts of Holywell, across the expanse of Halkyn Mountain and Rhosesmor. It has a south-western outcrop along the eastern side of the Vale of Clwyd around Tremeirchion. The band continues along the eastern side of the Alyn Valley to Loggerheads, Maeshafn and beyond. It is interrupted by a major fault but occurs again in the gorge of Nant-y-Ffrith, is prominent around Minera, then forms the famous scarp from

World's End along the Eglwyseg rocks to Llangollen. It continues as a band of varying width to the border between Montgomery and Shropshire at Llanymynech. The limestone is often rich in fossils – corals, brachiopods and bryozoans being especially important. There are many cave systems, some with bone fossils and human artefacts and many of the rivers running over the limestone disappear into swallow holes during the summer, leaving dry valleys.

The Millstone Grit occurs in our area, not as the well-known coarse grey grit of the Pennines, but as finer sandstones and even shales. It forms prominent ridges or domes, as at Hope, Nercwys and Ruabon Mountains, where it is usually seen as the Cefn-y-Fedw sandstone. The grey Holywell shales occur in the north and are associated with the Gwespyr sandstone, which is the uppermost bed of the Millstone Grit in our area.

The Coal Measures are a mixture of shales, thin sandstones, and even some limestones, interleaved with coal seams. They occur from Point of Ayr, along (and under) the Dee estuary as far as Queensferry and across the Mold and Buckley area. They are most extensive in the Wrexham, Ruabon and Oswestry areas and coalmining was a major industry well into the twentieth century.

The youngest rocks in our region belong to the Bunter sandstones and pebble beds of Triassic age. The Vale of Clwyd is a classic rift valley in which the original Carboniferous dome has down-faulted and the depression filled with sandy deposits below a great lake basin. Another branch of this vast lake covered the area we now know as Cheshire and the Wirral. A more detailed account of the geology of North East Wales may be found in Smith & George (1961).

Topography
The region is bounded on the west by the deep valley of the river Conwy. The land rises steeply to the east to form an undulating plateau which rises gradually to the expansive Denbigh Moors, whose height ranges between 400 and 500 metres, reaching 532 m in the central part of the moor, but 664 m at the extreme south-western tip of our area at Carnedd y Filiast. The moors gradually descend through the Clocaenog Forest to the sudden drop to the rift valley of the Vale of Clwyd, with its nearly flat bottom.

To the south of the region the Berwyns rise in rounded masses from the Dee valley on their west and north to form a high plateau between 500 and 600 m, reaching their maximum height, 827 m at Moel Sych on our southern border. The massif gradually slopes down eastwards to the Shropshire Plain outside our area.

To the east of the Vale of Clwyd the slopes of the Clwydian Hills are broken here and there by limestone escarpments, as at Y Graig near Tremeirchion. The rounded summits of the Clwydian hills reach 554 m at Moel Famau and several of the tops are crowned with Iron Age hill forts.

To the south the high ground north of the Dee forms the dome of Ruabon Mountain, rising to 511 m, with the outlier of Llantysilio Mountain reaching 578 m. Nearby is the impressive limestone escarpment of the Eglwyseg Rocks. East of the Clwydian Hills the valley of the Alyn follows the change from Silurian shales to Carboniferous limestone and the river forms a series of small gorges between Loggerheads and Rhydymwyn, where the river is forced to flow south-east by the low mass of Halkyn Mountain to the north. The steady down-slope from limestone to Coal Measures continues almost to sea level at the Dee estuary and the Welsh border marks the start of the Cheshire Plain. Several other rivers have made gorges as they have cut down through different rocks, notably the Clywedog, the Dee, the Elwy and the Ffrith. The only waterfalls of note, however, are the Conwy Falls, on the south-west edge of our region, the Pistyll Rhaeadr near Llanrhaeadr-ym-Mochnant, in the extreme south of our area, the Horseshoe Falls on the Dee at Llangollen and the Falls of Dyserth.

Introduction

Natural lakes of any size are rare but there are several lakes on the Denbigh Moors, some of which, notably Llyn Brenig and Llyn Alwen, have been enlarged by damming and are important reservoirs. Indeed, virtually all the larger lakes in the region have been dammed to some extent.

The coastal strip is marked by ancient sea-cliffs running from behind Colwyn Bay to Abergele, which are fronted by sand or shingle beaches. From Abergele to Prestatyn there is a large triangle of alluvial soil, once a major wetland but now drained, on which stand Rhuddlan and Rhyl, again fronted by sandy beaches. From Dyserth to Prestatyn and on to Ffynonngroyw, on the Dee estuary, the old sea-cliff returns. There are extensive sand dunes from Prestatyn to Point of Ayr, with an important area of saltmarsh at Talacre, sheltered by a shingle spit.

The hill land that is not forested is used as a grouse moor or for sheep grazing. Cattle are grazed at lower levels on reclaimed marshland or on grasslands to the east of the hills. Arable land is confined to the coastal fringe, the Vale of Clwyd and the Coal Measure and Triassic regions of the east. In many places the limestone has been quarried for building stone, road-stone or for the manufacture of cement. The sandstones of the Millstone Grit have also been quarried for building stone. There has been extensive mining for lead and other heavy metals in the area until the early twentieth century. Coalmining persisted even longer. Flintshire is the most rapidly developing county in Wales at the present time so more land is likely to be lost to industry and housing in the near future.

Soils

Soils affect fungal communities both directly and indirectly. The mycelia of many species live in the soil and are influenced by particle size, pore space, water, humus and mineral content and acidity. The vegetation on which parasitic, symbiotic and saprophytic fungi depend is itself strongly influenced by pH, available calcium content, drainage and moisture retention. Thus the character of the soil may play a significant part in the composition of the mycota.

The soils in the region are mostly derived from drift but the impact of the underlying rocks, especially in the effect on emergent water sources, is also very important. Overlying the limestone are small areas of shallow brown earths which carry typical species-rich limestone grassland, with its attendant fungal populations. Over much of the lowlands and on other areas on the limestone lie typical brown earths. These are fertile and well-drained and carry typical woodland and lowland grassland vegetation. Over much of the high ground, on acid shales and greywackes, there is extensive blanket peat and this may be covered by typical podzols, with or without deposits of iron oxides in pans.

The coastal fringe from Abergele to Point of Ayr has a narrow strip of blown sand, often enriched with molluscan shell fragments and thus base-rich, which forms dune systems. Some stretches have shingle beaches with minimal vegetation.

The river valleys have a variety of soils, including brown earths, brown sands and alluvial soils of various kinds, including gravels, sands and muds. There are several developments, especially in the Wheeler valley, of tufa marl deposits, as at Ddol Uchaf, which are strongly calcareous. In the Vale of Clwyd and along the coast and estuary between Holywell and the Cheshire border are non-calcareous stagnogleys, rich in clay minerals and prone to waterlogging. The Dee estuary itself is largely silted up with fine muds carrying saltmarsh vegetation and extensive sand banks.

Climate

North East Wales falls into the Eu-oceanic climatic zone, characterised by relatively mild winters and cool summers (Bendelow & Hartnup, 1980). However, the factors which affect the distribution of fungi operate on a more local level and include temperature – especially the frequency and duration of frosts, sunshine, rainfall, snow cover and wind.

The hottest months are July and August with an average daily maximum of 19°C at sea level and 16°C at 400 m. Above this height temperatures are much lower on north and east facing slopes but may be raised on south and west facing slopes. The summer warmth followed by autumn rain is the usual trigger for the mass fruiting of larger fungi in the autumn although many species are in fruit in the spring or throughout the year. The lowest temperatures are recorded in February with an average daily minimum at sea level between 1°C and 2°C and -1°C at 400 m, and again, even lower at higher altitudes. Temperatures tend to be higher on the coast because the sea acts as a thermal reservoir.

Both distance from the sea and altitude affect the frequency and severity of frosts, which greatly reduce fungal activity. A frosty morning usually means a poor crop of fresh fungi on a foray. Over the period 1961–1980 the number of frosty days per year ranged from fewer than thirty on the coast to forty at sea level, but away from the coast, to seventy or more above 200 m. As with temperature readings generally the higher the altitude the greater the number of frost days.

Direct sunshine affects our local fungi in two distinct ecosystems, grasslands and sand dunes. The heating of the soil, especially where the turf is grazed and thus short, is beneficial to many of the larger fungi which are thermophilic. These include several species of waxcaps, *Hygrocybe*, and puffballs, especially species of *Bovista*. Many of these fungi live in forests at lower latitudes but need the warmth of sunlit open habitats this far north. South and west facing slopes tend, therefore, to be richer in these thermophiles.

Rainfall is similarly linked to altitude with about 600 mm per year at sea level and more than 1100 mm above 400 m. Rain is distinctly seasonal with the period from March to July being drier than August to February. This is based on the period from 1941 to 1970 but recent years have shown a shift, not only in the amount of rainfall, but also when it falls. June 2007 was the wettest on record and in early July several species of larger fungi, including mycorrhizal agarics, were seen in fruit in the region, at least two months early.

The impact of rainfall affects not only the vegetation with which the fungi are associated but also the balance of physical factors in the soil. Waterlogging and the consequent reduction in available oxygen to mycelia is damaging to most fungal organisms. In contrast, the drought conditions experienced in many summers since 1976 have had a deleterious effect on fungal fruiting. In that year, for example, many early autumn species of *Amanita* and *Russula* did not appear at all as the rain did not arrive until October.

Humidity is an important factor, especially in woodlands, and the region has a Meyer's Precipitation-Saturation Deficit Quotient of around 300, which is midway between the driest parts of eastern England and the wettest parts of the west coasts of the British Isles. This is reflected in the differences between the bryophyte-rich woods of Snowdonia and the herb-rich woods of North East Wales and their associated fungi.

Snow cover affects vegetation and, indirectly, fungi, by covering the ground with an insulating but light-reducing blanket. The longer the snow lies the more obvious the changes to the vegetation and mycota. Even on the highest parts of our region snow cover lasts for less than the three months needed to produce distinctive snow-patch vegetation and fungal communities. Even in Snowdonia such vegetation is confined to a small area below the summit of Snowdon and a few patches in the Carneddau. Although the snow remains for a short time the actual snowfall, with drifting, may amount to several metres, but the impact on plants and fungi is limited to the addition of meltwater to the soil.

Introduction

The coastal fringe, especially around Point of Ayr, and the tops of the hills, experience very strong winds on occasion and this may reduce the growth of trees, thus indirectly affecting the fungi. The wind also has a drying effect on grassland and woodland edges which may also affect fungal growth. The dune system at Point of Ayr is continually on the move as a result of strong onshore winds with serious consequences for fungal mycelia, which may become exposed to desiccation.

In the last few decades the pattern of our climate has appeared to shift, for whatever reason. Winters are milder, there is more rain at unexpected times with consequent flooding, there is a greater frequency of gales and summer temperatures are higher than ever. Fungi are quick to take advantage of warmer and moister conditions and the lack of frost and it is not surprising that several southern species, some from the Mediterranean, eastern Europe and the warmer parts of the United States, have arrived and are settling into our region. It will be important to monitor their survival and spread during the next decade or so.

1.3 Vegetation and fungal habitats

The climax vegetation for most of the region would be broad-leaved woodland but thousands of years of human activity have modified the woodland cover so that it is now much reduced and often dominated by non-native species. Woodland has been replaced on the acid soils by moorland, and on the better soils by grassland or arable land.

Woodland
Native, semi-natural, woodland in the region either resembles the oak-dominated Atlantic woodland to the west, as in the Conwy valley, or the oak-elm-hazel woodland on the clay soils of the east, as typified by the remnants of ancient woodland in the Alyn valley near Mold. On limestone ash is the natural climax tree but is almost universally in-planted with beech, which is not native in North Wales. Along rivers and in some ill-drained areas alder may be dominant and on drier soils birches are primary colonisers. Much of the native woodland has been replaced with amenity plantings, either as shelter belts or as modified parkland, as at Bodelwyddan, Gwysaney, near Mold, and Hawarden, or with extensive conifer plantations such as those at Clocaenog, Brenig, Clwyd, Ceiriog Forests, Nercwys Mountain, Cwm Woods, near Dyserth, and the Berwyn plantations.

Within these woodlands, whether semi-natural or planted, fungi are key components of the ecosystem. Primarily, they form the essential symbiotic relationship with tree roots without which the tree cannot obtain nutrients or even sufficient water. These are the mycorrhizal fungi which either form a sheath around the root hairs – ectomycorrhizal – or live, as small cells, within the root – the vesicular-arbuscular mycorrhizae. The fungi involved as ectomycorrhizal partners include species of *Amanita*, boletes, chanterelles, *Cortinarius*, *Hebeloma*, *Inocybe*, *Laccaria*, *Lactarius* and *Russula*. Some species have subterranean fruit bodies, as the truffles and false truffles, and their spores are dispersed by beetles.

Another important strategy for woodland fungi is litter decomposition and without them the dead leaves produced by generations of trees would not be broken down and their valuable nutrients recycled. Many species of *Collybia*, *Marasmius* and *Mycena*, some restricted to conifer, are found as decomposers in our area. Fallen wood of all kinds is quickly attacked by wood-rotting fungi, from agarics to brackets and white-wash-like corticioid species; again many of these are restricted to particular substrates. Finally a few species are parasitic on living trees, notably species of *Armillaria*, *Inonotus* and *Piptoporus*, and many of these, too, are confined to a single species. In our region the woodlands on limestone, especially those with planted beech, are by far the richest habitat for fungi of all groups.

Dwarf shrub heath
Heather moorland is widespread on the higher ground of the Berwyn range, the Denbigh and Llandegla moors and the Clwydian Range. In areas used to raise grouse the moors are burned at regular intervals to encourage young heather growth. This results in dry heather moor with few plant species other than common heather or ling. In damper or unburned areas a richer flora containing bilberry and other plants, including more lichens and mosses, produces more diverse vegetation. These areas are not very rich in fungi but a few agarics, such as *Hygrophoropsis* and some species of *Hygrocybe*, are frequent, but these types of vegetation are much less frequented by mycologists and are therefore less studied. Where there are areas of wet peat or *Sphagnum* bogs a few characteristic species of *Galerina, Hypholoma, Mitrula* and *Tephrocybe* may occur. Because of the relatively low altitude attained in our region there is no typical montane vegetation and therefore no mountain fungi have been recorded.

Grasslands
Grass-dominated vegetation on acid soils, or grass heaths, is widespread in areas subject to heavy sheep grazing, where hard grass, *Nardus stricta*, is the main species. Although some species of *Hygrocybe* do occur in acid grassland this is not generally a rich fungal environment. On base-rich soils that have not been ploughed or treated with inorganic fertilisers or agrochemicals – the unimproved grasslands – a rich and varied fungal population with species of *Hygrocybe sensu lato, Entoloma, Geoglossum* and clavarioid fungi may develop. These are known as 'waxcap grasslands' by conservationists. Sympathetically managed ancient lawns associated with stately homes may also be good waxcap grasslands. By contrast, improved grasslands have only common litter-decomposing fungi, such as species of *Panaeolus, Psilocybe* and *Stropharia*, and may also be poor in *Agaricus* species. Some past industrial sites, such as lead mines and limestone quarries may develop rich grassland after suitable treatment to bind heavy metal residues or after several decades of natural change.

Wetlands
Neutral and base-rich wetlands occur rather sparingly across the region and fungi characteristic of such habitats are correspondingly rare. Of the remaining areas of raised bog in the border region only the Fenn's Moss, part of the Fenn's, Whixall and Bettisfield Mosses National Nature Reserve, retains important bog vegetation. Open water habitats such as lakes, ponds and reservoirs are mostly artificial and many are in acidic catchments so their margins do not have mycologically suitable habitats. Rivers in the region, on the other hand, are a good source of Ingoldian fungi, or aquatic hyphomycetes, whose elaborate spores are trapped in the natural foam which collects behind projecting tree roots or bends of the river bank. These fungi break down leaves which have fallen into the stream and are a vital link in the carbon economy of these aquatic ecosystems. These fungi are good indicators of water quality and have been well studied in the region.

Coastal habitats
We have few cliffs at the sea's edge but there are some shingle beaches from Towyn eastwards to Point of Ayr, but few fungi are associated with this habitat in North Wales. Saltmarshes used to occur along the estuary of the River Clwyd but these have largely been reclaimed. At Point of Ayr, however, a species-rich saltmarsh remains behind the shingle ridge and supports a few uncommon plant pathogens. The Dee estuary is silted up for much of its length and carries species-poor saltmarsh vegetation dominated by *Spartina anglica*. No fungi have been detected in this area. Sand dunes are found from Prestatyn through Gronant to Point of Ayr but only in the east are they well enough developed, or protected, to carry a rich fungus flora. Here there are *Melanoleuca cinerifolia, Psathyrella ammophila* and the nationally rare *Cyathus*

stercoreus in the mobile marram dunes, and a rich assemblage of grassland species in the dune pasture as well as a range of pathogens on the vegetation in the wet dune slacks.

Man-made habitats

Arable fields, with crop pathogens and plant litter decomposers, pastures with field mushrooms, hedges with a range of distinctive species and even railway embankments add to the range of suitable sites in the rural environment. In urban situations parks, gardens, playing fields, cemeteries, roadsides, shrub borders in car parks, especially when mulched with forest bark, all provide an abundance of opportunistic fungi.

There is no published up-to-date account of the flora and vegetation of Denbighshire, except for a valuable species list (Green, 2006). Flintshire is extremely well covered by the modern, detailed flora by Goronwy Wynne (1993). Throughout the region the range of vegetation is mirrored by the variety of fungal species associated with it. This is detailed under each species account. The richness of the semi-natural environment means that some sites are more fungus-friendly than others and these special sites are described in the next section.

1.4 Mycologically important sites

A number of sites have been found to be especially rich in a wide range of fungi. Many of these sites are also valuable for other biological groups and several are Sites of Special Scientific Interest (SSSI) or Areas of Conservation Importance. Within the species accounts these most important sites are listed by the code letters used in this section (in bold and also included in the list of abbreviations on p. viii). When a species is known only from a single site the site name is given in full. Where no details of ownership or management are given it may be assumed that the site is in private hands and is not associated with any countryside organisation.

Ownership abbreviations:

AONB	Area of Outstanding Natural Beauty
DCC	Denbighshire County Council
DEFRA	Department of Environment, Food and Rural Affairs
FCC	Flintshire County Council
FE	Forest Enterprise
LNR	Local Nature Reserve
NEWW	North East Wales Wildlife
NT	National Trust
NWWT	North Wales Wildlife Trust
P	Private ownership (not listed)
RSPB	Royal Society for the Protection of Birds
SNP	Snowdonia National Park
SSSI	Site of Special Scientific Interest
v.c.	Vice-county or vice-counties
WCBC	Wrexham County Borough Council
WT	Woodland Trust

AV Alyn Valley woodlands [SSSI; owners: P (various) and DEFRA; managed by: NEWW (part)] v.c. 51.
[located in 1 km grid squares: 1863, 1864, 1865, 1866, 1963, 1965, 1966, 2066]
These woodlands clothe the sides of the Alyn Valley downstream from the Loggerheads Country Park (see below) as far as the main A541 road at Rhydymwyn. They include a range of old estate woodlands, including planted beech, native ash woods and mature conifer plantings, and have five main constituents: the Leete woodlands from Loggerheads, at the county boundary, to the Mold–Cilcain road at Pont Newydd; Coed Nant Gain, the private nature reserve at the bottom of Trial Hill; the Nant Alyn woodlands on the east side of the valley between Pont Newydd and Rhydymwyn; the woodlands around the Coed Du Hospital on the west side of the same section of the valley, and the relict woodlands, grasslands and new plantings at Rhydymwyn in the grounds of the World War II armaments factory, since 2004 managed as a nature reserve by North East Wales Wildlife for the Department of Environment, Food and Rural Affairs. Apart from the Rhydymwyn section, which is on Holywell Shales (Millstone Grit) and river alluvium, all components are on Carboniferous Limestone, on the steep sides of the valley, some parts of which form a gorge. This part of the River Alyn disappears into its limestone bed in late spring and flows underground until the autumn rains, so the woodlands are relatively dry. All groups of fungi are well represented, especially those species associated with beech. Particular specialities are *Coprinopsis picacea*, *Cortinarius infractus*, *C. violaceus*, *Leccinum roseofractum*, *Lycoperdon mammiforme*, *Marasmius hudsonii*, *Phyllotopsis nidulans* and species of *Geastrum*.

BO Bodelwyddan Castle Estate [owner: Bodelwyddan Trust] v.c. 51.
[9875, 9974, 9975, 0074, 0075]
This historically important mycological site was visited in the second half of the nineteenth century by the Rev. Miles Berkeley and his associates and several species new to science were described from here. The site includes the one-time military base of Kinmel Camp (Lowther Barracks), now mostly a new industrial estate, from where numerous records were made by Major Roger Brown in the 1970s. Much of the nineteenth-century material and that collected by Roger Brown is in the Mycological Herbarium at the Royal Botanic Gardens, Kew. In recent years the area has been forayed by members of several local wildlife organisations. In particular the meadow below the castle has been found to be one of the best sites for grassland fungi, especially species of *Hygrocybe sensu lato*, *Entoloma* and *Clavariaceae*, notably *Clavaria incarnata*, in North Wales. The woodland is mostly planted amenity woodland and is not in itself of ecological interest. However the large number of non-native species and the presence of an old orchard add to the interest and provide a wide range of fungal substrates. Interesting species found here are *Buchwaldoboletus lignicola*, *Gyroporus castaneus* and *Pluteus griseoluridus*. The site is on drift overlying Carboniferous limestone.

BP Bishop's Wood (Coed yr Esgob), Meliden [SSSI; owner: DCC] v.c. 51.
[0681, 0781, 0782]
This is a mixed site on Carboniferous limestone, with mixed woodland on the seaward scarp (probably once a sea-cliff) and species-rich grassland on the top. Mycologically the grassland is diverse, with species of *Hygrocybe sensu lato* and *Lepista* notably abundant. The woodland contains a number of uncommon agarics, including *Hygrophorus nemoreus*. The site is close to the important limestone site at Craig Fawr (see below).

Introduction

BW Big Wood, Hendre [owner: WT] v.c. 51.
[1867, 1967]
This block of ash-oak-beech woodland on limestone is floristically and mycologically rich, especially in small ascomycetes associated with moss cushions on the floor of the disused quarry. In some ways it is an extension of the Alyn Valley woodlands but it is not sited in the valley bottom and is therefore better drained.

CC Coed Coch, near Betws-yn-Rhos [P] v.c. 50.
[8773, 8873, 8874]
This is a mix of grassland, heathland and various types of woodland on neutral, basic and acidic soils on drift over Carboniferous limestone, Ludlow shales and grits (Silurian). The estate has changed hands many times since it was last forayed in 1880; before that it was intensively studied by the leading mycologists of the day. Several species were described as new from here and numerous species have been recorded nowhere else in North Wales, including national rarities. Unfortunately it has not been possible to investigate the area fully in recent years so the continued occurrence of those species cannot be assumed. However, much of the material collected in the nineteenth century has been preserved at Kew. Several species of microfungi were collected in 2008.

CE Celyn Woods, Northop [owners: P and Welsh College of Horticulture] v.c. 51.
[2368, 2369]
These woods run along the top and sides of Wat's Dyke, a medieval earthwork, on both sides of the old A55 road west of Northop and adjacent to the campus of the Welsh College at Celyn. The wet alder woods on the campus are included in the site. There are stands of conifers and hardwoods, both native and introduced, on boulder clay overlying Buckley Fire Clay of the Coal Measures. They have been extensively forayed by the Merseyside Naturalists' Association and the North West Fungus Group, among other organisations. A number of species are known in North East Wales only from this site, including *Cortinarius sanguineus* and *Leccinum aurantiacum*, which is a reflection of the amount of collecting that has occurred here. The site is near, and in some ways similar to Wepre Woods (see below) but is generally drier.

CF Clocaenog Forest [owner: FE] v.c. 50.
[05]
The extensive coniferous plantings on the Denbighshire Moors are on Ludlow and Wenlock grits and shales (Silurian) under high rainfall, at least for North East Wales. The woods contain mature stands of pines, spruces and Douglas fir which are rich in mycorrhizal fungi associated with conifers. There is an important mature beech plantation too, which also carries a rich and characteristic mycota. Rotten moss-covered logs support populations of rare myxomycetes normally associated with the ravines of Snowdonia (e.g. *Diderma ochraceum*, *Lepidoderma tigrinum*) and the very rare *Stemonitopsis reticulata* has its only Welsh (and second British) locality here. The area which has been most studied and is apparently the richest in the whole forest is around Bod Petrual [0351] which, unfortunately, is a few metres over the v.c. boundary in Merioneth (v.c. 48) but is still in modern Denbighshire. It is therefore retained as part of North East Wales for our present purpose and, in any case, was omitted by Aron (2005). A few kilometres to the north is Bontuchel/Cyffylliog (see below) which differs slightly in its geology.

CN Coed y Felin Nature Reserve [owner: Tarmac Ltd; managed by NWWT] v.c. 51.
[1867, 1868, 1967]
This is another limestone wood, with ash, oak and some planted beech and conifers on the edge of a worked-out quarry on the edge of the village of Hendre. It is on the north side of the A541 and is quite different from the nearby Big Wood on the south side. It is steep, more open and more varied in terms of tree species. There is a considerable area of wetland along the southern edge, adjacent to the old railway line. Rare species found here include *Gyromitra esculenta*. This is not to be confused with the Woodland Trust reserve by the same name at 1764.

DU Ddol Uchaf Nature Reserve, Ysceifiog [SSSI; owner: NWWT] v.c. 51.
[1471]
This site is unusual in that it was worked up to fifty years ago as a calcareous marl quarry. It lies at the junction of the limestone, to the north, and the Wenlock shales and is built on a fan of tufa deposited by the Afon Pantgym, which flows through the reserve. It is a patchwork of sallow carr, hawthorn and blackthorn scrub, patchy grassland and a range of pools and wetlands; there is also an extensive area of tall herb vegetation. It is rich in smaller fungi, such as myxomycetes and ascomycetes, and is well known from the abundance of *Morchella esculenta* in the spring. Rare species found here include *Amanita echinocephala*, *Peziza vacinii* and *Hypoxylon mammatum*. It is close to the mycologically interesting limestone pavement woodlands at Ysceifiog (see below).

ER Erddig Estate, near Wrexham [SSSI; owner: NT (Hafod Wood managed by NWWT)] v.c. 50.
[3247, 3248, 3347, 3348, 3447, 3448]
This is one of the finest mycological sites in North Wales, with a range of woodlands, pasture and wetlands. Big Wood, behind the house, is an estate woodland with abundant beech, sycamore, oak and some conifer. It is rich in fungi of all groups, including *Hericium erinaceum*, in its only site in our region, and *Cortinarius bibulus*, *Lepiota grangei*, *Lycoperdon echinatum* and *Melanophyllum haematospermum*.

Hafod Wood is a damp wood, mostly of beech on the slopes, but with abundant alder along the Black Brook, which is a tributary of the River Clywedog, which runs through the park. This wood is notable for *Coprinopsis picacea*. Two rare myxomycetes have been found, *Stemonitis smithii* and *Licea erddigensis*, which was described from the bark of a sycamore. Other woodlands, mainly of oak, occur on the perimeter of the park and there are numerous old pollard oaks in the parkland, which is grazed by cattle. The site lies entirely on the Coal Measures and has been subject to some subsidence. It has been investigated since the early part of the nineteenth century but has been especially well studied in the last fifty years.

GW Gladstone Estate woodlands, Hawarden [owner: Gladstone Estates] v.c. 51.
[3163, 3164, 3165]
The site is centred on Bilberry Wood, which lies on a small ridge of Cefn-y-Fedw sandstone (Millstone Grit), but includes the woodland along Tinker's Dell, on both sides of the Hawarden–Wrexham road. The main body of the woodlands consists of beech and conifer plantings which are rich in larger fungi and have been studied for more than a century by local naturalists. The valley woodlands are more varied, having abundant alder. Uncommon fungi found here include the rare myxomycete, *Physarum conglomeratum*, in its only Welsh site, and *Pluteus hispidulus*.

Introduction

HC Halkyn Common [SSSI; owner: Grosvenor Estates]
[1871, 1872, 1873, 1970, 1971, 1972, 2069, 2020, 2071]
The area is generally known as Halkyn Mountain and is dominated by the remains of worked-out lead mines and current limestone quarries. However, much of the area is covered by sheep and rabbit-grazed grassland, heath and gorse scrub. There are several areas of developing scrub woodland and there are small plantations and areas of amenity woodland around the periphery of the site. The major mycological interest is in the waxcap grassland which carries over twenty species each of *Hygrocybe sensu lato* and *Entoloma*, including several nationally rare species, together with members of the Clavariaceae. The area featured strongly in the national waxcap grassland survey of 2003–4.

HM Waun-y-Llyn Country Park, Hope Mountain [owner: FCC] v.c. 51.
[2858]
This hill, of Cefn-y-Fedw sandstone, is prominent in the landscape, especially when seen from the east, and consists largely of heather moorland, grass heath and *Sphagnum* bog. There are derelict quarries and a rapidly disappearing lake, which is becoming covered with bog vegetation. The area is grazed by cattle, sheep and rabbits and dung fungi are well represented. Many typical moorland and bog species have been recorded, of which *Exobasidium arescens* is nationally rare. The sandstone has some calcareous intrusions and the grassland is not all acidic and carries *Porpolomopsis calyptriformis*. The site is interesting as it is a fine example of the vegetation on this type of sandstone in the absence of trees and can be compared with the nearby Nercwys Mountain (see below).

LH Llyn Helyg woodlands [SSSI; owner: Mostyn Estates] v.c. 51.
[1076, 1077, 1176, 1177]
Llyn Helyg is a small artificial lake surrounded by remnants of damp, mossy oakwood. There is a large larch plantation to the east. The site is on boulder clay overlying Carboniferous limestone but there is little evidence of basic influence in the vegetation or the fungi. The woodlands are rich in all groups of fungi and this is one of the most productive sites in Flintshire. Rarities include *Diderma testaceum*, which is otherwise known in Wales only from Merioneth, *Buchwaldoboletus lignicola*, *Cortinarius triumphans* and *Gyroporus castaneus*.

LO Loggerheads Country Park [SSSI; owner: DCC] v.c. 50 and 51.
[1962, 1963, 2062, 2063]
Loggerheads is currently the most mycologically productive site in North East Wales. For many years it was privately owned amenity woodland but has been a popular and well-managed country park since the 1970s. There are Carboniferous limestone cliffs on the east side of the River Alyn and the slopes on the west are covered in drift overlying sands and gravels, near to the fault which separates the limestone from the Silurian shales. The soils on the west bank are varied with some base-rich lozenges, picked out by *Allium ursinum* and *Mercurialis perennis*, and more acid soils without these indicators. Up to the early years of the twentieth century the valley was mined for lead and is of considerable industrial archaeological importance. The woodland is mostly floristically rich ancient ash/wych elm/oak wood which is slowly re-colonising the spoil heaps from the mine shafts. There is much planted beech and sycamore and small plantations of hybrid poplars on the flood plain and conifers on the upper limestone slopes. There are small areas of limestone pavement and grassland to the east of the site; much of the pavement is now wooded. The area has been intensively studied for over fifty years and is rich in all groups of fungi. Rarities include *Boletus appendiculatus*, *Gomphidius maculatus*, *G. roseus*, *Hygrophorus arbustivus*, *Leccinum roseotinctum*, *Leucoagaricus pilatianus*, *Lycoperdon mammiforme*, *Mycena kuhneriana*, *Phyllotopsis nidulans*, *Ramaria botrytis*, *R. flava* and seven species of truffles. Among the rare myxomycetes are

Lycogala conicum, *Physarum psittacinum* and *Stemonitis foliicola*; the first two are indicators of ancient woodland.

The old county boundary, which is also the boundary between v.c. 50 and 51, runs along the footpath which leaves the upper entrance by the boundary stone (recently moved a few metres east) on the A491 and rises to the top of the cliff. East of this path is the area of the Cadole lead mine, which is now limestone grassland and planted woodland (in v.c. 51). The boundary runs more or less along the main track from the cliff-top to the steps which zigzag back down to the river to the wall which is the old boundary. The woodland to the east of the path is in v.c. 51, that to the west is in v.c. 50.

Through the gate in the boundary wall by the river is the Alyn Woodlands site, along the Leete Walk, all in v.c. 51.

MA Maeshafn [Big Covert – owner: Tarmac Ltd; Aberduna Nature Reserve – owner: Hanson Aggregates, managed by NWWT] v.c. 50.
[1960, 1961, 2060, 2061]
This site, on Carboniferous limestone, is in two parts on opposite sides of the village. The larger area is Big Covert, a plantation of beech and conifer on the edge of Burley Hill Quarry. The wood has areas of limestone pavement and is rich in characteristic beechwood fungi including several species of *Cortinarius*, *Lycoperdon echinatum*, *L. mammiforme*, *Suillus collinitus* and several species of truffle. The rare, mainly Mediterranean, myxomycete *Didymium sturgisii*, has been found on the bark of living beeches.

The smaller area is the Aberduna Nature Reserve which includes parts of Coed-y-Fedw and lies on a steep north-west facing scarp, with some limestone pavement, grassland and woodland. The trees are mainly ash but there is some planted beech and a good deal of hazel and holly. The grassland areas to the east of the wood are rich in waxcaps and other uncommon fungi, such as *Leucoagaricus pilatianus*. These two areas may be regarded as a southern extension of the Alyn Valley and the Loggerheads woodlands but differ in aspect and management history.

MF Moel Famau [AONB; DCC and FE] v.c. 50 and 51.
[16]
This, the highest hill in the Clwydian range at 554 m, straddles the v.c. boundary. It is composed entirely of Silurian Ludlow shales and is covered with various types of grass heath and dwarf shrub heath, with extensive conifer plantations, dating from the 1950s, on its lower slopes – the Coed Clwyd. The area above the forest is a country park. The forest has little native woodland left but there are margins with broad-leaved trees and there is abundant birch and rowan throughout. There is a wide range of mycorrhizal and litter agarics associated with pines, larches, spruces, silver fir and other conifers. Some of the pinewood fungi are northern species and show evidence of southward migration into these maturing plantations. The moorland is relatively poor in fungi but a few acidophilous agarics, notably waxcaps, have been noted. The rare myxomycete *Lycogala exiguum* has been found on fallen beech trunks.

ML Glan-yr-Afon, Marli, Abergele [P] v.c. 50.
[9972]
This limestone site borders the River Elwy and consists of high grade unimproved grassland, rich in waxcaps and other important indicator species, and various small blocks of woodland, mainly of beech or conifers. The area surrounds the home of Major Roger Brown who has collected here intensively for over forty years. His material is mostly at Kew and was identified by the late Dr Derek Reid. Numerous rare agarics have been found here, several being the only records for North Wales, or even for Britain. Some of the more interesting are: *Amanita franchetii*, *Cortinarius callisteus*, *Cystolepiota hetieri*, *Hygrocybe citrinovirens*, *Hygrophorus persoonii*,

Introduction

Lepiota fuscovinacea, L. locquinii, L. xanthophylla, Leucoagaricus marriagei, L. melanotrichus, L. tener, Limacella delicata, Lycoperdon mammiforme, Pluteus pellitus, Ramaria biformis and *Volvariella murinella*. The site, just south of Bodelwyddan, lacks its historical significance, but is far richer, especially in woodland species.

MO Mold area v.c. 51. [P (various)]
[26]
This is the most studied area of North East Wales, largely because it was the home of the author, but also because it contains a very wide range of accessible habitats! The bulk of the records have been made within the confines of the town itself and include species from gardens, parks, cemeteries, churchyards, playing fields, roadsides, hedgerows and riversides. The River Alyn flows through the edge of the town. Several lanes and footpaths run east and west from the main axis, linking the town with farmland and small copses, some of ancient woodland. Two major private estates – Leeswood to the south [2561] and Gwysaney to the north [2266] – have extensive ancient woodland in their policies and these have been well studied. Although there is a wide range of common species of larger fungi, it is the wealth of plant pathogens that makes the area so interesting. The farmland adjacent to the town has been surveyed for dung fungi, especially on rabbit dung. Two rare myxomycetes have been found on farmland around the town – *Oligonema schweinitzii* and *Physarum didermoides*. The whole area lies on Productive Coal Measures but is overlain with boulder clay, gravel and alluvium.

NF Nant-y-Ffrith [owner: FE] v.c. 50 and 51.
[2653, 2654, 2754]
This ravine of the River Ffrith cuts through limestone, sandstone and Coal Measures and marks the southern boundary between v.c. 50 and 51. The sometimes vertical walls of the gorge are clothed in a mixture of semi-natural woodland and planted conifers and there is an abundance of rhododendron. The valley was once intensively mined for lead. Birch is the commonest broad-leaved tree and there are many associated mycorrhizal species, such as *Cortinarius umbrinolens*. There are some representatives of the Atlantic ravine myxomycete association and this is most easterly site for them in Wales.

NM Nercwys Mountain [owner: FE] v.c. 51.
[2157, 2158, 2159, 2257, 2258, 2259]
This prominent sandstone ridge reaches 378 m and in the 1960s was moorland. Now it carries a thriving conifer forest with the first crop clear felled and replanted. Because of its altitude, it receives above average rainfall for the district and snow often lies for several weeks in the winter. Birch and sallow accompany the conifers and there are a few pockets of acid sessile oak woodland along the southern margin. Mycorrhizal fungi are abundant. A largely overgrown lake is now covered with *Sphagnum* moss and wet heath and the area has a good selection of fungi on bare peat. The hill is part of the same formation as Hope Mountain – the Cefn-y-Fedw sandstone – but is higher and is well contrasted by the forest.

PA Point of Ayr [SSSI; owner: Hamilton Oil; part managed by RSPB] v.c. 51.
[1084, 1184, 1185, 1284, 1285]
This is the major sand dune system on the north-east coast of Wales and the site includes the extensive Talacre Warren and a small area of saltmarsh enclosed by a shingle spit. The large area of dune grassland is rich in grassland and coastal agarics, such as *Camarophyllus schulzeri*, and the wet slacks have interesting and uncommon ascomycetes. The yellow dunes are especially interesting. The first British record of *Hohenbuehelia culmicola*, the second British record of *Cyathus stercoreus* and the only recent Welsh collection of *Psathyrella flexispora* were

made here. There is a large population of *Melanoleuca cinerifolia* at the front of the dunes and the uncommon rust, *Puccinia dioica* var. *extensicola*, is in the saltmarsh.

PP Nant Mill Country Park, Coedpoeth [owners: WCBC and WT] v.c. 50.
[2949]
The woodlands in the valley of the River Clywedog downstream of Nant Mill include the former grounds of Plas Power and are mycologically important. The woodlands lie on the Coal Measures and include beech, oak and a wide variety of conifers. The ravine-like valley maintains a high humidity and the leaf litter is deep. Important species are *Amanita virosa*, *Ganoderma lucidum*, *Strobilomyces strobilaceus* in its only Welsh locality and *Tylopilus felleus*. The site is upstream from the Erddig Park estate.

WW Wepre Woods Country Park [SSSI; FCC] v.c. 51.
[2867, 2967, 2968]
The country park is a long strip of mixed woodland and grassland on both sides of the Wepre Brook, downstream of the ruins of Ewloe Castle, and extending into the urban area of Connah's Quay. The stream and its tributaries cut deeply into the Coal Measures, including some massively bedded sandstone. This rock has some calcareous content and is responsible for some areas of base-rich soil in the woodlands. The upper reaches, below the castle, are oak-dominated ancient woodland, with a rich ground flora. Beech and other planted trees inhabit the lower reaches, which include the amenity woodland around the remains of Wepre Hall. There are rough grasslands with playing fields and a rough golf course within the park but there are no unimproved grasslands and waxcaps are not well represented. Among the site's specialities are *Leucoagaricus georginae* and the myxomycete *Metatrichia vesparium* in one of its two sites in North East Wales.

The above sites are recorded in the species accounts by their code letters and may be regarded as the most important mycological sites in North East Wales. However, there are several others which show potential and, with further study, may deserve to be placed in the first rank. The descriptions of these are shorter and important species are only mentioned when they are the only records in the region.

v.c. 50

Alyn Waters Country Park, Gwersyllt [WCBC; LNR] [3154, 3254]
A mixture of developing grassland and relict woodland, partly on a former landfill site. The River Alyn flows along the eastern boundary and is lined with old, damp woodland. There is a wealth of rotting logs and extensive stands of Rosebay Willowherb, brambles and nettles, with great mycological potential. The site, which used to be Gwersyllt Park, is on the Coal Measures.

Bodnant Gardens [NT] [7972]
These famous gardens are mature enough to have attracted an interesting woodland mycota but are perhaps of greatest importance for the wide range of hosts for plant pathogens. The site lies on Wenlock grits on the east side of the Conwy Valley.

Ceiriog Forest [FE] [around 1638]
Mature, moist conifer plantations on the Ordovician shales of the Berwyn Mountains, with a good range of fungi in all groups, under moderately high rainfall.

Introduction

Chirk Castle Estate [NT] [2638, 2639, 2737, 2738, 2837]
The grounds of the castle contain various plots of low quality woodland which are nevertheless rich in common fungi. Some of the pastures, which are grazed by horses, are rich in clavarioid fungi and species of *Agaricus, Hygrocybe sensu lato* and other significant grassland species. The site lies on sandstones of the Millstone Grit but is close to limestone so the drift is base-rich in parts.

Cilygroeslwyd Nature Reserve, Llanfair Dyffryn Clwyd [SSSI; NWWT] [1255]
The woodland is on Carboniferous limestone pavement and is famous for its rich flowering plant flora. There are areas of short grassland in an old quarry. The fungus flora is diverse and is especially good for morels and earthstars.

Coed Hafod, near Betws-y-Coed [SNP] [8056, 8057, 8058]
This mixed woodland was visited twice by the British Mycological Society, in 1924 and 1988, and long lists of fungi were produced. It lies on Ordovician mudstones and Wenlock grits on the east side of the Conwy Valley and is in that small area of the Snowdonia National Park that comes into our area.

Cyffylliog valley, Bontuchel [P] [0657, 0757, 0857]
The valley, north of Clocaenog Forest, lies to the west of the Vale of Clwyd, into which it drains. It has been well studied by the Merseyside Naturalists who have produced valuable lists of common species. The site, on the River Clywedog, is upstream of the next site.

Lady Bagot's Drive, Rhewl [P] [0859, 0958, 0959, 1059]
This mixed woodland is more or less continuous downstream from the previous site but differs in being on Carboniferous limestone. The valley is wider here, there is more level ground and the soil is damper; the woodland is more varied.

Llyn Brenig [FE] [9654, 9655, 9656]
This high level conifer forest on the Ludlow shales of the Denbigh Moors is in an area of high rainfall and has an abundance of litter decomposing and mycorrhizal fungi associated with conifers. Grassy areas alongside the forest tracks allow a few grassland agarics to thrive. The forest is younger than the nearby Clocaenog Forest and is not yet as rich mycologically.

Llyn Syberi [P] [7869]
An area of conifer and beech plantation around an acid lake on a ledge above the Conwy valley. The site, on the Wenlock grits, is rich in agarics.

Marford Quarry Nature Reserve [SSSI; NWWT] [3555]
This disused sand quarry in the flood plain of the River Alyn has developing vegetation ranging from birch and conifer scrub to grassland and moss beds on pure sand. The workings of the sand pit have left dune-like mounds and several species of fungi, e.g. *Geoglossum cookeianum*, which are characteristic of coastal dunes, have been found here.

Mostyn Ucha, Llansannan [P] [9367, 9467]
Beech and conifer plantations, over Ludlow shales, on the north side of the valley of the River Aled is rich in agarics in all groups, especially chanterelles.

Tŷ Mawr Country Park, Acrefair [WCBC] [2741, 2841]
An area of grassland, scrub and small copses alongside the River Dee and near to the Pontcysyllte Aqueduct of the Llangollen Canal. The site is close to the junction between

Silurian and Carboniferous systems and consists of shales and sandstones but is influenced by the nearby limestone, below a good depth of rich, alluvial soil. Some of the land has been taken out of agriculture and is nutrient-rich, which will encourage a range of grassland fungi, although waxcaps are unlikely to be important.

World's End [FE] [2347]
A favourite picnic area at the northern end of the limestone cliffs of Eglwyseg Mountain, along the line of Offa's Dyke Path, this site has an interesting mixture of limestone woodland and grassland, together with woodland and moorland on the nearby Millstone Grit and blanket peat. The site has been a regular site for fungus forays since 1910.

Wynnstay Park [Wynnstay Estates] [3041, 3042]
The woodlands in Nant-y-Belan, on the western flank of Wynnstay Park, are oak-dominated ancient woodland, rich in fungi of all groups. Of special interest is the rare rust, *Ochropsora ariae*, on wood anemone. The site is entirely on the Coal Measures and is a few kilometres south of Erddig Park.

v.c. 51

Big Pool Wood, Gronant [NWWT] [1084]
A dense, damp broad-leaved woodland on coastal drift surrounding a small lake. Although the site is small it has a rich and varied fungus flora.

Coed Bach-y-Graig, Tremeirchion [NWWT] [0771]
A small area of ancient woodland on the edge of the flood plain of the River Clwyd, on alluvium overlying Triassic Bunter sandstone, below the limestone scarp on the east side of the Vale of Clwyd. The site is well known for its wild service trees and has a wide range of fungi, many associated with planted beech.

Etna Country Park, Buckley [FCC] [2865]
Developed on a reclaimed landfill pit, in a Buckley Fire Clay quarry, this site has a large area of recently sown grassland, as yet of little ecological interest, surrounded by older woodland of oak, sallow and birch, some on spoil heaps. The woodland supports a rich fungus flora including the first (of two) sites in North East Wales for *Cortinarius violaceus*, the first Welsh record of the rare myxomycete *Lepidoderma chailletii*, and good colonies of *Macrotyphula fistulosa* var. *fistuosla*.

Graig Fawr, Meliden [NT] [0580, 0680]
This botanically rich area of limestone grassland and scrub is directly south of the Bishop's Wood site. It has already produced a good range of limestone fungi.

Y Graig Nature Reserve, Tremeirchion [NWWT] [0872]
Limestone grassland and woodland on the steep scarp on the eastern edge of the Vale of Clwyd provide a good range of uncommon fungi in all groups.

Ysceifiog woodlands [P] [1471]
Beech and ash woodland on limestone pavement, close to the Fisheries Lake and Ddol Uchaf Nature Reserve, with some cattle-grazed grassland in between. The woodland is rich in agarics, including *Agaricus moelleri*, *Cystolepiota moelleri* and *Melanophyllum eyrei*.

Introduction

The forty-six sites more or less mirror the relative distribution of limestone and non-limestone areas within the region and include most of the sites which have hosted day forays during the last fifty years.

1.5 Fungal recording in North East Wales

The earliest records are to be found in Thomas Pennant's account of the history of the Whitford area (Pennant, 1796) and a few species of common fungi are included, as well as some lichenised species. This period saw the dawn of mycology in the UK but it was not until the following century that anything in the way of organised recording began.

John Bowman was a banker in Wrexham and also a fine botanist. He described the myxomycete genus *Enerthenema* in 1830 and recorded the single British species from Erddig. During the second half of the nineteenth century the most skilled and influential mycologists of the day, led by the Rev. Miles Berkeley, the 'father' of British mycology, met regularly in the area between Colwyn Bay and Bodelwyddan and collected and recorded assiduously. Much of the field work centred on the Coed Coch estate and several species were named in honour of the Wynne family who at that time owned the estate. Many rare and poorly known species were found and, fortunately for us, material has been preserved at Kew. The fungus flora of the area was written up by Kew's chief mycologist at the time, Mordecai Cubitt Cooke (Cooke, 1880). Berkeley and his collaborator Broome published many new species first found in our area, in a series of Notices of British Fungi, published in the *Annals and Magazine of Natural History* from 1838 to the 1860s.

The next group of records is provided by the work of the Liverpool Field Naturalists' Club who included fungi in their studies of Welsh sites (Green, 1902). In 1910 the British Mycological Society held both its spring and autumn forays in the region. During the spring foray, based in Chester, they visited Erddig Park, and in the autumn, based in Wrexham, they forayed again at Erddig and also at Erbistock, Wynnstay Park and World's End (Rea, 1911a, 1911b). In autumn 1924 the Society visited Betws-y-Coed and forayed in two sites on the east side of the River Conwy, and therefore in v.c. 50, namely Coed Hafod and the Fairy Glen. The latter site is on the county boundary but the main track, and the only one likely to be used by elderly mycologists in 1924, is on the north bank. The site is included in Caernarvonshire (v.c. 49) by Aron (2005) but at least the records are not forgotten!

The next recording began in the 1950s when a distinguished amateur mycologist from Cheshire, J. Terry Palmer, an expert in gasteroid basidiomycetes and Sclerotiniaceae (ascomycetes) spent many years exploring the Alyn Valley and the Clwydian hills. His material is at Kew and he has kindly made his notebooks available for this book. Local interest from the natural history societies in the area continued at modest level and a useful paper was produced by the Dyserth and District Field Club (Higgins, 1962).

In 1969 the present author made a brief, but life-changing, visit to Loggerheads and recorded many myxomycete species. In 1971 he began a long career as Lecturer in Biology at Chester College of Education (now the University of Chester). In 1972 he and his family came to live in Mold. In 1972 the British Mycological Society, based at Liverpool, visited the Hawarden Woods and Loggerheads and produced the first broad-based list of species from Flintshire (Thomas, 1973). The area around Loggerheads became the focus of important research into the ascomycete fungi associated with sycamore carried out at the University of Liverpool (Bevan & Greenhalgh, 1983). From about the same period much useful collecting of members of the Xylariaceae (also ascomycetes) was started by Professor A. Whalley of Liverpool John Moores University, who also came to live in Flintshire.

The 1970s saw a dramatic increase in interest in fungi in the area. The North Wales Wildlife Trust branches began to organise regular fungus forays which have persisted up to the present day and over the years the following sites have been visited by groups of members, usually led by the author: Alyn Waters Country Park; Big Pool Wood Nature Reserve, Gronant; Big Wood, Hendre; Bishop's Wood, Meliden; Bodrhyddan Hall Estate; Chirk Castle; Cilygroeslwyd Nature Reserve; Clocaenog Forest; Coed Bach-y-Graig Nature Reserve; Coed Clwyd, Moel Famau; Coed y Felin Nature Reserve; Coed y Garth; Graig Fawr; Ddol Uchaf Nature Reserve; Erddig Estate; Foxhall, Denbighshire; Glan-yr-Afon, Marli; Hawarden Woods; Lady Bagot's Drive, Rhewl; Llyn Helyg; Loggerheads Country Park; Marford Quarry Nature Reserve; Minera Lead Mines and Country Park; Nant Gwyn, Denbighshire; Nant-y-Ffrith; Nant Mill Country Park; Pant-yr-Ochain, Gresford; Pen Parc Llwyd Nature Reserve, Henllan; Point of Ayr Nature Reserve; Rhyd-y-Foel, near Abergele; Coed-y-Gopa, Abergele; Vivod Estate, Llangollen; World's End; Wynnstay Hall Estate, Y Graig Nature Reserve and Rhydymwyn Nature Reserve. Several of these were joint meetings with the Dyserth and District Field Club.

Since the 1970s Major Roger Brown, who lives at Marli, has forayed with great skill and success in the area around his home, at Llyn Syberi, in Snowdonia, at Bodelwyddan and along the coast at Point of Ayr. He has found many rare and interesting species. He has sent much of his material to Kew, where it was identified by the late Dr Derek Reid, then Head of Mycology at Kew. Dr Reid visited the area and added several species. The Brown collections are an important resource for all students of fungi in North Wales.

A succession of the author's students at Chester carried out field studies in the area on bark myxomycetes, dictyostelids and aquatic hyphomycetes as part of their degree requirements. These studies produced information that could not be obtained during traditional fungus forays. In 1985 the British Mycological Society returned to Chester and visited Ddol Uchaf, Erddig, Loggerheads, Moel Famau and Point of Ayr. Numerous species were added to the local lists, including several rare species, such as *Cyathus stercoreus*. The records are published on the website of the British Mycological Society's Foray Database.

This period saw the emergence of several Country Parks in the region and the Rangers were keen to introduce the general public to the world of fungi so regular forays were arranged, usually led by the author, at Loggerheads, Nant Mill, Wepre, Tŷ Mawr, Greenfield Valley, Alyn Waters and the Etna Country Parks. These forays added yet more species and to some extent allowed these sites to be monitored. Other forays have been organised for different groups at Bodelwyddan Estate, the Legacy Estate, near Mold, Llyn Gweryd, Llanarmon Dyffryn Ceiriog, Llyn Brenig, Ffrwd Quarry Nature Reserve and Coed Nant Gain Reserve. At the same time the Merseyside Naturalists and the North West Fungus Group forayed at Bontuchel, Celyn Wood, Coed y Felin Reserve, Connah's Quay Reserve, Erddig, Halkyn Mountain, Hawarden Woods and Loggerheads, and their records are incorporated into the species accounts.

In 1988 the British Mycological Society returned to Betws-y-Coed for its Autumn foray, based at the Rhyd-y-Creuau Field Study Centre. Collections were made at Rhyd-y-Creuau, Coed Hafod, Capel Garmon, Clocaenog Forest, Bodnant Gardens, Llyn Syberi and, for a few members en route for Chester, at Bodelwyddan. Again the records are on the database.

In 2003 and 2004 a detailed survey of waxcap grasslands was carried out at selected sites and the experienced field mycologist Liz Holden made many important finds, including several species new to Wales and three species probably new to the United Kingdom.

The present author has been collecting and recording fungi in North Wales since 1969 and a large proportion of the records are his. He is especially interested in myxomycetes, gasteroid basidiomycetes, rusts, smuts, exobasidia, taphrinas and powdery mildews but all groups of fungi have been studied and, when necessary, sent to specialists for identification. The Clwyd Mycological Institute, which the author set up as a consultancy at Mold, housed a large herbarium in which representative material from the region is deposited. This would have

Introduction

been sent to a Mycological Herbarium at the National Botanic Garden of Wales, but the project to develop a mycological centre at the garden has not yet materialised. The material will be sent to Edinburgh if it cannot be housed in Wales. The author left Wales to live in north-west Scotland in 2010 but continued to record in North East Wales until 2019.

Four local naturalists have contributed a great deal to the present account of North East Wales fungi. Joe Phillips of Llanferres has provided numerous records and samples of plant pathogens from all over the region as well as records of larger fungi, for over forty years. Graham Rose of Carmel, near Holywell, an expert photographer, has also been most active, especially with agarics, and has provided extensive lists from sites not visited by the author. Dave Sykes of Mynydd Isa, near Mold, another expert photographer, has an interest in a wide range of fungal groups and has added many new species to the list in recent years. Charles Aron, of Anglesey, began studying fungi in North East Wales in 2006 and in the following years made a number of important discoveries of rare fungi at the remarkable Rhydymwyn Nature Reserve in the Alyn Valley. Together with Roger Brown and Liz Holden these colleagues have made such an impact on our knowledge of the fungi of the region that this book would have been much poorer without their input. In more recent years a local group of keen amateur mycologists, the Cheshire and Clwyd Fungus Group, has been recording in many of the sites named above and added greatly to our knowledge. They are listed in the main list of recorders.

The number of species of each group recorded in North East Wales and in v.c. 50 and 51 is shown in the following table.

Table 1. Species numbers in North East Wales and in vice-counties 50 and 51

Group	No. of species in North East Wales	No. of species in v.c. 50	No. of species in v.c. 51
Myxomycota	186	179	159
Plasmodiophoromycota	2	0	2
Peronosporomycota	51	34	45
Chytridiomycota	4	2	3
Glomeromycota	1	0	1
Zygomycota	21	9	18
Ascomycota	811	505	687
Basidiomycota – Agaricomycetes, etc.	1,281	1,090	1,017
Pucciniomycetes	147	113	129
Ustilaginomycetes	30	17	27
Total	2,534	1,949	2,088

In general the microfungi have been less well recorded in Denbighshire than in Flintshire.

1.6 Sources of records

The bulk of the records have been made during the extensive forays described in the previous section; lists of identified species have been kept and checked. Where material could not be named with confidence with the literature and herbarium specimens available (see 1.8), material has been sent to appropriate experts for naming (see the list of specialists on p. vi). Records have been made by other mycologists who are listed above. The herbarium collections at the Royal Botanic Gardens, Kew have been studied and these records are clearly of high quality. Records from the literature (see 1.11) have been treated with caution in the absence of specimens. In many cases they have been rejected as unacceptable but some, of historic interest, are included with a suitable comment.

1.7 List of individual collectors

The following persons have contributed individual records and material for study during the course of the preparation of this fungus flora, viz. 1969 to 2020:

> Charlotte Anderson, Charles Aron, Henry Beker, Dave Bennett, Rod Bevan, Roger Brown, Arthur Chater, Keith Davies, Paul Day, Elizabeth Dearden, R.W.G. Dennis, Debbie Evans, Sarah Furse, John Gambles, Clive Garnett, Angela Gorlon, Andrew Graham, Janet Graham, Ted Green, Geoff Greenhalgh, Liz Holden, Bruce Ing, Eleanor Ing, Ken Jordan, Betty Lee, Gareth Lewis-Ing, Roy Mantle, Chris McCracken, Don McNeil, Angela Medlam, Emily Meilleur, James Morris, Lara Morris, Adrian Newton, Julian Newton, Terry Palmer, Joe Phillips, Michael Pilkington, Ron Poole, R.D. Pryce, John Ratcliffe, Derek Reid, Graham Rose, Julie Rose, Shan Runagall, Stuart Smith, Geoffrey Spencer, Dave Sykes, David Toyne, Dave Higginson-Tranter, John Watt, Anthony Whalley, D. Winnard, Roger Wood J.P. Woodman and Ray Woods. In addition numerous members of the North Wales Wildlife Trust, the National Trust, the British Trust for Conservation Volunteers, the Merseyside Naturalists' Association, the North West Fungus Group, the Dyserth Field Club, the British Mycological Society and the general public have contributed specimens at forays on innumerable occasions.

1.8 Classification and identification of fungi

Fungi are a heterogeneous collection of organisms which share only their mode of nutrition – heterotrophism – in which they secrete enzymes which digest their food, mostly dead plant material, outside their bodies and they then absorb the nutrients in solution. They range from single-celled amoeboid or flagellate cells to complex multicellular fruit bodies which may originate from a vast network of tubes – the hyphae – which may form a mycelium occupying several square kilometres of soil and live for centuries. They occupy soil of all kinds, dead and living plants and animals, other fungi, freshwater and the sea. They may even penetrate rock and their spores are dispersed by the wind, by animal life and by water. The wide range of reproductive organs and the details of their microscopic structure and biochemistry has led to the realisation that the organisms we call fungi belong to a number of separate evolutionary lines and should be classified in several different kingdoms, none of which includes plants, with which fungi have traditionally been grouped.

Modern studies using the electron microscope have elucidated the differences and similarities of structure, while molecular studies, including DNA analyses, have brought about a better understanding of the relationships between different groups at all levels of classification.

All the organisms in this book have been treated as fungi in the past but can now be classified into at least six distinct kingdoms. The first, Amoebozoa, typified by *Amoeba*, includes groups of slime moulds, which have an amoeboid cell at some stage in their life cycle, in the Phylum Myxomycota, also known as Mycetozoa, including the cellular and plasmodial slime moulds (Protosteliomycetes, Dictysteliomycetes, Ceratiomyxomycetes and Myxomycetes). The club-root fungi are placed in the Phylum Plasmodiophoromycota, which may belong to

Introduction

this kingdom. The acrasid slime moulds are not closely related to the others and are placed in the Phylum Heterolobosea, which may represent a third kingdom.

The Phylum Peronosporomycota includes the blights, downy mildews and some primitive aquatic fungoid organisms which have cellulose cell walls and reproduction via large egg cells and small, swimming male cells; they have recently been transferred from the kingdom Chromista, which includes a range of coloured algal organisms, to the kingdom Chromoalveolata (also known as Straminopila). These are often called Pseudofungi or Oomycetes. The groups so far considered are studied mostly by mycologists, using mycological methods, and may be regarded as 'honorary' fungi. The groups of fungi are now named according to the rules governing nomenclature of the kingdoms they inhabit. However, because they were all originally described as fungi, the traditional fungal groups and names are used here. To transfer nomenclature to the Animal Kingdom would involve the renaming of a large number of genera in both animal and plant codes!

The kingdom Ciliofungi contains 'true' fungi, in that they have chitin cell walls, but they have ciliated cells and little development of hyphae or a mycelium. They comprise the Phylum Chytridiomycota, most of which are either parasites of diatoms and other planktonic algae, or of terrestrial flowering plants.

The true fungi, with chitin walls to the hyphal system and no flagellated cells in the life cycle, include all the remaining organisms which are recognisable as fungi by the non-specialist, and make up the kingdom Fungi.

The Phylum Zygomycota includes the familiar bread moulds, various dung fungi and several groups of unusual symbiotic fungi found in the gut of arthropods as well as important insect parasites. All are characterised by a stage in the life cycle in which two hyphal tips conjugate and form a thick-walled resting spore. The other spore stages are produced in thin-walled sporangia.

The recently separated Phylum Glomeromycota includes a small number of vastly important species which are mostly single-celled symbionts inside the root cells of terrestrial plants, forming the vesicular-arbuscular mycorrhizas, without which their hosts could not absorb water or most nutrients.

The Ascomycota is the largest phylum of fungi and its classification has undergone major revisions in recent years and is still unsettled. The system used here is based on the 9th edition of the *Dictionary of the fungi* (Kirk, Cannon, David & Stalpers, 2001) which is the most accessible source to non-specialists. Over half of all ascus-forming fungi are lichenised (i.e. are lichens) but these have not been studied intensively in the region and have not been included in the present account. Ascomycetes occupy all ecosystems where fungi are found and are decomposers, symbionts and parasites of a wide range of materials.

The basidia-forming fungi, or Basidiomycota, have been the most intensively studied by non-professional mycologists and account for the majority of records made in the area. The Pucciniomycetes include not only the rust fungi but some parasites of other fungi and also some species which traditionally have been regarded as flower smuts, in the order Microbotryales. These have been shown to be genetically closer to the rusts than the rest of the smuts. The Ustilaginomycetes, as well as embracing the rest of the smut fungi, also include the Exobasidiales, which gall plants of the heather family in our area, and a number of yeast-like fungi.

The Basidiomycetes have been the subject of much revision based on DNA relationships, with surprising and, to the layman, unlikely results. The classification followed here, is based on the 9th edition of the *Dictionary of the Fungi* and *Funga Nordica* (Knudsen & Vesterholt, 2008). More recent molecular studies have brought about great changes in generic concepts, especially in the boletes and the waxcaps. The current (new) names are used here. The presence of gills, pores or other spore-bearing surfaces is no longer a sign of common ancestry.

Most major groups of basidiomycetes have diversified along similar structural lines from different genetic starting points. The traditional division between mushrooms, brackets, puffballs and corals is no longer maintained as a result of these molecular investigations. The classification is changing at a greater rate than in the past. The views shown here may change!

Mitosporic or anamorphic fungi, also known as imperfect fungi, are those for which there is a sexually produced spore stage that may be rare, or absent, in Nature. Modern techniques enable these fungi to be assigned to a sexual genus, family, order, class or sub-phylum and they are so placed in this list. They are mostly ascomycetes but a few anamorphic basidiomycetes are also included. They may continue to be known by their anamorph (imperfect) names in the absence of the perfect stage. In the appropriate sections of the species account these fungi are listed by their anamorph names, after the entries for their teleomorph (sexual spore-bearing) relatives.

Nomenclature
The names given to fungi are governed by strict rules under the *International Code of Botanical Nomenclature*. In recent years there have been considerable changes to the names given to familiar fungi. This is partly due to a revision of the genera or families in which they are placed but also many familiar names are technically incorrect because they are not the earliest valid name published for the species. In a few cases well established names, when they are incorrect, may be conserved to save the confusion and irritation of a change. The lists followed here are given under the major groups.

Myxomycetes: Ing, 1999; Lado, 2001.
Dictyosteliomycetes: Hagiwara, 1989; Raper, 1984.
Peronosporomycota: Chater et al., 2020; Preece, 2002.
Chytridiomycota: Ellis & Ellis, 1985.
Glomeromycota: Pegler, Spooner & Young, 1993.
Zygomycota: Ellis & Ellis, 1988; Waterhouse & Brady, 1982.
Ascomycota: Cannon, Hawksworth & Sherwood-Pike, 1985; Ellis & Ellis, 1985 and 1988.
Basidiomycota: Legon & Henrici, 2005 plus online updates; Knudsen & Vesterholt, 2008.
Anamorphic fungi: Ellis & Ellis, 1985 and 1988.

Most of these lists are kept updated on the British Mycological Society's website and also on www.britishfungi.info. Current names, and synonyms, may also be checked by consulting *Index fungorum* (online).

Identification
In order to identify fungi with confidence it is usually necessary to examine a number of anatomical structures under the microscope, often using special stains. It is rarely possible to reach an accurate identification in the field or solely by the use of pictures. Among the numerous field guides now available only three are comprehensive enough to include here, together with valuable series with good descriptions and illustrations. The following have been consulted:

General works: Bon, 1987; Courtecuise & Duhem, 1995; Kibby, 2017; Phillips, 2006;
Richardson & Watling, 1997; Laessoe & Petersen, 2019.
Series: *Fungi of Switzerland* – 6 vols (Breitenbach & Kränzlin, 1984-2006);
Nordic Macromycetes – 3 vols (Hansen & Knudsen, 1992–2000);
Flora Agaricina Neerlandica – 5 vols (Bas, Kuyper, Noordeloos & Vellinga, 1988–2001).
Funga Nordica (Knudsen & Vesterholt, 2008).
Fungi of Temperate Europe (Laessoe & Petersen, 2019).

Introduction

Single works: Mushrooms and Toadstools of Britain and Europe (Kibby, 2017, 2010).
Myxomycetes: Ing, 1999.
Dictyosteliomycetes: Raper, 1984; Hagiwara, 1989.
Peronosporomycota: Ellis & Ellis, 1985; Preece, 2002.
Chytridiomycota: Ellis & Ellis, 1985.
Glomeromycota: Pegler, Spooner & Young, 1993.
Zygomycota: Ellis & Ellis, 1988; Waterhouse & Brady, 1982.
Ascomycota: Ellis & Ellis, 1985 and 1988; Braun, 1995; Braun & Cook, 2012; Dennis, 1960; Pegler, Spooner & Young, 1993.
Anamorphic fungi: Ellis & Ellis, 1985 and 1988; Ingold, 1975.
Basidiomycota – Pucciniomycetes: Wilson & Henderson, 1966.
- *Ustilaginomycetes*: Mordue & Ainsworth, 1984; Nannfeldt, 1981;
- Spooner & Legon, 2006.
- *Basidiomycetes – Agaricales*: Antonin & Noordeloos, 1993, 1997; Boertman, 1995; van Waveren, 1985; Orton, 1986; Orton & Watling, 1979; Watling, 1982; Watling & Gregory, 1987, 1989, 1993; Kuyper, 1986; Senn-Irlet, 1995; Vesterholt, 2005.
- *Boletales*: Watling & Hills, 2005.
- *Cantharellales etc*: Pegler, Roberts & Spooner, 1997; Watling & Turnbull, 1998.
- *Gasteroid spp.*: Pegler, Laessoe & Spooner, 1995.
- *Polyporales*: Ryvarden & Gilbertson, 1993–1994.
- *Russulales*: Heilmann-Clausen, Verbeken & Vesterholt, 1998; Rayner, 1985; Hjortstam, Larsson & Ryvarden, 1987.

In addition numerous articles in mycological journals have been consulted to check species newly arrived in the United Kingdom.

For full details of all these references consult the Bibliography in 1.11.

1.9 Conservation of fungi

Fungi play such a vital role in the functioning of terrestrial ecosystems and the survival of such systems often depends on the continuing good health of the associated fungal populations. This is especially true of forest ecosystems where the reduction in the populations of mycorrhizal fungi threatens the long-term health of the entire forest. For instance the outbreaks in recent decades of forest death or *Waldsterben* in montane forest in Central Europe, triggered by photochemical smog and ozone generated by local road vehicles, can be traced to a much earlier incidence of acid deposition in rain and snow, notably when nitric and sulphuric acids are involved. This changed the soil environment adversely and initiated the decline of mycorrhizal species such as *Cantharellus* and *Cortinarius* species. This may have severely reduced the ability of the trees to resist later severe pollution episodes. This phenomenon has been observed in our area above 300 m on Nercwys Mountain near Mold, in *Tsuga heterophylla* plantings, but does not appear to have been reported elsewhere in North Wales. Climate change is also affecting the occurrence of fungi, by changing the habitat and vegetation, but the arrival of species from warmer lands is not yet causing problems.

Because of their ecological importance, fungi are now covered by wildlife conservation legislation and included in Biodiversity Action Plans at national and local level. Only one of our North East Wales species, *Hericium erinaceum*, is protected under the Wildlife and Countryside Act and it is an offence to collect specimens without a licence. A revised Red Data List of endangered British fungi is now available (Evans, Henrici, Ing & Rotheroe, 2007). A Red List for selected species of basidiomycetes was prepared by Smith, Suz & Ainsworth in 2016. More recently lists for rusts and smuts have been prepared specifically for Wales by

Woods et al. (2015, 2018). A national Red List for myxomycetes is in press (Ing, 2020). Those species recorded in North East Wales which appear in these recent lists are indicated with the appropriate threat levels. The abbreviations used are as follows:

RDL – Red Data List
EX – probably extinct (not recorded anywhere for fifty years)
RE – regionally extinct (not recorded in Wales for fifty years)
CR – critically endangered (known only from one or two sites in the UK)
EN – endangered (currently known from five or fewer sites in the UK, having been more widespread in the past)
VU – currently known from five or fewer sites, either having been more widespread in the past, or known from few sites, but without evidence of decline; vulnerable
NT – known from few sites but not in serious decline, near threatened
LC – common and widespread species giving least concern
DD – data deficient and requiring more research to assess their threat levels

It is imperative that species which are featured on the Red Lists should not be deliberately collected and their presence should be used to increase the protection given to sites where they occur, including appropriate management. Most of the species are easy to identify so do not need to be picked in any quantity.

1.10 Layout of entries in the species account

In order to reduce space and words the species entries are set out in standard form. The generalised form is given below.

Currently accepted name of species with author citation *English names (if available)*
Synonyms with author citations
Habitat and/or substrates in North East Wales; general frequency in the British Isles.
Vice-counties from which the species is recorded. 10 km squares of the National Grid from which the species is recorded. Code letters for important sites in which the species is known to occur (see 1.4). (Total number of localities from which the species is known in the area.) Year of first record. (Whether the species is known from v.c. 48 (Merioneth), 49 (Caernarfon) or 52 (Anglesey.) [This information taken from Aron, 2005 and recent collecting.] **Red Data List/threat level** (where appropriate; see 1.9). Notes of special occurrence, decline or spread. Where there is only one known site for a species in the region the locality and date are given.

Anamorphic forms of ascomycetes and basidiomycetes are listed in the same taxonomic group as their sexual relatives but their names are placed in normal, rather than bold, type.

Notes:
1. The currently accepted name is taken from the most recent major work as listed in 1.8 and 1.11.
2. English names are those in the 'official' list edited by Holden (2005). Where the English name used here is in parentheses it is not in the official list but is likely to be included in future editions.
3. Synonyms included allow the reader to relate unfamiliar 'new' names to those in standard texts and field guides.
4. Habitats and substrates may be broader outside our region; the frequency details are very general as detailed distribution data is available for very few species.

5. The botanical names for associated native or naturalised plants are in accordance with Stace (2019).
6. The botanical names for associated cultivated plants follow the current editions of *Plant Finder*, published for the Royal Horticultural Society.

1.11 Bibliography

The list that follows includes the references cited in the various sections of the introduction and un-cited references that have been used as sources of records and other relevant information.

Antonin, V. & Noordeloos, M.E. 1993. *A Monograph of Marasmius, Collybia and Related Genera in Europe*, Pt. 1. München, IHW Verlag.
Antonin, V. & Noordeloos, M.E. 1997. *A Monograph of Marasmius, Collybia and Related Genera in Europe*, Pt. 2. München, IHW Verlag.
Aron, C. 2005. *The Fungi of North West Wales*. Pentraeth, Aron.
Bas, C., Kuyper, T.W., Noordeloos, M.E. & Vellinga, E.C. 1988. *Flora Agaricina Neerlandica*, Vol. 1: *Entolomataceae*. Rotterdam, Balkema.
Bas, C., Kuyper, T.W., Noordeloos, M.E. & Vellinga, E.C. 1990. *Flora Agaricina Neerlandica*, Vol. 2: *Pleurotaceae, Pluteaceae and Tricholomataceae*, Pt. 1. Rotterdam, Balkema.
Bas, C., Kuyper, T.W., Noordeloos, M.E. & Vellinga, E.C. 1995. *Flora Agaricina Neerlandica*, Vol. 3: *Tricholomataceae*, Pt. 2. Rotterdam, Balkema.
Bas, C., Kuyper, T.W., Noordeloos, M.E. & Vellinga, E.C. 1999. *Flora Agaricina Neerlandica*, Vol. 4: *Strophariaceae, Tricholomataceae*, Pt. 3. Rotterdam, Balkema.
Bendelow, V.C. & Hartnup, R. 1980. *Climatic Classification of England and Wales*. Soil Survey Technical Monograph **15**. Harpenden, Soil Survey.
Bevan, R.S. & Greenhalgh, G.N. 1983. Pyrenomycetes and loculoascomycetes on sycamore wood and bark in the northwest of England. *Transactions of the British Mycological Society* **80**, 83–89.
Boertmann, D. 1995. *Fungi of Northern Europe*, Vol. 1: *The Genus Hygrocybe*. Copenhagen, Danish Mycological Society.
Bon, M. 1987. *The Mushrooms and Toadstools of Britain and North West Europe*. London, Hodder & Stoughton.
Bowman, J.E. 1830. Account of a new plant of the gastromycetous order of fungi. *Transactions of the Linnean Society of London* **16**, 151–154.
Braun, U. 1995. *The Powdery Mildews (Erysiphales) of Europe*. Jena, Fischer Verlag.
Braun, U. & Cook, R.T.A. 2012. *Taxonomic Manual of the Erysiphales (Powdery Mildews)*. Utrecht, Netherlands, CBS.
Breitenbach, J. & Kränzlin, F. 1984. *Fungi of Switzerland*, Vol. 1: *Ascomycetes*. Luzern, Mycological Society of Lucerne.
Breitenbach, J. & Kränzlin, F. 1986. *Fungi of Switzerland*, Vol. 2: *Non-gilled Fungi*. Luzern, Mycological Society of Lucerne.
Breitenbach, J. & Kränzlin, F. 1991. *Fungi of Switzerland*, Vol. 3: *Boletes and Agarics*, Pt. 1. Luzern, Mycological Society of Lucerne.
Breitenbach, J. & Kränzlin, F. 1995. *Fungi of Switzerland*, Vol. 4: *Agarics*, Pt. 2. Luzern, Mycological Society of Lucerne.
Breitenbach, J. & Kränzlin, F. 2000. *Fungi of Switzerland*, Vol. 5: *Cortinariaceae*. Luzern, Mycological Society of Lucerne.
Breitenbach, J. & Kränzlin, F. 2006. *Fungi of Switzerland*, Vol. 5: *Russulaceae*. Luzern, Mycological Society of Lucerne.

Brown, R. 2007. First British record of yellow *Sarcoscypha*. *Field Mycology* **8**, 9–12.
Butterfill, G.B. & Spooner, B.M. 1995. *Sarcoscypha* (Pezizales) in Britain. *Mycologist* **9**, 20–26.
Cannon, P.F., Hawksworth, D.L. & Sherwood-Pike, M.A. 1988. *The British Ascomycotina – An Annotated Checklist*. Farnham Royal, Commonwealth Mycological Institute.
Chater, A.O., Woods, R.G., Stringer, R.N., Evans, D.A. & Smith, P. 2020. *Downy Mildews and White Blister-rusts of Wales*. Aberystwyth, Chater.
Cooke, M.C. 1880. Coed Coch and Colwyn fungi. *Grevillea* **9**, 75–79.
Courtecuise, R. & Duhem, B. 1995. *The Mushrooms and Toadstools of Britain and Europe*. London, Harper Collins.
Dallman, A.A. 1938. Halophytic Uredineae. *North-western Naturalist* **10**, 268.
Dennis, R.W.G. 1960. *British Cup Fungi and their Allies*. London, Ray Society.
Dennis, R.W.G. 1968. *British Ascomycetes*. Vaduz, Cramer.
Ellis, M.B. & Ellis, J.P. 1985. *Microfungi on Land Plants*. London, Croom Helm (reprinted by Richmond Publishing, Slough).
Ellis, M.B. & Ellis, J.P. 1988. *Microfungi on Miscellaneous Substrates*. London, Croom Helm (reprinted by Richmond Publishing, Slough).
Ellis, M.B. & Ellis, J.P. 1990. *Fungi Without Gills*. London, Chapman & Hall.
Evans, D. 2006. *Results of a Mycological Survey of Chirk Castle Grasslands*. Unpublished report to the Countryside Council for Wales.
Evans, S., Henrici, A., Ing, B. & Rotheroe, M. 2007. Red Data List of Threatened British Fungi. Available online at https://www.britmycolsoc.org.uk/mycology/conservation/red-data-list.
Green, C.T. 1902. A preliminary index of local fungi, mainly from Wirral. *Proceedings of the Liverpool Field Naturalists' Club*, **1901**, 19–28.
Green, J.A. 2006. *The Flowering Plants and Ferns of Denbighshire*. Tremeirchion, Green.
Grove, W.B. 1935. *British Leaf and Stem Fungi*. 2 vols. Cambridge, Cambridge University Press. Reprinted in 1967 by J. Cramer, Lehre.
Hagiwara, H. 1989. *The Taxonomic Study of Japanese Dictyostelid Cellular Slime Moulds*. Tokyo, National Science Museum.
Hansen, L. & Knudsen, H. 1992. *Nordic Macromycetes*, Vol. 2: *Polyporales, Boletales, Agaricales, Russulales*. Copenhagen, Nordsvamp.
Hansen, L. & Knudsen, H. 1997. *Nordic Macromycetes*, Vol. 3: *Heterobasioid, Aphyllophoroid and Gastromycetoid Basidiomycetes*. Copenhagen, Nordsvamp.
Hansen, L. & Knudsen, H. 2000. *Nordic Macromycetes*, Vol. 1: *Ascomycetes*. Copenhagen, Nordsvamp.
Heilmann-Clausen, J., Verbeken, A. & Vesterholt, J. 1988. *Fungi of Northern Europe* Vol. 2: *The Genus Lactarius*. Copenhagen, Danish Mycological Society.
Higgins, L. 1962. Fungi. *Proceedings of the Dyserth and District Field Club* **1961**, 53–54.
Hills, A.E. 2008. The genus *Xerocomus*. *Field Mycology* **9**, 77–96.
Hjortstam, K., Larsson K.-H. & Ryvarden, L. 1987. *The Corticiaceae of Northern Europe*, Vols 1–8. Oslo, Fungiflora.
Holden, E. 2005. *Recommended English Names for Fungi in the United Kingdom*. Salisbury, Plant Life International.
Hughes, J. 1992. Botanical recording at Iscoyd Park. *Proceedings of the Dyserth and District Field Club* **1991**, 63–68.
Hughes, J. & Phillips, J. 1992. Bachygraig woodland trail, Tremeirchion. *Proceedings of the Dyserth and District Field Club* **1991**, 41–46.
Ing, B. 1978. New and interesting fungi in north Wales. *Nature in Wales* **16**, 91–94.
Ing, B. 1980. New and interesting fungi in north Wales – 2. *Nature in Wales* **17**, 104–106.

Ing, B. 1999. *The Myxomycetes of Britain and Ireland*. Slough, Richmond Publishing. (reprinted 2020).

Ing, B. & Phillips, J. 1991. Fungus foray in Bishopwood (Coed-yr-Esgob), Prestatyn, October 1990. *Proceedings of the Dyserth and District Field Club* **1990**, 25–28.

Ing, B. & Spooner, B. 2002. The horse chestnut powdery mildew Uncinula flexuosa in Europe. *Mycologist* **16**, 112–113.

Ingold, C.T. 1975. *An Illustrated Guide to Aquatic and Water-Borne Hyphomycetes (Fungi Imperfecti) With Notes on Their Biology*. Scientific Publication No. 30. Ambleside, Freshwater Biological Association.

Kibby, G. 2000. A user-friendly key to the genus *Leccinum* in Great Britain. *Field Mycology* **1**, 20–29.

Kibby, G. 2001. Synoptic keys to the British species of the genus *Russula*. Unpublished keys – available from the author.

Kibby, G. 2006. *Leccinum* revisited – a new synoptic key to species. *Field Mycology* **7**, 113–122. This supersedes Kibby, 2000.

Kibby, G. 2007. The genus *Russula* in Great Britain with synoptic key to species. Unpublished Keys – available from the author; this supersedes Kibby, 2001.

Kibby, G. 2017, 2020. *Mushrooms and Toadstools of Britain & Europe*. Vol. 1, ed. 2. London, Kibby.

Kirk, P.M., Cannon, P.F., David, J.C. & Stalpers, J.A. 2001. *Dictionary of the Fungi*, ed. 9. Wallingford, CAB International.

Knudsen, H. & Vesterholt, J. (eds.) 2008. *Funga Nordica*. Copenhagen, Nordsvamp.

Kuyper, T.W. 1986. A revision of the genus *Inocybe* in Europe I. *Persoonia*, Supp. **3**, 1–247.

Lado, C. 2001. Nomenmyx – a nomenclatural taxabase of myxomycetes. *Cuardenos de trabaja de Flora Micologica Iberica* **16**, Madrid, Real Jardín Botánico.

Laessoe, T. & Petersen, J.H. 2019. *Fungi of Temperate Europe*. 2 vols. Princeton, Princeton University Press.

Legon, N.W. & Henrici, A. 2005. *Checklist of the British and Irish Basidiomycota*. Kew, Royal Botanic Gardens.

Lister, G. 1926. Mycetozoa found during the Bettws-y-coed foray. *Transactions of the British Mycological Society* **10**, 240–242.

McNeil, D. 2012. *Clathrus archeri* from the North West's area... Just! *Newsletter, North West Fungus Group*, **April 2012**, 14–15.

Moore, W.C. 1959. *British Parasitic Fungi*. Cambridge, Cambridge University Press.

Mordue, J.E.M. & Ainsworth, G.C. 1984. *Ustilaginales of the British Isles*. Kew, Commonwealth Mycological Institute.

Nannfeldt, J.A. 1981. *Exobasidium* – a taxonomic reassessment applied to the European species. *Symbolae Botanicae Upsaliensis* **23**, 1–71.

Noordeloos, M.E., Kuyper, T.W. & Vellinga, E.C. 2001. *Flora Agaricina Neerlandica* Vol 5: *Agartcaceae*. Lisse, Balkema.

Orton, P.D. 1986. *British Fungus Flora* Vol. 4: *Pluteaceae*. Edinburgh, Royal Botanic Garden.

Orton, P.D. & Watling, R. 1979. *British Fungus Flora* Vol. 1: *Coprinus*. Edinburgh, HMSO.

Pegler, D.N., Laessoe, T. & Spooner, B.M. 1995. *British Puffballs, Earthstars and Stinkhorns*. Kew, Royal Botanic Gardens.

Pegler, D.N., Roberts, P.J. & Spooner, B.M. 1997. *British Chanterelles and Tooth Fungi*. Kew, Royal Botanic Gardens.

Pegler, D.N., Spooner, B.M. & Young, T.W.K. 1993. *British Truffles*. Kew, Royal Botanic Gardens.

Pennant, T. 1796. *The History of the Parishes of Whiteford and Holywell*. B. and J. White, London.

Phillips, J. 1990. Fungus foray at Loggerheads Country Park, October 1989. *Proceedings of the Dyserth and District Field Club* **1989**, 28–30.

Phillips, J. 1992. Fungus foray: Coed Felin-Blwm and Coed Garth, Ffynongroew, October 1991. *Proceedings of the Dyserth and District Field Club* **1991**, 247–249.

Phillips, J. 1994a. Fungus foray: Coed Felin Blwm and Coed Garth, Ffynnongroew, October 1991. *Proceedings of the Dyserth and District Field Club* **1991**, 71–72.

Phillips, J. 1994b. Fungus foray: Bachygraig Woodland Trail, Tremeirchion, October 1992. *Proceedings of the Dyserth and District Field Club* **1993**, 73–75.

Phillips, J. 1995. Fungus foray: Lady Bagot's Drive, Rhewl, near Ruthin, October 1994. *Proceedings of the Dyserth and District Field Club* **1994**, 29–34.

Phillips, J. 1996. Fungus foray: Talacre sand dunes, October 1995. *Proceedings of the Dyserth and District Field Club* **1995**, 53–56.

Phillips, J. 1997. Fungus foray: Bodelwydden Castle Grounds, November 1996. *Proceedings of the Dyserth and District Field Club* **1996**, 74–79.

Phillips, R. 2006. *Mushrooms*. London, Macmillan.

Preece, T.F. 2002. *A Checklist of the Downy Mildews (Peronosporaceae) of the British Isles*. Kew, British Mycological Society.

Raper, K.B. 1984. *The Dictyostelids*. Princeton, Princeton University Press.

Rayner, R.W. 1985. *Keys to the British Species of Russula*. Stourbridge, British Mycological Society.

Rea, C. 1911a. The Chester Spring Foray. *Transactions of the British Mycological Society* **3**, 233–238.

Rea, C. 1911b. The Wrexham Foray. *Transactions of the British Mycological Society* **3**, 239–249.

Richardson, M.J. & Watling, R. 1997. *Keys to Fungi on Dung*. Stourbridge, British Mycological Society.

Roberts, P. & Spooner, B. 2000. *Nodulisporium cecidiogenes*: a gall-causing fungus new to Britain. *Mycologist* **14**, 177–178.

Ryvarden, L. & Gilbertson, R.L. 1993–1994. *European Polypores* Vols, 1 & 2. Oslo, Fungiflora.

Senn-Irlet, B. 1995. The genus *Crepidotus* (Fr.) Staude in Europe. *Persoonia* **16**, 11–80.

Smith, B. & George, T.N. 1961. *British Regional Geology – North Wales*, ed. 3. London, HMSO.

Smith, J.H., Suz, L.M. & Ainsworth, M. 2016. Red List of fungi for Great Britain: Bankeraceae, Cantharellaceae, Geastraceae, Hericiaceae and selected genera of Agaricaceae and Fomitopsidaceae. Natural England.
http://fungi.myspecies.info/sites/fungi.myspecies.info/files/Smith%20et%20al.%20%282015%29.pdf

Spooner, B.M. & Kemp, S.L. 2005. *Epichloë* in Britain. *Mycologist* **19**, 82–87.

Spooner, B.M. & Legon, N.W. 2006. Additions and amendments to the list of British smut fungi. *Mycologist* **20**, 90–96.

Stace, C. 2019. *New Flora of the British Isles*, ed. 4. Stowmarket, C & M Floristics.

Thomas, A. 1973. Autumn foray, Liverpool. *Bulletin of the British Mycological Society* **7**, 52–58.

Thompson, T. & Highes, J. 1990. Conservation on the farm: Pant-yr-Ochain. *Proceedings of the Dyserth and District Field Club* **1989**, 35–37.

Van Waveren, E. Kits, 1985. The Dutch, French and British species of *Psathyrella*. *Persoonia*, Supplement **2**, 1–306.

Vesterholt, J. 2005. *Fungi of Northern Europe*, Vol. 3: *The Genus Hebeloma*. Copenhagen, Danish Mycological Society.

Wakefield, E.M. 1926. The Bettws-y-coed foray. *Transactions of the British Mycological Society* **10**, 233–239.

Waterhouse, G.M. & Brady, B.L. 1982. Key to the species of *Entomophthora* sensu lato. *Bulletin of the British Mycological Society* **16**, 113–143.

Watling, R. 1982. *British Fungus Flora*, Vol. 3: *Bolbitiaceae*. Edinburgh, HMSO.

Introduction

Watling, R. & Gregory, N.M. 1987. *British Fungus Flora,* Vol. 5: *Strophariaceae and Coprinaceae p.p.* Edinburgh, Royal Botanic Garden.

Watling, R. & Gregory, N.M. 1989. *British Fungus Flora,* Vol. 6: *Crepidotaceae, Pleurotaceae and other Pleurotoid Agarics.* Edinburgh, Royal Botanic Garden.

Watling, R. & Gregory, N.M. 1993. *British Fungus Flora,* Vol. 7: *Coretinariaceae p.p.* Edinburgh, Royal Botanic Garden.

Watling, R. & Hills, A.E. 2005. *British Fungus Flora,* Vol. 1: *Boletes and their Allies,* ed. 2. Edinburgh, Royal Botanic Garden.

Watling, R. & Turnbull, E. 1998. *British Fungus Flora,* Vol. 8: *Cantharellaceae, Gomphaceae etc.* Edinburgh, Royal Botanic Garden.

Watson, H.C. 1847–59. *Cybele Britannica.* London, Quaritch.

Wild, M. 1991. The natural history of Dyserth churchyard. *Proceedings of the Dyserth and District Field Club* **1990**, 49–56.

Wilson, M. & Henderson, D.M. 1966. *British Rust Fungi.* Cambridge, Cambridge University Press.

Woods, R.G., Stringer, N., Evans, D.A. & Chater, A.O. 2015. *Rust Fungus Red Data List and Census Catalogue for Wales.* Aberystwyth, Chater.

Woods, R.G., Chater, A.O., Smith, P.A., Stringer, N. & Evans, D.A, 2018. *Smut and Allied Fungi of Wales. A Guide, Red Data List and Census Catalogue.* Aberystwyth, Chater.

Wynne, G. 1993. *Flora of Flintshire.* Denbigh, Gee.

PART 2

SPECIES ACCOUNTS

THE FUNGI OF NORTH EAST WALES

Phylum HETEROLOBOSEA
Class ACRASIOMYCETES
Order ACRASIDALES
Family Guttulinaceae

Pocheina rosea (Cienk.) A.R. Loebl, & Tappan
Guttulina rosea Cienk.
Frequent on the bark of living trees, especially those affected by acid pollution or with naturally acid bark.
50, 51. SH 77, 86, 95–97; SJ 04, 13–16, 23–26, 34, 43. BO, ER, LO, MF, MO, NF. (19) 1977. (48, 49, 52)

Phylum MYXOMYCOTA
Class PROTOSTELIOMYCETES
Order PROTOSTELIALES
Family Protosteliaceae

Protostelium arachisporum Olive & Stoian.
On the lichen *Parmelia sulcata* Taylor, on birch bark in moist chamber culture; rare.
50. Recorded only in our area from Pot Hole, near Eryrys (SJ 26) in 1978.

Class CERATIOMYXOMYCETES
Order CERATIOMYXALES
Family Ceratiomyxaceae

Ceratiomyxa fruticulosa (O.F. Müll.) T. Macbr.
Famintzinia fruticulosa (O.F. Müll.) Lado
Very common on rotten wood, especially of conifers, from early summer to autumn.
50, 51. SH 76, 87, 95; SJ 02, 05, 07, 08, 13, 15-18, 23-26, 34, 36. AV, BP, CC, CF, ER, GW, LH, LO, MF, MO, NF, WW. (32) 1910. (48, 49, 52) **RDL/LC**

Class DICTYOSTELIOMYCETES
Order DICTYOSTELIALES
Family Dictyosteliaceae

Dictyostelium brefeldianum Hagiwara
Dictyostelium mucoroides sensu Norberg
Isolated from soil and cow dung; common and widespread.
50, 51. SH 95; SJ 05, 06, 15-17, 25, 26. CE, CF, HM, LO, MF, WW. (13) 1978.

Dictyostelium giganteum Singh
Isolated from soil; frequent in temperate regions.
50, 51. SJ 05, 16, 26. CF, MF. (4) 1987.

Dictyostelium minutum Raper
Isolated from soil; frequent in forest soils.
50, 51. SH 95; SJ 05, 06, 15-17, 26. CE, CF, LO, MF, WW. (12) 1987.

Dictyostelium mucoroides Brefeld
Dictyostelium sphaerocephalum (Oud.) Sacc. & March.
Isolated from soil; common and widespread.
50, 51. SH 95; SJ 05, 06, 15-17, 26. CE, CF, LO, MF, WW. (12) 1987. (48)

Dictyostelium rosarium Raper & Cavender
Isolated from soil; widespread, perhaps more common in drier soils.
50, 51. SH 95; SJ 05, 06, 16, 17, 26. CE, CF, MF. (9) 1987.

Polysphondylium pallidum Olive
Isolated from soil; frequent in temperate forests.
50, 51. SH 95; SJ 16, 17. MF. (4) 1987.

Species Accounts

Polysphondylium violaceum Brefeld
Isolated from soil, common and widespread.
50, 51. SH 95; SJ 05, 16, 26. CF, MF. (6) 1987.

Class MYXOMYCETES
Order ECHINOSTELIALES
Family Clastodermataceae

Clastoderma pachypus Nann.-Bremek.
On the bark of *Arbutus menziesii*; uncommon.
50. Recorded only from Bodnant Gardens (SH 77) in 2010. (49, 52) This is the third Welsh record. **RDL/LC but VU in Wales**

Family Echinosteliaceae

Echinostelium apitectum K.D. Whitney
On the bark of *Arbutus menziesii*; rare.
50. Recorded only from Bodnant Gardens (SH 77) in 2010. (48, 49) **RDL/LC but VU in Wales**

Echinostelium brooksii K.D. Whitney
On the bark of living trees, in moist chamber culture; common on acid-barked trees.
50, 51. SH 76, 86, 87, 95–97; SJ 07, 13–16, 23–26, 34, 37, 43. AV, BO, ER, LO, MO, NF. (24) 1990. (48, 49, 52) **RDL/LC**

Echinostelium colliculosum K.D. Whitney & H.W. Keller
On the bark of living trees, in moist chamber culture; common and widespread.
50, 51. SH 97; SJ 06, 13–16, 23–27, 34. BO, ER, LO, MA, MF, MO, NM. (18) 1976. (48, 49, 52) **RDL/LC**

Echinostelium corynophorum K.D. Whitney
On the bark of living trees, in moist chamber culture; widespread but less common than the other species in the genus.
50, 51. SH 97; SJ 22, 24, 26, 36. BO, MO. (5) 1997. (48, 49, 52) **RDL/LC**

Echinostelium fragile Nann.-Bremek.
On the bark of living trees, in moist chamber culture; frequent and widespread on isolated trees.
50, 51. SH 76, 97; SJ 06, 07, 13–16, 23, 24, 26, 34, 43. AV, BO, ER, LO, MF, MO. (21) 1973. (48, 49, 52) **RDL/LC**

Echinostelium minutum de Bary
On the bark of living trees, in moist chamber culture; common and widespread.
50, 51. SH 77, 95, 97; SJ 02, 04, 05, 13–17, 23–27, 34, 36, 37. AV, BO, CF, DU, ER, GW, LO, MF, MO. (24) 1972. (48, 49, 52) **RDL/LC**

Order LICEALES
Family Liceaceae

Licea biforis Morgan
On the bark of living trees and shrubs; becoming more common and widespread.
50, 51. SH 77, 97; SJ 04, 07, 12, 16, 24, 26, 35, 37, 43. BO, LO, MO. (11) 2004. (48, 49, 52) **RDL/LC**

Licea bryophila Nann.-Bremek.
On liverworts on the bark of living trees, in moist chamber culture; uncommon.
50, 51. SJ 02, 16, 25. LO. (3) 1997. (48, 49, 52) **RDL/LC**

Licea castanea G. Lister
On the bark of living trees, in moist chamber culture; uncommon.
50. 51. SH 97; SJ 16, 26. BO, MF, WW. (3) 1978. **RDL/LC but NT in Wales**

Licea clarkii Ing
On dead bramble stems; probably ubiquitous but difficult to detect.
50, 51. SJ 16, 23, 24, 26, 34, 35, 54. LO, MA, MO. (8) 1988. (48, 49, 52) **RDL/LC**

Licea denudescens T.E. Brooks & H.W. Keller
On the bark of living trees, in moist chamber culture; uncommon.
50, 51. SH 76; SJ 07, 16, 35, 37. AV, LO. (6) 1997. (48, 49, 52) **RDL/LC**

Licea eleanorae Ing
On the bark of living trees, in moist chamber culture; uncommon.
50. SJ 35, 45. (2) 2007. This species, originally described from Switzerland, is named after the author's wife. **RDL/LC but NT in Wales**

Licea erddigensis Ing
On the bark of living trees, in moist chamber culture; rare, known from a few sites in England and Scotland and also in Glamorgan.
50. SJ 24, 34, 35. ER. (3) 1999. Described from Erddig Park (Ing, 1999). **RDL/LC but NT in Wales**

Licea gloeoderma Döbbeler & Nann.-Bremek.
On the liverwort *Frullania*, on the bark of living trees; uncommon.
50. Recorded only from Chirk Castle (SJ 23) in 2011. (48) **RDL/LC but NT in Wales**

Licea inconspicua T.E. Brooks & H.W. Keller
On the bark of living trees, in moist chamber culture; uncommon.
50, 51. SJ 14, 16, 22, 26. LO, MO. (4) 1989. (48, 49, 52) **RDL/LC**

Licea kleistobolus G.W. Martin
Common on acid bark of living trees, in moist chamber culture.
50, 51. SH 76, 77, 84, 86–88, 94–97; SJ 02, 04, 06–08, 13–17, 23–27, 33–35, 37, 43–45. AV, BO, ER, LO, MF, MO, NF. (45) 1977. (48, 49, 52) **RDL/LC**

Species Accounts

Licea longa Flatau
On the bark of living trees, in moist chamber culture; rare.
51. Recorded only from a sycamore at Bodelwyddan (SH 97) in 2008. This is the first record from Wales. **RDL/LC but DD in Wales**

Licea marginata Nann.-Bremek.
On the bark of living trees, in moist chamber culture; frequent and widespread.
50, 51. SJ 07, 14, 16, 27, 37. AV. (6) 1993. (48, 48, 52) **RDL/LC**

Licea microscopica D.W. Mitchell
Widespread on the bark of living elder trees, associated with dark growths of cyanobacteria, in moist chamber culture.
50, 51. SH 76, 87, 88, 97; SJ 02, 06, 07, 13–17, 23–27, 34–36, 44, 54. BO, CC, DU, ER, GW, LO, MO. (30) 2000. (48, 49, 52) **RDL/LC**

Licea minima Fr.
On thick bark of living trees, in moist chamber culture; widespread.
50, 51. SJ 15, 16, 24, 25, 34–36. ER, GW, MF. (7) 1974. (48, 49, 52) **RDL/LC**

Licea operculata (Wing.) G.W. Martin
On the bark of old trees, in moist chamber culture, becoming more frequent.
50, 51. SH 87, 97; SJ 06, 07, 34, 35, 43. BO, CC, ER. (8) 2000. (48, 49, 52) **RDL/LC**

Licea parasitica (Zukal) G.W. Martin
Very common on the bark of living trees, in moist chamber culture.
50, 51. SH 76, 77, 84, 86–88, 94, 96, 97; SJ 02, 04–08, 12–17, 22–27, 34–37, 43–45, 53, 54. AV, BO, CC, CF, DU, ER, GW, LO, MA, MO, NF, NM, WW. (66) 1973. (48, 49, 52) **RDL/LC**

Licea pedicellata (H.C. Gilb.) H.C. Gilb.
On the bark of living trees, in moist chamber culture; uncommon.
50, 51. SH 88, 97; SJ 23, 26, 34. ER, LO. (5) 1978. **RDL/LC but NT in Wales**

Licea perexigua T.E. Brooks & H.W. Keller
On the bark of living trees, in moist chamber culture; uncommon.
50, 51. SH 88; SJ 08, 34. ER. (3) 2003. **RDL/LC but NT in Wales**

Licea pusilla Schrad.
On acid bark of living trees, in moist chamber culture, also on rotten conifer logs; common.
50, 51. SH 95; SJ 05, 13, 26, 27. CF, MO. (5) 1977. (48, 49, 52) **RDL/LC**

Licea pygmaea (Meylan) Ing
On acid bark of living trees, in moist chamber culture; frequent.
50, 51. SH 88; SJ 04, 07, 26, 27, 53. MO. (6) 2004. **RDL/LC**

Licea sambucina D.W. Mitchell
On the bark of living elder; uncommon.
51. Recorded only from Flint (SJ 27) in 2007. New to Wales. **RDL/LC but DD in Wales**

Licea scyphoides T.E. Brooks & H.W. Keller
On mossy bark of living trees, in moist chamber culture; more common in the west.
50, 51. SH 97; SJ 02, 16, 35, 36, 45. AV, BO, GW. (6) 1973. (48, 49, 52) **RDL/LC**

Licea synsporos Nann.-Bremek.
On tips of moss leaves on the bark of living trees, in moist chamber; uncommon.
50, 51. SJ 16, 22, 26. LO, WW. (3) 2005. (48) **RDL/LC**

Licea testudinacea Nann.-Bremek.
On the bark of living trees; uncommon.
50, 51. SH 97; SJ 24, BO. (3) 2007. (48) **RDL/LC**

Licea variabilis Schrad.
On decorticated pine branches on the forest floor; only common in older plantations.
50, 51. SJ 05, 13, 26. CF, LO. (3) 1969. (48, 49, 52) **RDL/LC**

Family Dictydiaethaliaceae

Dictydiaethalium plumbeum (Schum.) Rostaf.
On decaying beech logs; frequent, especially in southern Britain.
50, 51. SJ 15, 16, 26, 34. ER, LO, MO. (8) 1910. (48, 49, 52) **RDL/LC**

Family Reticulariaceae

Lycogala confusum Nann.-Bremek.
On rotten beech log; rare.
50. Recorded only from Pant-yr-Ochain (SJ 35) in 2005. (49) **RDL/LC but VU in Wales**

Lycogala conicum Pers.
On rotten trunks, characteristic of ancient woodland; uncommon.
50. Recorded only from a rotting beech log in the Loggerheads Country Park (SJ 16) from 1994 onwards. (49) **RDL/LC but NT in Wales**

Lycogala epidendrum *sensu lato*
On dead wood of all kinds; very common everywhere. This very familiar slime mould is now considered to comprise two common species. Unfortunately, older records cannot be assigned and there are few herbarium specimens to check.
The aggregate species is recorded from **50, 51**. SH 76, 87, 97; SJ 02, 05–08, 13, 15–18, 23, 25, 26, 34–36. AV, BP, BW, CF, CN, DU, ER, GW, LH, LO, MA, MF, ML, MO, NF, NM, PP, WW. (63) 1910. (48, 49, 52) **RDL/LC**

Lycogala epidendrum (L.) Fr. *sensu stricto*
On dead wood; characterised in the field by its scarlet plasmodium, grey/buff spore mass, smaller and darker aethalia and by a number of microscopic characters. It is less common than the next species.
50, 51. SH 95, 97; SJ 25, 16. LO, MA, ML, NM. (5) 1994. (48, 49, 52) **RDL/LC**

Species Accounts

Lycogala terrestre Fr.
On dead wood, and occasionally on terrestrial mosses; characterised by its by pink plasmodium, pink spore mass, larger and paler aethalia and microscopic characters.
50, 51. SH 87, 97; SJ 02, 05–07, 13, 15–18, 23, 25, 26, 34–36. CN, ER, GW, LO, MF, ML, MO, WW (37) 1994. (48, 49, 52) **RDL/LC**

Lycogala exiguum Morgan
On rotten wood, especially of beech; rare.
50. Recorded only from a beech log in Coed Clwyd on Moel Famau (SJ 16) in 1989. **RDL/LC but VU in Wales**

Lycogala flavofuscum (Ehrenb.) Rostaf.
Inside hollow trees or on dead standing trees; rare.
50. Recorded only from Coed Coch (SH 87) in 1880. **RDL/RE in Wales**

Reticularia jurana Meylan
Enteridium splendens (Morg.) T. Macbr. var. *juranum* (Meylan) Härkönen
On fallen oak branches in summer and autumn; common, especially in the west.
50, 51. SH 97; SJ 16, 17, 26, 36. GW, LO, MA, MO. (6) 1972. (48, 49, 52) **RDL/LC**

Reticularia lobata Lister
Enteridium lobatum (Lister) M.L. Farr
Under the bark at the base of pine stumps; uncommon, but difficult to detect.
50, 51. SJ 25. NF. (2) 1910. (49, 52) **RDL/DD**

Reticularia lycoperdon Bull.
Enteridium lycoperdon (Bull.) M.L. Farr
On dead wood, including door and window frames of houses, but especially on dead, standing alder trunks by rivers; very common in spring.
50, 51. SH 76, 87; SJ 06–08, 14–18, 23–27, 34–36, 43, 45. AV, BP, CC, CE, CN, DU, ER, GW, LH, LO, MF, MO, NF, WW. (54) 1880. (48, 49, 52) **RDL/LC**

Reticularia olivacea (Ehrenb.) Fr.
Enteridium olivaceum Ehrenb.
On small, fallen branches; uncommon.
51. Recorded only from the Leete Walk, Alyn Valley (SJ 16) in 1977. **RDL/LC but VU in Wales**

Tubifera ferruginosa (Batsch) J.F. Gmel.
Tubulifera arachnoidea Jacq.
On rotting conifer trunks and branches; very common.
50, 51. SH 76, 87, 95–97; SJ 02, 05, 07, 13, 15–17, 23, 25–27, 34, 36. AV, CC, CE, CF, CN, ER, GW, LH, LO, MF, MO, WW. (30) 1880. (48, 49, 52) **RDL/LC**

Order CRIBRARIALES
Family Cribrariaceae

Cribraria argillacea (Pers.) Pers.
On decaying conifer wood; common.
50, 51. SH 76, 85, 87, 95; SJ 02, 07, 13, 15-17, 23-26, 34. CC, ER, LO, MF, NF. (18) 1910. (48, 49, 52) **RDL/LC**

Cribraria aurantiaca Schrad.
On decaying conifer wood; common.
50, 51. SH 95; SJ 02, 05, 07, 13, 16, 23, 26, 36. CF, GW, LO. (9) 1969. (48, 49, 52) **RDL/LC**

Cribraria cancellata (Batsch) Nann.-Bremek.
On decaying conifer wood; common.
50, 51. SH 95; SJ 02, 05, 07, 13, 15-17, 23, 26, 34. CF, DU, ER, LO. (11) 1969. (48, 49, 52) **RDL/LC**

Cribraria persoonii Nann.-Bremek.
On decaying conifer wood; uncommon.
50, 51. SJ 16, 26, 36. AV, GW, LO. (3) 1972. (48, 49, 52) **RDL/LC**

Cribraria pyriformis Schrad.
On decaying conifer wood; uncommon.
50. SJ 05, 16. CF, LO. (2) 1972. (48, 49) **RDL/LC but NT in Wales**

Cribraria rufa (Roth) Rostaf.
On decaying conifer wood; common.
50, 51. SH 95; SJ 13, 16, 26. LO. (4) 1976. (48, 49, 52) **RDL/LC**

Order TRICHIALES
Family Dianemataceae

Calomyxa metallica (Berk.) Nieuw.
On the bark of old, living trees, especially elder, in moist chamber culture; frequent.
50, 51. SH 76, 77, 87, 88, 97; SJ 02, 06-08, 13-17, 23-27, 34-37, 44, 54. BO, CC, DU, ER, GW, LO, MO, NF. (35) 1910. (48, 49, 52) **RDL/LC**

Dianema corticatum Lister
On rotten, standing pine trunk; rare.
50. Recorded only from Pistyll Rhaeadr (SJ 02) in 2008. This is the first record for Wales, but since found in v.c. 48, Merioneth. **RDL/NT but EN in Wales**

Family Arcyriaceae

Arcyria cinerea (Bull.) Pers.
On mossy, dead wood and the bark of living trees; common.
50, 51. SH 85-87, 97; SJ 05, 06, 08, 13-18, 23-27, 34, 36. AV, BO, CC, CE, CF, CN, DU, ER, GW, LH, LO, MA, NF. (40) 1910. (48, 49, 52) **RDL/LC**

Species Accounts

Arcyria denudata (L.) Wettst.
On dead wood; common.
50, 51. SH 85, 87, 97; SJ 02, 06, 15-18, 23-27, 34, 36. AV, CC, DU, ER, GW, LO, MA, MO, NF, PP, WW. (34) 1780. (48, 49, 52) **RDL/LC**

Arcyria ferruginea Sauter
On dead wood; uncommon and mostly found during the winter period.
50, 51. SH 85; SJ 16, 17, 26. LO. (4) 1924. (49, 52) **RDL/LC but NT in Wales**

Arcyria incarnata (Pers.) Pers.
On dead wood, especially fallen oak branches; common.
50, 51. SH 76, 85, 87; SJ 05, 13, 16-18, 23, 24, 26. BW, DU, ER, LH, LO. (16) 1910. (48, 49, 52) **RDL/LC**

Arcyria minuta Buchet
On dead wood; uncommon.
50, 51. SJ 16, 17, 34. AV, DU, ER. (3) 1977. **RDL/LC but VU in Wales**

Arcyria obvelata (Oeder) Onsberg
On dead wood, often quite dry, in summer and early autumn; common.
50, 51. SH 76, 87; SJ 05, 15-17, 23, 25, 26, 34. CE, CF, ER, LO, MO, WW. (15) 1969. (48, 49, 52) **RDL/LC**

Arcyria oerstedtii Rostaf.
On rotten beech and pine stumps and logs; uncommon.
50, 51. SH 85; SJ 26. LO. (2) 1988. (49) **RDL/LC**

Arcyria pomiformis (Leers.) Rostaf.
On fallen branches and, more frequently, on the bark of living trees; common.
50, 51. SH 85-88, 97; SJ 02, 05, 06, 13-17, 23-26, 34, 36, 44, 45, 54. BO, CF, ER, GW, LO, MO, NF. (28) 1924. (48, 49, 52) **RDL/LC**

Perichaena chrysosperma (Currey) List.
On the bark of living trees, in moist chamber culture; common, especially in the west.
50, 51. SH 76, 84, 87, 88, 97; SJ 02, 04, 06-08, 13-17, 23-27, 34-37, 44, 53, 54. AV, BO, CC, DU, ER, GW, LO, MA, MO, NF. (40) 1977. (48, 49, 52) **RDL/LC**

Perichaena corticalis (Batsch) Rostaf.
On fallen bark, especially of ash; common.
50, 51. SJ 16, 17, 25, 26, 34. AV, ER, LO, MA, NF. (7) 1910. (48, 49, 52) **RDL/LC**

Perichaena depressa Libert
On fallen bark, especially of ash, rarely on the bark of living trees; uncommon except in the south.
50, 51. SJ 07, 08, 16, 18, 24, 26, 34. ER, LO. (9) 1910. (48, 52) **RDL/LC**

Perichaena vermicularis (Schwein.) Rostaf.
In leaf litter, especially of sycamore; uncommon.
51. Recorded only from Mold (SJ 26) in 1975. **RDL/LC but NT in Wales**

Family Trichiaceae

Hemitrichia abietina (Wigand) G. Lister
Hyporhamma abietina (Wigand) Lado
On the bark of living trees, rare.
50. Recorded only from an elm in Old Colwyn (SH 87) in 2003. (48, 49) **RDL/NT but VU in Wales**

Hemitrichia calyculata (Speg.) M.L. Farr
Hyporhamma calyculata (Speg.) Lado
On rotten logs, especially of beech; frequent.
50, 51. SH 87; SJ 02, 15–17, 25, 26, 35. DU, LO, MO, NF. (9) 1972. (48, 49, 52) **RDL/LC**

Hemitrichia leiotricha (Lister) G. Lister
Hyporhamma leiotricha (Lister) Lado
On heather litter in acid woodland; uncommon.
50. Recorded only from Nant Mill (SJ 24) in 1991; elsewhere in Wales recorded only from Caernarvonshire. (49) **RDL/LC but EN in Wales**

Hemitrichia minor G. Lister
Hyporhamma minor (G. Lister) Lado
On liverworts on the bark of living trees; uncommon.
50, 51. SJ 16. AV, LO. (2) 1973. (48, 49) **RDL/LC**

Hemitrichia pardina (Minakata) Ing
Hyporhamma pardina (Minakata) Lado
On mosses and liverworts, and occasionally on bare bark of living trees; uncommon.
50. SJ 16, 34. ER, LO. (2) 1999. **RDL/LC but NT in Wales**

Metatrichia floriformis (Schwein.) Nann.-Bremek.
On rotten wood; common.
50, 51. SH 87; SJ 02, 05, 08, 15–18, 23, 25, 26, 34, 36. AV, BP, CF, DU, ER, GW, LH, LO, MO, NF, NM, WW. (24) 1969. (48, 49, 52) **RDL/LC**

Metatrichia vesparium (Batsch) Nann.-Bremek.
On rotten beech trunk, scarce, rare in the north of Britain.
50, 51. SJ 26, 34. ER, WW. (2) 2006. (48, 49) **RDL/LC but NT in Wales**

Oligonema schweinitzii (Berk.) G.W. Martin
On sticks in ditches; rare.
50. Recorded only from Mold (SJ 26) in 1978; this remains the only record from Wales. **RDL/LC but EN in Wales**

Prototrichia metallica (Berk.) Massee
On small twiggy litter; rare.
50, 51. SJ 17, 25. DU, NF. (2) 1910. (48, 49) **RDL/LC but NT in Wales**

Trichia affinis de Bary
On very rotten, often mossy, logs and stumps; common.
50, 51. SH 87, 95; SJ 13, 15–17, 25–27, 34. CC, DU, ER, LO, MA, MO. (18) 1880. (48, 49, 52) **RDL/LC**

Species Accounts

Trichia botrytis (J.F. Gmel.) Pers.
On fallen branches, especially of oak or conifer; common.
50, 51. SH 85, 95; SJ 02, 05, 13–17, 23–26, 34. AV, CF, ER, LO, MF. (20) 1910. (48, 49, 52) **RDL/LC**

Trichia contorta (Pers.) Ditm. var. **contorta**
On fallen branches; frequent.
50, 51. SJ 24, 26, 34. ER, WW. (4) 1910. (48) **RDL/LC**

Trichia contorta var. **inconspicua** (Rostaf.) Lister
On herbaceous litter; frequent.
50. SJ 24, 34. ER, WW. (2) 1910. **RDL/LC**

Trichia decipiens (Pers.) T. Macbr.
On fallen branches, especially of broad-leaved trees; common. Some of the records may refer to the related *Trichia meylanii* but in the absence of herbarium specimens this cannot now be checked.
50, 51. SH 85, 87, 97; SJ 05, 07, 15–18, 23, 25–27, 34–36. AV, CC, DU, ER, GW, LO, MA, MF, MO, NF, WW. (34) 1880. (48, 49, 52) **RDL/LC**

Trichia meylanii Ing
Trichia decipiens var. *olivacea* Meylan
On fallen branches; probably common but only recently separated at specific level.
50, 51. SJ 16, 26. LO, MA. (3) 2003. **RDL/LC**

Trichia munda (Lister) Meylan
In mosses on the bark of living trees; frequent, especially in the west.
50, 51. SJ 24, 26, 34. ER, LO. (3) 2003. (48) **RDL/LC**

Trichia persimilis P. Karst.
On rotten trunks and branches, much less rotten than for *T. affinis*, and not usually moss-covered; common.
50, 51. SH 85; SJ 02, 15–17, 23, 26, 34, 36. AV, DU, ER, GW, LO, MA, MO, WW. (20) 1910. (48, 49, 52) **RDL/LC**

Trichia scabra Rostaf.
On rotten trunks, especially of beech or elm; frequent.
50, 51. SJ 15–17, 24, 26, 34, 36. AV, BW, ER, GW, LO, WW. (12) 1910. (48, 49, 52) **RDL/LC**

Trichia varia (Pers.) Pers.
On soggy, rotten wood; very common.
50, 51. SH 85, 97; SJ 02, 06, 07, 13, 15–17, 23–27, 34–36. AV, BO, BW, CE, DU, ER, GW, LH, LO, MA, MO, NF, WW. (39) 1910. (48, 49, 52) **RDL/LC**

Trichia verrucosa Berk.
On rotten conifer trunks; rare.
50, 51. SH 95; SJ 05, 36. CF, GW. (3) 1972. (48, 49, 52) **RDL/LC but NT in Wales**

Order STEMONITALES
Family Stemonitadaceae

Amaurochaete atra (Alb. & Schwein.) Rostaf.
Lachnobolus ater (Alb. & Schwein.) Lado
On newly felled conifer trunks; uncommon.
50, 51. SJ 16, 17. LH, MF. (2) 1979. (48, 52) **RDL/LC**

Brefeldia maxima (Fr.) Rostaf.
On stumps of broad-leaves; frequent, possibly becoming more common. This is the largest species of myxomycete and the emerging plasmodium, which resembles cold porridge, may extend for as much as a square metre.
50, 51. SJ 05, 15–17, 23, 25, 26, 34. AV, DU, ER, LO, MA, WW. (13) 1976. (52) **RDL/LC**

Collaria elegans (Racib.) Ing
On fallen conifer branches; frequent.
50, 51. SH 95; SJ 13, 16, 25. AV, MF, NM. (5) 1974. (48, 49, 52) **RDL/LC**

Collaria lurida (Lister) Nann.-Bremek.
On leaf litter; rare.
50. Recorded only from Hafod Wood, near Betws-y-Coed (SH 85) in 1924; this is the only Welsh record. **RDL/NT but RE in Wales**

Collaria rubens (Lister) Nann.-Bremek.
On leaf litter; rare.
50, 51. SJ 26, 34. ER, MO. (2) 1975. (49) **RDL/LC but VU in Wales**

Colloderma oculatum (Lippert) G. List.
On alga-covered bark of living trees; frequent, especially in the west.
50, 51. SJ 16, 25, 26. LO, MF, NF. (3) 1974. (48, 49, 52) **RDL/LC**

Comatricha alta Preuss
On fallen trunks; uncommon.
50, 51. SJ 16, 34. AV, ER. (2) 1990. (49) **RDL/LC but NT in Wales**

Comatricha laxa Rostaf.
On fallen branches; frequent.
50, 51. SH 85; SJ 26. LO. (2) 1924. (48, 49) **RDL/LC**

Comatricha nigra (Pers.) Schroet.
On dead wood of all kinds; very common.
50, 51. SH 76, 85, 87, 97; SJ 02, 05–08, 13, 15–18, 23–27, 34–36. AV, BO, BP, BW, CC, DU, ER, GW, LH, LO, MA, MF, MO, NF, NM, WW. (47) 1910. (48, 49, 52) **RDL/LC**

Comatricha pulchella (C. Bab.) Rostaf.
On leaf litter, especially of holly, and also on decaying fern fronds; common.
50, 51. SH 87; SJ 02, 07, 16, 17, 23, 25, 27, 34, 36. CC, ER, GW, LH, LO, MA, NF. (12) 1910. (48, 49, 52) **RDL/LC**

Species Accounts

Comatricha rigidireta Nann.-Bremek.
On acidic bark of living trees; uncommon.
50. Recorded only from a cedar at Hanmer (SJ 43) in 2008. (49) **RDL/LC but NT in Wales**

Comatricha tenerrima (M.A. Curt.) G. List.
On dead herbaceous stems, especially in marshy places; uncommon.
50, 51. SJ 17, 24, 26, 34. DU, ER, LH, MO. (5) 1910. (48, 52) **RDL/LC**

Enerthenema papillatum (Pers.) Rostaf.
On fallen branches and the bark of living trees; common.
50, 51. SH 86, 95, 97: SJ 02, 05–07, 13–17, 23–25, 26, 34, 36. AV, BO, CF, DU, ER, GW, LH, LO, MO. (26) 1828. (48, 49, 52) **RDL/LC**

Lamproderma columbinum (Pers.) Rostaf.
On mosses on rotten trunks and on damp rocks in ravines; uncommon.
50, 51. SJ 02, 05, 16, 25. CF, LO, NF. (5) 1978. (48, 49, 52) **RDL/LC**

Lamproderma nigrescens (Rostaf.) Rostaf.
Lamproderma arcyrioides (Sommerf.) Rostaf.
On leaf litter, especially of ivy; uncommon.
50, 51. SJ 17, 25, 26, 35. DU, MO, NF. (5) 1975. (48, 49) **RDL/LC**

Lamproderma scintillans (Berk. & Br.) Morgan
On leaf litter, especially of holly, and on decaying fern fronds; frequent.
50, 51. SH 87; SJ 06, 16, 17, 26, 27, 35, 36. CC, GW, LH, MA, MO. (13) 1971. (48, 49, 52) **RDL/LC**

Macbrideola cornea (G. List. & Cran) Alexop.
On mosses on the bark of living trees; common in the west.
50, 51. SH 76, 87; SJ 02, 05, 12–16, 23, 24, 26, 34, 35. AV, CC, CF, ER, LO, MA. (16) 1976. (48, 49, 52) **RDL/LC**

Macbrideola macrospora (Nann.-Bremek.) Ing
On the bark of living alder; rare.
50. Recorded only from Alyn Waters Country Park (SJ 35) in 2007. (48, 49) **RDL/LC**

Paradiacheopsis acanthodes (Alexop.) Nann.-Bremek.
On the bark of living trees; rare.
50. Recorded only from an oak in Clocaenog Forest (SJ 05) in 1985. This, and an earlier record from Merioneth are the only British records of this rare but distinctive species. (48) **RDL/VU**

Paradiacheopsis cribrata Nann.-Bremek.
On the bark of living trees; uncommon.
50, 51. SH 76, 77, 97; SJ 13, 14, 16, 26, 34. LO, MO. (8) 1997. (48, 49, 52) **RDL/LC**

Paradiacheopsis fimbriata (G. List. & Cran) Hertel
On the acid bark of living trees, including polluted bark; very common.
50, 51. SH 76, 77, 86, 87, 95–97; SJ 02, 06, 07, 13–17, 23–27, 33–37, 43, 44, 53, 54. AV, BO, ER, GW, LO, MF, NF. (38) 1976. (48, 49, 52) **RDL/LC**

Paradiacheopsis rigida (Brandza) Nann.-Bremek.
On the bark of living trees; rare.
50. Recorded only from a sycamore in the Loggerheads Country Park (SJ 16) in 2005. (48, 49) **RDL/LC**

Paradiacheopsis solitaria (Nann.-Bremek.) Nann.-Bremek.
On the bark of living trees, in moist chamber culture; common.
50, 51. SH 76, 77, 84, 86, 87, 94, 97; SJ 02, 05, 12-17, 23-25, 34, 35, 54. BO, CC, CF, ER, LO, MF, NF. (25) 1979. (48, 49, 52) **RDL/LC**

Stemonitis axifera (Bull.) T. Macbr.
On dead wood; common.
50, 51. SH 97; SJ 16, 23, 25, 26, 34. BO, CE, ER, LO, NF, NM. (9) 1910. (48, 49, 52) **RDL/LC**

Stemonitis flavogenita E. Jahn
On rotten wood; frequent.
50, 51. SH 76, 97; SJ 07, 15-17, 24-26, 34, 35. CE, DU, ER, MA, MF, ML, NF. (16) 1910. (48, 49, 52) **RDL/LC**

Stemonitis foliicola Ing
On leaf litter; rare.
51. Recorded only from Loggerheads Country Park (SJ 26) in 1976; this remains the only Welsh record. **RDL/VU but CR in Wales**

Stemonitis fusca Roth
On dead wood; very common.
50, 51. SH 76, 85, 87, 97; SJ 02, 06-08, 15-17, 23-27, 34-36, 43. AV, BP, CC, CE, DU, ER, GW, LH, LO, MA, ML, MO, NF, NM, WW. (40) 1880. (48, 49, 52) **RDL/LC**

Stemonitis herbatica Peck
In leaf litter and on decaying herbaceous vegetation on the ground; uncommon.
50, 51. SH 97; SJ 16, 34. AV, ER, ML. (3) 1985. (49, 52) **RDL/LC**

Stemonitis smithii T. Macbr.
On twigs in wet woodland; rare.
50. Recorded only from the Erddig estate (SJ 34) in 1974; this is the only record for Wales. **RDL/NT but CR in Wales**

Stemonitopsis amoena (Nann.-Bremek.) Nann.-Bremek.
On the bark of living horse chestnut; rare.
50. Recorded only from Llanrwst (SH 76) in 2008. **RDL/LC**

Stemonitopsis hyperopta (Meylan) Nann.-Bremek.
On very wet conifer wood; frequent.
50, 51. SH 95; SJ 13, 16, 25. AV, LO. (6) 1975. (48, 49, 52) **RDL/LC**

Stemonitopsis reticulata (H.C. Gilb.) Nann.-Bremek. & Y. Yamam.
On fallen conifer branch; rare.
50. Recorded only from Clocaenog Forest (SJ 05) in 2002; this is the first Welsh and third British record. **RDL/VU but CR in Wales**

Species Accounts

Stemonitopsis subcaespitosa (Peck) Nann.-Bremek.
On the bark of living trees, in moist chamber culture; rare.
51. Recorded only from a willow at Queensferry (SJ 36) in 2001; this is the only record from Wales. **RDL/NT but CR in Wales**

Stemonitopsis typhina (Wiggers) Nann.-Bremek.
On wet, rotten wood, mainly of broad-leaved trees; common.
50, 51. SJ 15–17, 23, 24, 26, 34, 36. AV, ER, GW, LO, WW. (13) 1910. (48, 49, 52) **RDL/LC**

Symphytocarpus amaurochaetoides Nann.-Bremek.
On stumps; rare.
50. SJ 16, 34. ER, LO. (3) 1969. (48, 49) **RDL/NT**

Symphytocarpus flaccidus (Lister) Ing & Nann.-Bremek.
On dead, standing conifer trunks, uncommon.
50, 51. SJ 23, 26. MO. (2) 1972. (48, 49, 52) **RDL/LC**

Symphytocarpus impexus Ing & Nann.-Bremek.
On woody litter on the forest floor; uncommon.
51. Recorded only from Loggerheads Country Park (SJ 26) in 1988; this is the only record for Wales. **RDL/LC but CR in Wales**

Order PHYSARALES
Family Physaraceae

Badhamia affinis Rostaf.
On mossy bark of living trees; frequent in the west.
50, 51. SH 88; SJ 02, 04, 27, 35, 54. (6) 2007. (48, 49, 52) **RDL/LC**

Badhamia capsulifera (Bull.) Berk.
On branches and trunks of living trees; uncommon.
50, 51. SJ 26, 34. ER, MO. (2) 1975. (52) **RDL/LC but NT in Wales**

Badhamia foliicola Lister
On moss on decaying trunks and dead grass in dunes; frequent.
50, 51. SJ 16, 18. LO, PA. (2) 2011. (48, 52) **RDL/LC**

Badhamia lilacina (Fr.) Rostaf.
On *Sphagnum* in bogs; uncommon, mostly western.
50, 51. SH 76; SJ 25, 53. HM. (3) 1978. (48, 49, 52) **RDL/LC**

Badhamia panicea (Fr.) Rostaf.
On the bark of fallen trunks, especially of beech; frequent.
50, 51. SH 95; SJ 06, 16, 17, 26, 34, 36. ER, GW, LO, MO. (10) 1910. (48, 49, 52) **RDL/LC**

Badhamia utricularis (Bull.) Berk.
On bracket fungi on fallen trunks; frequent, especially in the winter months.
50, 51. SJ 16, 17, 25, 26, 34, 36. ER, GW, LO, NF, WW. (6) 1910. (48, 49, 52) **RDL/LC**

Craterium aureum (Schum.) Rostaf.
In leaf litter, especially of beech; uncommon.
50, 51. SJ 16, 17, 26, 34. ER, LH, LO, MF. (4) 1910. (48, 52) **RDL/LC**

Craterium leucocephalum (Pers.) Ditmar
In leaf litter; frequent.
50, 51. SJ 16-18, 26, 34. AV, LH, MO, PA. (5) 1910. (48, 49) **RDL/LC**

Craterium minutum (Leers) Fr.
In leaf litter and on stems of living herbaceous plants; common.
50, 51. SH 87; SJ 06, 07, 15-18, 24-27, 34-36. AV, CC, DU, ER, GW, LH, LO, MA, MF, MO, NF, PA. (29) 1910. (48, 49, 52) **RDL/LC**

Fuligo candida Pers.
On dead wood; uncommon.
51. Recorded only from Coed Talon (SJ 25) in 1975. (52) **RDL/LC but NT in Wales**

Fuligo cinerea (Schwein.) Morgan
On straw and manure heaps; probably extinct in Britain.
50. Recorded only from Llangollen (SJ 24) in 1910; the only Welsh record. **RDL/ RE**

Fuligo muscorum Alb. & Schwein.
On damp mosses on woodland floors; common only in the west of Britain.
50, 51. SH 85, 95; SJ 02, 05, 24, 25. CF, NF. (6) 1910. (48, 49) **RDL/LC**

Fuligo rufa Pers.
On very rotten, often dry, stumps; uncommon.
51. SJ 07, 08, 18, 26, 36. MO. (4) 1985. **RDL/LC but VU in Wales**

Fuligo septica (L.) Wiggers
On dead wood; very common, usually as the var. *flava* (Pers.) Lazaro Ibiza which has all the calcareous material inside the aethalium yellow, rather than the var. *septica* where this material is yellow and white.
50, 51. SH 76, 85, 87, 95, 97; SJ 02, 05-08, 13-18, 24-27, 34-36. AV, BP, BW, CC, CE, CF, CN, DU, ER, GW, LH, LO, MF, ML, NF, NM, WW. (39) 1871. (48, 49, 52) **RDL/LC**

Leocarpus fragilis (Dicks.) Rostaf.
On leaf litter, especially of conifers, and also on living stems of herbaceous plants; common.
50, 51. SH 87; SJ 05-07, 15-17, 23-27, 34-36. AV, CC, CF, DU, ER, GW, LH, LO, MA, MF, NF, NM. (15) 1892. (48, 49, 52) **RDL/LC**

Physarum album (Bull.) Chevall.
Physarum nutans Pers.
On dead wood; very common.
50, 51. SH 76, 87; SJ 02, 05-08, 14-17, 23-27, 34-36, 43. AV, BP, BW, CC, CF, DU, ER, GW, LH, LO. (31) 1880. (48, 49, 52) **RDL/LC**

Physarum auriscalpium Cooke
On mosses and lichens on the bark of living trees; frequent.
50, 51. SJ 05, 25. CF, NF. (2) 2011. (48) **RDL/LC**

Species Accounts

Physarum bitectum G. List.
On litter and old bramble stems; uncommon.
50, 51. SJ 17, 25, 26, 34, 36. DU, ER, GW, LH, MO, NF. (7) 1910. **RDL/LC**

Physarum bivalve Pers.
On leaf litter; frequent.
50, 51. SJ 15–17, 23, 25, 26, 34, 36. DU, ER, GW, LH, LO, MA, MF, NF. (13) 1972. (48, 49, 52) **RDL/LC**

Physarum cinereum (Batsch) Pers.
On leaf litter and on living lawn grass; frequent.
50, 51. SJ 16, 23–26, 34, 36. AV, ER, GW, MO, NF. (8) 1910. (48, 49, 52) **RDL/LC**

Physarum compressum Alb. & Schwein.
On mosses and herbaceous material in woodland and on a compost heap; frequent.
50, 51. SJ 16, 26. LO, MF, MO. (3) 1975. (48, 52) **RDL/LC**

Physarum conglomeratum (Fr.) Rostaf
On leaf litter in damp woods; rare.
51. Recorded only from Hawarden Woods (SJ 36) in 1972; this remains the only Welsh record.
RDL/NT but CR in Wales

Physarum contextum (Fr.) Pers.
On mossy litter in damps woods; uncommon.
50. SJ 17, 34. ER. (2) 1910. (48) **RDL/NT**

Physarum crateriforme Petch
On mossy bark of living horse chestnut; uncommon.
50. Recorded only from Holt (SJ 45) in 2007. (48) **RDL/LC**

Physarum didermoides (Pers.) Rostaf.
On old straw bales; rare.
51. Recorded only from Mold (SJ 26) in 1976. (52) **RDL/NT but VU in Wales**

Physarum leucophaeum Fr.
On dead wood, occasionally on the bark of living trees; common.
50, 51. SH 85; SJ 08, 15–18, 23, 24, 26, 34–36. AV, DU, ER, GW, LH, LO, MO, PP. (17) 1910. (48, 49, 52) **RDL/LC**

Physarum psittacinum Ditmar
On rotten stumps and logs in ancient woodland; rare.
50, 51. SJ 16, 34. ER, LO. (2) 1985. At Loggerheads it has appeared at the same spot in the second week of July every year since 1994. (48) **RDL/LC but NT in Wales**

Physarum pusillum (Berk. & Curt.) G. List.
On grass litter, and, occasionally, on the bark of living trees; uncommon.
50, 51. SJ 12, 16, 18, 23, 26. LO, MO, PA. (5) 1976. (49, 52) **RDL/LC**

Physarum robustum (Lister) Nann.-Bremek.
On fallen branches; frequent.
50, 51. SJ 16, 17, 34. DU, ER, LO. (3) 2003. (48, 49, 52) **RDL/LC**

Physarum straminipes Lister
On straw and grass litter; very rare.
50. Recorded only from Nant-y-Ffrith (SJ 25) in 1910; the only Welsh record. **RDL/NT but RE in Wales**

Physarum vernum Fr.
On grass litter; uncommon.
50. Recorded only from Moel Famau (SJ 16) in 1984. (48) **RDL/NT but VU in Wales**

Physarum virescens Ditmar
On terrestrial mosses in damp woods; uncommon.
50. SJ 16, 17. MF. (2) 1979. (48, 49, 52) **RDL/LC**

Physarum viride (Bull.) Pers.
On dead wood, especially conifer brashings; frequent.
50, 51. SH 76, 87; SJ 05, 07, 16, 17, 23, 24, 26, 34. CC, CF, ER, LH, LO. (12) 1880. (48, 49, 52) **RDL/LC**

Willkommlangea reticulata (Alb. & Schwein.) Kuntze
On sticks in damp woodland; uncommon.
50. Recorded only from Hafod Wood, near Betws-y-Coed (SH 85) in 1924. This is the only Welsh record. **RDL/LC but CR in Wales**

Family Didymiaceae

Diachea leucopodia (Bull.) Rostaf.
On leaf litter and on the living stems of herbaceous plants; uncommon.
50, 51. SJ 15, 17, 23, 26, 34, 36. CE, DU, ER, GW, LH, WW. (9) 1910. (48, 49, 52) **RDL/LC**

Diderma chondrioderma (de Bary & Rostaf.) G. Lister
On mossy bark of living trees, in moist chamber culture; frequent.
50, 51. SJ 04, 08, 15, 35, 54. (5) 1986. (48, 49, 52) **RDL/LC**

Diderma deplanatum Fr.
On terrestrial mosses in damp woods; uncommon.
50, 51. SH 85; SJ 17, 25. DU, NF. (3) 1924. (48, 49) **RDL/LC**

Diderma effusum (Schwein.) Morg.
On leaf litter, especially of beech, rarely on the bark of living trees; frequent.
50, 51. SJ 06, 15, 16, 26. AV, LO, MA. (6) 1975. (48, 49, 52) **RDL/LC**

Diderma floriforme (Bull.) Pers.
On rotten logs in ancient woodland; uncommon.
50. Recorded only from Coed Coch (SH 87) in 1880. This is the only Welsh record. **RDL/LC but RE in Wales**

Diderma globosum Pers.
On litter in damp places; uncommon.
50, 51. SH 85; SJ 17, 18, 25, 26, 34, 36. GW, LH, LO, MA, PA. (8) 1910. (48, 49, 52) **RDL/LC**

Species Accounts

Diderma hemisphaericum (Bull.) Hornem.
On leaf litter in damp places, rarely on the bark of living trees; frequent.
50, 51. SJ 14, 16, 17, 25, 27, 34–36. AV, DU, ER, GW, LO, MA. (12) 1910. (48, 49, 52) **RDL/LC**

Diderma ochraceum Hoffm.
On mosses on wet rocks or very soggy logs; uncommon, mostly in the west.
50. Recorded only from the Clocaenog Forest (SJ 05) in 2002. (48, 49) **RDL/LC**

Diderma spumarioides (Fr.) Fr.
On mosses and short turf in sand dunes and in leaf litter over limestone; frequent.
50, 51. SH 97; SJ 15, 18, 25, 26, 34, 35. ER, MA, ML, PA. (7) 1910. (48, 49, 52) **RDL/LC**

Diderma testaceum (Schrad.) Pers.
On leaf litter; rare.
51. Recorded only from Llyn Helig (SJ 17) in 1977. (48) **RDL/NT but VU in Wales**

Diderma umbilicatum Pers.
On dead bramble stems; uncommon.
51. Recorded only from Coed Bach-y-Graig Nature Reserve (SJ 07) in 1992. (48) **RDL/LC but VU in Wales**

Didymium anellus Morgan
On leaf litter; uncommon.
51. Recorded only from Northop (SJ 26) in 1991. (49) **RDL/LC but VU in Wales**

Didymium bahiense Gottsberger
On herbaceous litter; frequent.
50, 51. SJ 16, 25, 26, 34–36. GW, LO, MA, MF, MO, WW. (11) 1971. (48, 49, 52) **RDL/LC**

Didymium clavus (Alb. & Schwein.) Rabenh.
In leaf litter; frequent.
50, 51. SJ 05, 16–18, 25, 26, 36. DU, GW, MF, PA, WW. (7) 1975. (48, 49, 52) **RDL/LC**

Didymium crustaceum Fr.
On litter in hedge bottoms; rare.
50, 51. SJ 16, 34. (2) 1910. **RDL/VU**

Didymium difforme (Pers.) S.F. Gray
On plant litter of all kinds, on living roots of shrubs in soil and on bulbs in water culture; very common.
50, 51. SH 85, 87, 97; SJ 06–08, 15–18, 23–27, 34–36. AV, BP, CC, DU, ER, GW, LH, LO, MF, MO, NF, NM, WW. (40) 1910. (48, 49, 52) **RDL/LC**

Didymium ilicinum Ing
On leaf litter, especially of holly; frequent.
50, 51. SJ 16, 26. LO, MA. (2) 2006. (49, 52) This is a segregate from the common and widespread *D. squamulosum*. **RDL/LC**

Didymium megalosporum Berk. & M.A. Curt.
On plant litter and, rarely, on the bark of living elder; frequent.
51. SJ 17, 26, 27. LH, MO. (3) 1976. **RDL/LC but VU in Wales**

Didymium melanospermum (Pers.) Macbr.
On leaf litter, especially of conifers; common.
50, 51. SH 85; SJ 05, 07, 13, 15, 16, 24, 26, 36. CF, GW, LO, MF, WW. (10) 1910. (48, 49, 52) **RDL/LC**

Didymium minus (List.) Morg.
On dead herbaceous stems, especially of Rosebay Willowherb; uncommon.
50, 51. SH 85; SJ 26, 34. ER, MO. (4) 1910. (48, 49, 52) **RDL/LC**

Didymium nigripes (Link) Fr.
On leaf litter, especially of holly; common.
50, 51. SH 85, 87; SJ 06, 15–17, 25–27, 34, 36. CC, DU, GW, LH, LO, MA, MF, MO. (19) 1910. (48, 49, 52) **RDL/LC**

Didymium serpula Fr.
On fallen beech leaves and bark of a living shrub; uncommon.
50, 51. SJ 26, 34. ER, MO. (2) 2004. (48, 49, 52) **RDL/LC but NT in Wales**

Didymium squamulosum (Alb. & Schwein.) Fr.
On leaf litter of all kinds; very common and very variable.
50, 51. SH 85, 87; SJ 06–08, 15–18, 23–27, 34–36. AV, BP, CC, DU, ER, GW, LH, LO, MA, MF, MO, NF, NM, WW. (40) 1910. (48, 49, 52) **RDL/LC**

Didymium sturgisii Hagelst.
Trabrooksia applanata H.W. Keller
On the bark of living trees; rare.
50. Recorded only from a sycamore at Maeshafn (SJ 26) in 1992. (49) This species is characteristic of the bark of evergreen and other oaks in the Mediterranean. **RDL/NT but VU in Wales**

Didymium tubulatum E. Jahn
On leaf litter; rare.
51. Recorded only from Mold (SJ 26) in 1976; the only record from Wales. **RDL/EN but CR in Wales**

Didymium verrucosporum Welden
On straw and herbaceous litter; rare.
51. Recorded only from Mold (SJ 26) in 1976; the only record from Wales. **RDL/LC but CR in Wales**

Lepidoderma chailletii Rostaf.
On birch leaf litter, sometimes on heather and occasionally, in Scotland, on vegetation at the snowline in the spring; uncommon.
51. Recorded only from the Etna Country Park, Buckley (SJ 26) in 2002; the first record for Wales.
RDL/NT but CR in Wales

Lepidoderma tigrinum (Schrad.) Rostaf.
On mosses on damp rocks; frequent in the west.
50, 51. SH 95; SJ 05, 25. CF, NF. (4) 1978. This is the farthest east that this member of the ravine association of myxomycetes has been found in Wales. (48, 49) **RDL/LC**

Mucilago crustacea Wiggers
On living grass stems, especially on limestone soils; common.
50, 51. SH 76, 87, 88, 95, 97; SJ 06–08, 15–18, 23–27, 34–37. AV, BO, DU, ER, GW, LO, MA, MF, MO, NF, PA, WW. (46) 1973. (48, 49, 52) **RDL/LC**

Phylum PLASMODIOPHOROMYCOTA
Class PLASMODIOPHOROMYCETES
Order PLASMODIOPHORALES
Family Plasmodiophoraceae

Plasmodiophora brassicae Woronin
The cause of club-root disease in Brassicaceae, especially in poorly drained base-poor soils. Although common this is rarely reported to mycologists.
51. Recorded from Mold (SJ 26) since 1978, but undoubtedly widespread.

Spongospora subterranea (Wallr.) Lagerh.
The cause of corky scab disease of potato tubers. Widespread and sometimes common, especially on older varieties of potato, some of which are no longer permitted to be grown. As with the last species it is far more common than the single record suggests.
51. Recorded from Mold (SJ 26) since 1978.

Species Accounts

Phylum PERONOSPOROMYCOTA
Class PERONOSPOROMYCETES
Order PERONOSPORALES
Family Albuginaceae

Albugo candida (Pers.) O. Kuntze
White blister rust on leaves and stems of Brassicaceae, including *Aurinia saxatilis, Arabis caucasica, Aubreta deltoides, Brassica rapa, Capsella bursa-pastoris, Iberis sempervirens* and *Lunaria annua*; common.
50, 51. SH 77; SJ 07, 08, 16, 18, 24, 26, 34–37. ER, MO, PA, WW. (19) 1975. (48, 49, 52)

Albugo labaichii Thines & Y.-J. Choi
White blister rust on *Arabidopsis thaliana*; uncommon.
51. Recorded only from Mold (SJ 26) in 1975.

Pustula lepigoni (de Bary) O. Kuntze
White blister rust on leaves of *Spergularia rubra* and *Stellaria media*, uncommon.
51. SJ 26, 27. MO. (2) 1978.

Pustula obtusata (Link) C. Rost.
Pustula tragopogonis (Pers.) Thines in part; *Albugo tragopogonis* (Pers.) Gray
White blister rust on leaves of *Senecio squalidus* and *S. vulgaris*; uncommon.
50, 51. SJ 15, 17, 26, 36, 45. MO. (5) 1977. (49, 52)

Family Peronosporaceae

Bremia centaureae Styd.
Bremia lactucae in part
Downy mildew on *Centaurea nigra*; uncommon.
50. Recorded only from Betws Gwerfil Goch (SJ 04) in 1969.

Bremia lactucae Regel.
Downy mildew on leaves of *Lactuca sativa*; common.
50. Recorded only from Loggerheads (SJ 16) in 1971. (49, 52)

Bremia lapsanae Syd.
Bremia lactucae in part
Downy mildew on *Lapsana communis*; frequent.
50, 51. Recorded from Mold (SJ 26) since 1978 and recorded from Denbighshire, without locality, by Chater et al. (2020).

Bremia tulasnei (Hoffm.) Syd.
Bremia lactucae in part
Downy mildew on *Jacobaea vulgaris*; uncommon.
51. Recorded only from Hawarden (SJ 36) in 2015. (49)

Hyaloperonospora brassicae (Gäum.) Göker, Riethm., M. Weiss & Oberw.
Peronospora parasitica (Pers.) Constant. in part
Downy mildew on *Brassica*; common.
50, 51. Recorded from Mold (SJ 26) in 2000 and from Denbighshire, without locality, by Preece (2002). (49)

Hyaloperonospora galligena (S. Blujmer) Göker, Riethm., Voglmayr & Oberw.
Peronospora parasitica in part
Downy mildew on *Aurinia saxatilis*; uncommon.
51. Recorded only from Mold (SJ 26) in 1978.

Hyaloperonospora nasturtii-aquatici (Gäum.) Voglmayr
Peronospora parasitica in part
Downy mildew on *Cardamine*; uncommon.
50, 51. SJ 12, 18, 36. GW, PA. (3) 1917. (49)

Hyaloperonospora niessliana (Berl.) Constant.
Peronospora niessliana Berl.
Downy mildew on leaves of *Alliaria petiolata*; common.
59, 51. SJ 16, 17, 26, 34–36. ER, LO, MO, WW. (9) 1976. (49)

Hyaloperonospora sisymbrii-loeselii (Gäum.) Göker, Riethm., Voglmayr, M. Weiss & Oberw.
Peronospora parasitica in part
Downy mildew on *Sisymbrium*; uncommon.
51, 51. Recorded from Mold (SJ 26) and from Denbighshire, without locality, by Chater et al. (2020). (49)

Peronospora agrestis Gaüm.
Downy mildew on leaves of *Veronica chamaedrys*; uncommon.
50. Recorded only from Loggerheads (SJ 16) in 2012. (49)

Peronospora alsinearum Casp.
Downy mildew on leaves of *Stellaria media*; uncommon.
50, 51. SJ 16, 26. LO, MO. (2) 2000. (49)

Peronospora alta Fuckel
Downy mildew on leaves of *Plantago lanceolata* and *P. major*; frequent.
50, 51. SJ 16, 17, 26, 36. DU, LO, M. (4) 1972. (49, 52)

Peronospora aparines (de Bary) Gäum.
Downy mildew on leaves of *Galium aparine*; common.
50, 51. SJ 18, 26, 35. ER, LO, MO, PA, WW. (6) 1972. (49, 52)

Peronospora arenariae (Berk.) Tul.
Downy mildew on leaves of *Moehringia trinervia*; uncommon.
51. Recorded only from Llyn Helyg (SJ 17) in 1978.

Peronospora arvensis Gäum.
Downy mildew on *Veronica hederifolia*; common.
50, 51. SJ 16, 26, 34, 36. ER, GW, LO, MO. (5) 2014. (48, 49)

Species Accounts

Peronospora calotheca de Bary
Downy mildew on leaves of *Galium odoratum*; uncommon.
50. Recorded only from Loggerheads Country Park (SJ 16) in 1978.

Peronospora chenopodii Schltdl.
Downy mildew on leaves of *Chenopodium album*; uncommon.
50, 51. Recorded from Bagillt (SJ 17) in 1978 and from Denbighshire, without locality, by Chater et al. (2020). (49, 52)

Peronospora conferta (Unger) Unger
Downy mildew on leaves of *Cerastium fontanum* ssp. *holosteoides*; frequent.
50, 51. SJ 16, 26, 36. LO, MO. (3) 2012. (49, 52)

Peronospora destructor (Berk.) Casp.
Downy mildew on leaves of *Allium cepa* and *A. ursinum*; uncommon.
50, 51. 16, 26. AV. (2) 1978 and recorded from Denbighshire, without locality, by Preece (2002).

Peronospora digitalidis Gäum.
Downy mildew on leaves of *Digitalis purpurea*; uncommon.
51. Recorded only from Llyn Helyg (SJ 17) in 1978. (49, 52)

Peronospora ficariae Tul.
Downy mildew on leaves of *Ficaria verna*, causing the leaves to swell and stand upright; common.
50, 51. SH 87, 97; SJ 14, 16, 17, 23, 24, 26, 34, 36. AV, ER, GW, LO, MO, WW. (14) 1978. (49, 52)

Peronospora gei H. Syd.
Downy mildew on leaves of *Geum urbanum*; rare.
51. Recorded only from Loggerheads Country Park (SJ 26) since 1976; this is the only Welsh site.

Peronospora grisea (Unger) Unger
Downy mildew on leaves of *Veronica serpyllifolia*; frequent.
50. Recorded only from near Chirk Castle (SJ 23) in 1972. (48, 49, 52)

Peronospora lamii A. Braun
Downy mildew on leaves of *Lamium purpureum*; uncommon.
51. SJ 26, 36. GW, WW. (3) 2012. (52)

Peronospora meconopsidis Major
Downy mildew on leaves of *Papaver cambrica*; uncommon.
50, 51. SH 77; SJ 26. MO. (2) 1958. (48, 49)

Peronospora minor (Casp.) Gäum.
Downy mildew on leaves of *Atriplex prostrata*; uncommon.
50. Recorded only from Llanferres (SJ 16) in 1976. (48, 49, 52)

Peronospora myosotidis de Bary
Downy mildew on leaves of *Myosotis sylvatica*; uncommon.
51. Recorded only from Mold (SJ 26) in 2009. (49)

Peronospora oerteliana Kuhn
Downy mildew on leaves of *Primula veris* and *P. vulgaris*; uncommon.
50, 51. SJ 04, 16, 17, 23, 36. LO. (6) 1866. (48)

Peronospora potentillae-reptantis Gäum.
Downy mildew on leaves of *Potentilla reptans*; rare.
51. Recorded only from Mold (SJ 26) in 1978. (49)

Peronospora ranunculi Gäum.
Downy mildew on leaves of *Ranunculus repens*; common.
50, 51. Recorded only from Mold (SJ 26) in 1978. and from Denbighshire, without locality, by Chater et al. (2020.) (48, 49, 52)

Peronospora romanica Saud. & Rayss
Downy mildew on leaves of *Medicago lupulina*; uncommon.
50. Recorded only from Erddig (SJ 34) in 1985.

Peronospora rubi Rabenh.
Downy mildew on leaves of *Rubus idaeus*; uncommon.
50, 51. SH 97; SJ 26, 34, 36. ER, MO. (4) 1978. (49, 52)

Peronospora sepium Gäum.
Downy mildew on leaves of *Vicia sepium*; uncommon.
50, 51. SJ 26, 36. MO. (2) 1978. (49)

Peronospora sordida Berk. & Br.
Downy mildew on leaves of *Scrophularia nodosa*; frequent.
50, 51. SJ 16, 34. AL, ER. (2) 1925. (49, 52)

Peronospora symphyti Gäum.
Downy mildew on leaves of *Symphytum officinale*; uncommon.
51. Recorded only from Mold (SJ 26) in 1978. This is the only Welsh record.

Peronospora trifolii-hybridi Gäum.
Powdery mildew on *Trifolium hybridum* and *T. pratense*; uncommon.
50, 51. SJ 06, 16, 35. LO. (4) 1985.

Peronospora trifoliorum de Bary
Downy mildew on leaves of *Trifolium dubium, T. hybridum, T. medium* and *T. repens*; common.
51. SJ 16, 18. LO, PA. (2) 1978.

Plasmopara angelicae (Casp.) Trotter
Plasmopara umbelliferarum (Casp.) J. Schröt. & Wartenw. in part
Downy mildew on *Angelica*; common.
50, 51. Recorded from Ysceifiog (SJ 17) in 1976 and from Denbighshire, without locality, by Chater et al. (2020). (48, 49, 52)

Species Accounts

Plasmopara chaerophylli (Casp.) Trotter
Plasmopara umbelliferarum in part
Downy mildew on *Anthriscus*; common.
50, 51. SJ 16, 17, 26. DU, LO, MA, MF, WW. (5) 1983. (49, 52)

Plasmopara densa (Rab.) J. Schröt.
Downy mildew on leaves of *Odontites verna* and *Rhinanthus minor*; frequent.
50, 51. SJ 18, 26, 34. ER, MO, PA. (3) 1978. (48, 49, 52)

Plasmopara nivea J. Schröt.
Plasmopara umbelliferarum in part
Downy mildew on *Aegopodium*; common.
50, 51. Recorded from the Alyn Valley (SJ 16) and from Denbighshire, without locality, by Chater et al. (2020). (48, 49, 52)

Plasmopara pusilla (de Bary) J. Schröt.
Downy mildew on leaves of *Geranium pyrenaicum*; frequent.
51. SJ 26. MO, WW. (2) 1978. (52)

Plasmoverna pygmaea (Unger) Constant., Voglmayr, Fatehi & Thines
Plasmopara pygmaea (Unger) J. Schröt.
Downy mildew on leaves of *Anemone nemorosa*; frequent.
50, 51. SH 77; SJ 15–17, 26, 34. AV, LO, WW. (7) 1910. (48, 49, 52)

Pseudoperonospora urticae (Berk.) Salmon & Ware
Downy mildew on leaves of *Urtica dioica*; frequent.
51. SJ 08, 18, 26, 36. GW, MO. (4) 1978.

Order PYTHIALES
Family Pythiaceae

Phytophthora infestans (Mont.) de Bary
The cause of potato blight, common but rarely reported to mycologists.
50. Recorded from Llanferres (SJ 16) since 1978, but undoubtedly more widespread. (49)

Phytophthora ramorum Warres et al.
The cause of 'sudden oak death', recently arrived in Britain from the United States.
50, 51. SH 77; SJ 05. CF. (2) Reported from nursery stock in both counties, from Bodnant Gardens in 2009 and from a private larch plantation in Denbighshire in August 2010, but locality not released. In 2011 the species was reported from Clocaenog Forest, a worrying development.

Pythium ultimum Trow.
The cause of fruit rot of cucumber but also isolated from soil; frequent.
51. SJ 26. MO. (2) 1959.

Class SAPROLEGNIOMYCETES
Order LEPTOMITALES
Family Leptomitaceae

Leptomitus lacteus (Roth) Agardh
In streams contaminated with cow dung and sewage tanks; widespread.
51. Recorded only from Mold (SJ 26) in 1977.

Order SAPROLEGNIALES
Family Saprolegniaceae

Achlya racemosa Hildebr.
Water mould on algae in pond; frequent.
51. Recorded only from Ddol Uchaf Nature Reserve (SJ 17) in 1985.

Species Accounts

Phylum CHYTRIDIOMYCOTA
Class CHYTRIDIOMYCETES
Order CHYTRIDIALES
Family Synchytriaceae

Synchytrium aureum J. Schröt.
Crinkle galls on leaves of *Plantago media*; rare.
50. Recorded only from World's End (SJ 24) in 1910.

Synchytrium endobioticum (Schilb.) Perc.
Causes wart disease of potatoes; widespread.
51. Recorded from Mold (SJ 26) since 1978 but surely common throughout the area.

Synchytrium mercurialis (Lib.) Fuckel
Crinkle galls on leaves of *Mercurialis perennis*; frequent.
50, 51. SJ 15–17, 24, 27, 34. AV, CN, DU, ER, LO. (9) 1977. (49, 52)

Synchytrium taraxaci de Bary & Woron.
Crinkle galls on leaves of *Taraxacum*; frequent.
51. SJ 16, 17, 26, 36. AV, LO, WW. (6) 1976.

Phylum GLOMEROMYCOTA
Class GLOMEROMYCETES
Order GLOMALES
Family Glomaceae

Glomus macrocarpus Tul. & C. Tul.
Subterranean, on beech roots; frequent.
51. Recorded only from Loggerheads Country Park (SJ 26) in 1993.

Species Accounts

Phylum ZYGOMYCOTA
Class ZYGOMYCETES
Order ENTOMOPHTHORALES
Family Entomophthoraceae

Entomophthora muscae (Cohn.) Fres.
Parasitic on muscid flies; common.
50, 51. SJ 13, 16, 17, 23, 24, 26, 34. DU, ER, LO, MO. (8) 1975. (49, 52)

Erynia anglica (Petch) Ben-Ze'er
Parasitic on staphylinid beetles; frequent.
51. Recorded only from Loggerheads Country Park (SJ 26) in 1985.

Erynia coleopterorum (Petch) Humber & Ben-Ze'er
Parasitic on curculionid beetles; frequent.
51. Recorded only from Point of Ayr (SJ 18) in 1985.

Erynia conica (Now.) Rem. & Henn.
Parasitic on mayfly nymphs, spores collected in river foam; common.
50, 51. SJ 16, 26, 35. LO, MO. (3) 1985. Widespread in the River Alyn.

Zoophthora radicans (Bref.) Batko
On moribund insects; uncommon.
50. Recorded only from Erddig (SJ 34) in 1910.

Order MORTIERELLALES
Family Mortierellaceae

Mortierella baineri Costantin
Isolated from garden soil; frequent.
51. Recorded only from Mold (SJ 26) in 1976.

Mortierella isabellina Oud.
In bracken litter; frequent.
51. Recorded only from Hope Mountain (SJ 25) in 1978.

Order MUCORALES
Family Chaetocladiaceae

Chaetocladium brefeldi v. Tiegh. & Le Monn.
Isolated from garden soil; frequent.
51. Recorded only from Mold (SJ 26) in 1976.

Family Cunninghamellaceae

Cunninghamella elegans Lendn.
On the bark of living oak; uncommon.
50. Recorded only from Fenn's Bank (SJ 53) in 2008.

Family Mucoraceae

Mucor hiemalis Wehmer
Isolated from bark and plant litter; common.
50, 51. SJ 16, 25, 26, 35. LO, MO. (5) 1976.

Mucor mucedo (L.) Bref.
On mammalian dung, especially of dogs; common.
50, 51. SJ 08, 16, 18, 25, 26, 35. LO, MO, NF. (7) 1976. (49, 52)

Mucor pusillus Lindt
Isolated from garden compost heap; frequent.
51. Recorded only from Mold (SJ 26) in 1978.

Mucor racemosus Fres.
Isolated from garden soil; common.
51. Recorded only from Mold (SJ 26) in 1978.

Rhizopus stolonifer (Ehrenb.) Lindt
The 'bread mould'; common.
51. Recorded from Mold (SJ 26) in 1975, but surely more widespread.

Spinellus fusiger (Link) v. Tiegh.
Parasitic on mushrooms of the genus *Mycena*, including *M. amicta*, *M. epipterygia*, *M. galopus*, *M. metata*, *M. pura* and *M. sanguinolenta*; common.
50, 51. SH 85; SJ 07, 13, 15–17, 25, 34–36. DU, ER, GW, LH, MF, NF, NM. (15) 1924. (49, 52)

Syzygites megalocarpus Ehrenb.
Parasitic on *Hygrocybe* and other agarics; frequent.
50. Recorded only from the Aberduna Nature Reserve, Maeshafn (SJ 26) in 2008.

Family Mycotyphaceae

Mycotypha microspora Fenner
On rabbit dung; uncommon.
51. Recorded only from Mold (SJ 26) in 1978.

Family Pilobolaceae

Pilaira anomala J. Schröt.
On rabbit dung; frequent.
51. Recorded only from Mold (SJ 26) in 1976. (48)

Pilobolus crystallinus (F.H. Wigg.) Tode
On rabbit dung; common.
50, 51. SJ 15, 26. MO. (2) 1976. (48, 49, 52)

Species Accounts

Pilobolus kleinii v. Tiegh.
On rabbit dung; frequent.
51. Recorded only from Mold (SJ 26) in 1976. (48)

Order ZOOPAGALES
Family Piptocephalidaceae

Piptocephalus cylindrospora Bain.
Parasitic on *Mucor* on dung; uncommon.
51. Recorded only from Mold (SJ 26) in 1976.

The Fungi of North East Wales

Phylum ASCOMYCOTA
Sub-phylum PEZIZOMYCOTINA
Class DOTHIDEOMYCETES
Order CAPNODIALES
Family Capnodiaceae

Unassigned anamorphs of **Trichomerium:**

Tripospermum camelopardus Ingold, Dann & McDougall
In foam in fast flowing rivers; uncommon.
51. Recorded only from the River Alyn near Cilcain (SJ 16) in 1981. (48, 49)

Tripospermum myrti (Lind) S.J. Hughes
In river foam from River Alyn; uncommon.
50, 51. SJ 16, 26. LO, MO. (3) 1978. (48)

Family Davidiellaceae

Unassigned anamorphs of **Davidiella:**

Cladosporium cladosporioides (Fresen.) de Vries
'Sooty mould' on living leaves and stems; frequent.
50, 51. SH 97; SJ 15, 16, 34. AV, BO, ER. (5) 1976.

Cladosporium herbarum (Pers.) Link
On litter in damp places; common.
50, 51. SJ 15, 16, 25, 26, 34. AV, ER, HM, MO, NF. (7) 1910. (48, 52)

Cladosporium sphaerospermum Penzig
On dead stems of *Urtica dioica*; frequent.
51. Recorded only from Mold (SJ 26) since 1977.

Family Mycosphaerellaceae

Cymadothea trifolii (Pers.) Wolf *Clover Blotch*
Polythrincium trifolii Kunze
On leaves of *Trifolium pratense* and *T. repens*; common.
50, 51. SH 77; SJ 16, 17, 26. DU, LO, MO. (4) 1977. (49)

Mycosphaerella ascophylli Cotton
Thallus spots on *Pelvetia canaliculata*; common.
50. Recorded only from Rhos-on-Sea (SH 88) in 2006. (48, 52)

Mycosphaerella buxicola (DC.) Tomlin
Leaf spot on *Buxus sempervirens*; uncommon.
50. Recorded only from Erddig (SJ 34) in 1985.

Mycosphaerella chelidonii (Fautrey & Lambotte) Guyot
Leaf spot on *Chelidonium majus*; rare.
50. Recorded only from Maeshafn (SJ 16) in 2019.

Species Accounts

Mycosphaerella fagi (Auersw.) Lindau
Leaf spot on *Fagus sylvatica*; frequent.
51. Recorded only from Hawarden (SJ 36) in 2017.

Mycosphaerella fragariae (Tul.) Lindau
Leaf spot on *Fragaria* and *Potentilla*; common.
50, 51. SJ 15, 26, 35. MO. (3) SJ 26 in 1978, as the anamorph, *Ramularia grevilleana*, but surely more widespread.

Mycosphaerella hedericola Lindau
Leaf spot on ivy; common.
50, 51. SH 97; SJ 15– 17, 25, 26, 35, 36. BO, CN, DU, LO, MA, MO, NF. (11) 1910. Found usually as the anamorph, *Septoria hederae* Desm.

Mycosphaerella isariophora (Desm.) Johanson
Leaf spot on *Stellaria holostea* and *S. media*; uncommon.
50, 51. SJ 17, 26, 35. LH, MO, WW. (4) 1978. Found only as the anamorph, *Septoria stellariae* Rob. & Desm.

Mycosphaerella ligustri (Desm.) Lindau
Leaf spot on *Ligustrum ovalifolium*; uncommon.
50. Recorded only from Llanferres (SJ 16) in 1978.

Mycosphaerella lineolata (Desm.) J. Schröt.
Leaf spot on *Ammophila arenaria*; common.
51. Recorded only from Point of Ayr (SJ 18) in 1978.

Mycosphaerella mercurialis (Lamb.) Magnus
Leaf spot on *Mercurialis perennis*; frequent.
50, 51. SH 85; SJ 16, 26. LO, MA. (4) 1924. (49, 52)

Mycosphaerella plantaginis (Sollm.) Vestergr.
Leaf spot on *Plantago major*; frequent.
50. SJ 34, 45. ER. (2) 2016, as the *Ascochyta* anamorph.

Mycosphaerella podagrariae (Fr.) Petrak
Leaf spot of *Aegopodium podagraria*; frequent.
50, 51. SJ 15, 16, 26, 35. LO, MA, MO. (4) 1978. (49)

Mycosphaerella punctiformis (Pers.) Starb.
On oak leaf litter; common.
50, 51. SH 85; SJ 17, 36. GW. (3) 1924.

Mycosphaerella superflua (Auersw.) Petrak
Leaf spot of *Urtica dioica*; common.
50, 51. SH 85; SJ 26, 35. MO. (3) 1978. Found as the anamorph. *Ramularia urticae* Ces. (49, 52)

Mycosphaerella ulmi Kleb.
Leaf spot of *Ulmus glabra*; frequent.
50, 51. SJ 15, 16, 26. LO, MA. (4) 1977, as the anamorph, *Phleospora ulmi* (Fr.) Wallr.

Unassigned anamorphs of **Mycosphaerella:**

Cercospora armoraciae Sacc.
Leaf spot on *Armoracia rusticana*; frequent.
51. Recorded only from Mold (SJ 26) since 1977.

Cercospora beticola Sacc.
Leaf spot on beetroot; frequent.
51. Recorded only from Mold (SJ 26) since 1977.

Cercospora violae Sacc.
Leaf spot on *Viola riviniana*; common.
51. SJ 16, 26. AV, LO, MO. (3) 1977.

Fusidella depressa (Berk. & Broome) Videira & Crous
Passalora depressa (Berk. & Br.) Sacc.; *Cercosporidium depressum* (Berk. & Br.) Deighton
Leaf spot on *Angelica sylvestris*; frequent.
51. SJ 16, 26, 36. GW, LO, MA. (3) 1972. (48, 52)

Paracercosporidium microsorum (Sacc.) U. Braun, C. Nakash, Viderira & Crous
Leaf spot on *Corylus avellana*; frequent.
50. Recorded only from Wynnstay Park (SJ 34) in 2019.

Phleospora pseudoplatani (Rab. & Desm.) Bubak
Leaf spot on sycamore; uncommon.
50, 51. SJ 16, 34. AV, ER. (2) 1976.

Ramularia abscondita (Fautrey & F. Lamb) U. Braun
Leaf spot on *Arctium minus*; frequent.
50, 51. SJ 16, 26, 35. MA, MO. (3) 1978.

Ramularia adoxae (Rabenh.) Karst.
Leaf spot on *Adoxa moschatellina*; uncommon.
51. SJ 17, 36. (2) 1986.

Ramularia ajugae (Niessl.) Sacc.
Leaf spot on *Ajuga reptans*; frequent.
51. SJ 17, 26. MO. (2) 1978.

Ramularia armoraciae Fuckel
Leaf spot on *Armoracia rusticana*; frequent.
51. Recorded only from Mold (SJ 26) since 1976. (49)

Ramularia asplenii Jaap
Leaf spot on *Aspleniun ruta-muraria*; uncommon.
51. Recorded only from Holywell (SJ 17) in 2017.

Ramularia beticola Fautrey & Lamb.
Leaf spot on *Beta maritima*; frequent.
51. Recorded only from Point of Ayr (SJ 18) since 1978.

Ramularia cardamines Syd.
Leaf spot on *Cardamine pratensis*; frequent.
51. Recorded only from Mold (SJ 26) in 1978. (49, 52)

Ramularia carneola (Sacc.) Nannf.
Leaf spot on *Scrophularia nodosa*; frequent.
50, 51. SJ 16, 26, 35, 36. GW, LO, MA (6) 1972. (48, 49)

Ramularia chamaenerii Rostr.
Leaf spot on *Chamaenerion angustifolium*; common.
50, 51. SH 85; SJ 16, 26, 35. LO, MA, MO. (6) 1978.

Ramularia cynarae Sacc.
Ramularia cardui Karst.; *R. cirsii* Allesch.
Leaf spot on Cirsium; common.
50, 51. SJ 16, 26. MA, MO. (2) 1978.

Ramularia deusta (Fuckel) Karakulin
Leaf spot on *Lathyrus montanus*; uncommon.
51. Recorded only from Maes-y-Groes (SJ 17) in 1978.

Ramularia didyma Unger
Leaf spot on *Ranunculus acris* and *R. repens*; common.
50, 51. SH 85; SJ 16, 26. MA, MO. (4) 1924. (49, 52)

Ramularia digitalis (Fuckel) U. Braun
Leaf spot on *Digitalis purpurea*; common.
50, 51. SJ 16, 26. LO, MA, MO. (3) 1978. (49)

Ramularia epilobiana (Sacc. & Fautrey) B. Sutton & Piroz
Leaf spot on *Epilobium montanum*; uncommon.
51. Recorded only from Mold (SJ 26) in 1977. (49)

Ramularia filaris Fresen.
Leaf spot on *Jacobaea vulgaris*; frequent.
51. Recorded only from Mold (SJ 26) in 1978. (49)

Ramularia gei (A.G. Eliass.) Lindroth
Leaf spot on *Geum rivale*; uncommon.
50, 51. SJ 17, 34. DU, ER. (2) 1982. (48)

Ramularia geranii Fuckel
Leaf spot on *Geranium robertianum*; common.
50. SJ 16, 34. ER, MA. (2) 2016.

Ramularia glechomatis U. Braun
Leaf spot on *Glechoma hederacea*; common.
50, 51. SJ 17, 26, 34. ER, MO. (5) 1910.

Ramularia heraclei (Oud.) Sacc.
Leaf spot on *Heracleum sphondylium*; common.
51. SJ 16, 26. MA, MO. (3) 1977.

Ramularia inaequale (Preuss) U. Braun
Leaf spot on *Taraxacum*; common.
51. SJ 17, 26, 36. DU, MO. (4) 1978. (49)

Ramularia kriegeriana Bres.
Leaf spot on *Plantago major*; uncommon.
51. Recorded only from Mold (SJ 26) in 1978. (49)

Ramularia lactea (Desm.) Sacc.
Leaf spot on *Viola riviniana*; frequent.
51. SJ 17, 26. MO. (2) 1978. (49)

Ramularia lamii Fuckel
Leaf spot on *Lamium album*; uncommon.
51. Recorded only from Loggerheads Country Park (SJ 26) in 1978.

Ramularia lampsanae (Desm.) Sacc.
Leaf spot on *Lapsana communis*; frequent.
51. Recorded only from Mold (SJ 26) since 1978. (49)

Ramularia lychnicola Cooke
Leaf spot on *Silene dioica*; common.
50, 51. SH 97; SJ 15, 17, 26, 36. BO, GW, LO, MO. (6) 1972.

Ramularia parietariae Passer
Leaf spot on *Parietaria judaica*; uncommon.
51. Recorded only from Prestatyn (SJ 08) in 1992. (49)

Ramularia pratensis Sacc.
Leaf spot on *Rumex acetosa, R. acetosella* and *Rheum barbarum*; common.
50, 51. SJ 16, 26. LO, MO. (3) 1978. (49)

Ramularia primulae Thüm.
Leaf spot on *Primula polyanthus, P. veris* and *P. vulgaris*; common.
50, 51. SJ 17, 26. DU, MO. (3) 1975. (48)

Ramularia rhabdospora (Berk. & Br.) Nannf.
Leaf spot on *Plantago lanceolata* and *P. major*; common.
50, 51. SJ 16, 26, 35. CN, MA, MO. (5) 1978. (49)

Ramularia rosea (Fuckel) Sacc.
Leaf spot on *Salix cinerea*; common.
50. Recorded only from Wynnstay Park (SJ 34) in 2019.

Ramularia rubella (Bron.) Nannf.
Leaf spot on *Rumex crispus, R. nemoreus* and *R. obtusifolius*; common.
50, 51. SH 85; SJ 16, 17, 26, 27, 34–36. ER, LO, MA, MO. (14) 1910. (48, 49)

Ramularia sambucina Sacc.
Leaf spot on *Sambucus nigra*; common.
50. Recorded only from Maeshafn (SJ 16) in 2019.

Ramularia sphaeroidea Sacc.
Ovularia sphaeroidea (Sacc.) Sacc.
Leaf spot on *Lotus corniculatus*; uncommon.
51. Recorded only from Mold (SJ 26) in 1978. (49, 52)

Ramularia succisae Sacc.
Leaf spot on *Succisa pratensis*; common.
51. Recorded only from Loggerheads Country Park (SJ 26) in 2016.

Ramularia triboutiana (Sacc. & Letendre) Nannf.
Leaf spot on *Centaurea nigra*; common.
50, 51. SJ 16, 26, 35, 45. LO, NA. (4) 2016.

Ramularia ulmariae Cooke
Leaf spot on *Filipendula ulmaria*; frequent.
50, 51. SJ 15, 16, 26, 27. LO, MO. (4) 1978.

Septoria antirrhini Rob. & Desm.
Leaf spot on *Antirrhinum majus*; common.
51. Recorded only from Mold (SJ 26) since 1978.

Septoria apiicola Speg.
Leaf spot on *Apium nodiflorum*; uncommon.
51. Recorded only from Gwysaney Park, Mold (SJ 26) in 1978.

Septoria asperulae Baüml.
Leaf spot on *Galium odoratum*; rare.
50. Recorded only from Loggerheads (SJ 16) in 2015.

Septoria bellidis Desm. & Rob.
Leaf spot on *Bellis perennis*; frequent.
51. Recorded only from Mold (SJ 26) since 1978.

Septoria berberidis Niessl.
Leaf spot on *Berberis gagnepanii*; uncommon.
51. Recorded only from Mold (SJ 26) since 1977.

Septoria clematidis Rob. & Desm.
Leaf spot on *Clematis vitalba*; uncommon.
50. Recorded only from Tan-y-Gopa, Abergele (SH 97) in 1999.

Septoria convolvuli Desm.
Leaf spot on *Convolvulus arvensis*; uncommon.
50, 51. SJ 26, 35. MO. (2) 1978.

Septoria dianthi Desm.
Leaf spot on *Dianthus barbatus* and *D. plumarius*; frequent.
51. Recorded only from Mold (SJ 26) since 1978.

Septoria digitalis Pers.
Leaf spot on *Digitalis purpurea*; common.
51. Recorded only from Llyn Helyg (SJ 17) in 1977.

Septoria epilobii Westend.
Leaf spot on *Epilobium* spp.; common.
50. Recorded only from Alyn Waters Country Park (SJ 35) in 2016.

Septoria exotica Speg.
Leaf spot on *Veronica x franciscana*; uncommon.
51. Recorded only from Mold (SJ 26) in 1974. (49)

Septoria ficariae Desm.
Leaf spot on *Ranunculus ficaria*; frequent.
51. Recorded only from Ysceifiog (SJ 17) since 1978.

Septoria geranii Rib. & Desm.
Leaf spot on *Geranium* species; uncommon.
50. Recorded from Chirk (SJ 23) by Grove (1935). (49)

Septoria hippocanstani Berk. & Broome
Leaf spot on *Aesculus hippocastanum*; uncommon.
50. Recorded only from Wynnstay Park (SJ 34) in 2019.

Septoria hyperici Desm.
Leaf spot on *Hypericum* sp.; frequent.
51. Recorded only from Mold (SJ 26) in 2017.

Septoria lychnidis Desm.
Leaf spot on *Silene dioica*; frequent.
51. Recorded only from Loggerheads Country Park (SJ 26) in 2016.

Septoria sanguisorbae Jørst.
Leaf spot on *Poterium sanguisorba*; uncommon.
50. Recorded only from Maeshafn (SJ 16) in 2019.

Septoria stachydis Rob. & Desm.
Leaf spot on *Stachys sylvatica*; common.
50, 51. SJ 15, 17, 36. GW. (3) 1972. (52)

Septoria urticae Rob. & Desm.
Leaf spot on *Urtica dioica*; common.
50, 51. SJ 16, 26, 35, 45. LO, MA. (4) 2016.

Species Accounts

Septoria veronicae Rob. & Desm.
Leaf spot on *Veronica* species; uncommon.
50. Recorded from Denbighshire by Grove (1935) without locality.

Septoria violae-palustris Died.
Leaf spot on *Viola riviniana*: uncommon.
50. Recorded only from Hafod Wood (SH 85) in 1924.

Septoria wistariae Brun.
Leaf spot on *Wisteria sinensis*; common.
51. Recorded only from Mold (SJ 26) since 1977.

Spermosporina aricola U. Braum
Leaf spot on *Arum maculatum*; frequent.
50, 51. SJ 16, 26. AV, LO, WW. (3) 1977. (52)

Sphaerulian westendorpii Verkley, Quaedvl. & Crous
Sphaerulina rubi Demane & M.S. Wilcox; *Mycosphaerella rubi* Roark
Leaf spot on *Rubus* spp.; uncommon.
50, 51. SJ 16, 26, 35. CN, LO, MO. (4) 1976, as the *Septoria* anamorph.

Stigmidium eucline (Nyl.) Vězda
Parasitic on the lichen *Pertusaria lactea*; rare.
50. No details.

Unassigned anamorph of Mycosphaerellaceae:

Stigmina carpophila (Lév.) M.B. Ellis
On leaves and fruit of peach trees; uncommon.
51. SH 97; SJ 27. BO. (2) 1864.

Family Torulaceae

Unassigned anamorph of Torulaceae:

Torula herbarum (Pers.) Link
On herbaceous litter; common.
50, 51. SJ 26, 34. ER, MO. (2) 1977. (48, 49, 52)

Order DOTHIDEALES
Family Dothioraceae

Dothiora ribesia (Pers.) Barr
On fallen twigs of *Ribes rubrum*; uncommon.
50. Recorded only from Erddig (SJ 34) in 1985.

Family Saccotheciaceae

Unassigned anamorph of **Discosphaerina**:

Aureobasidium pullulans (de Bary) Arnaud
On bracken litter and in soil; common.
51. SJ 25, 26, 36. GW, HM, MO. (3) 1976.

Unassigned anamorph of Saccotheciaceae:

Pseudoseptoria donacis (Pers.) Sutton
Leaf spot on *Poa pratensis*; uncommon.
50. Recorded only from Llanferres (SJ 16) in 1976.

(Family uncertain)

Passeriniella obiones (P. Crouan & H. Crouan) K.D. Hyatt & Mouzouras
On stems of *Halimione portulacoides*; rare.
51. Recorded only from Point of Ayr (SJ 18) in 2017.

Order MYRIANGIALES
Family Myriangiaceae

Myriangium duriaei Mont. & Berk.
Parasitic on scale insects; uncommon.
50. Recorded only from Hafod Wood (SH 85) in 1924.

Order PLEOSPORALES
Family Didymellaceae

Unassigned anamorphs of **Didymella**:

Ascochyta aquilegiae (Roum. & Pat.) Sacc.
Actinonema aquilegiae (Roum. & Pat.) Grove
Leaf spot on *Aquilegia alpina* and *A. vulgaris*; frequent.
50, 51. SJ 16, 26. MO. (2) 1978.

Ascochyta avenae (Petrak) Sprague & Johnson
Leaf spot on *Phleum pratense*; frequent.
51. Recorded only from Mold (SJ 26) in 1978.

Ascochyta dahliicola (Brun) Petrak
Leaf spot of dahlias; frequent.
51. Recorded only from Mold (SJ 26) on 1977.

Ascochyta metulispora Berk. & Br.
Leaf spot of ash; frequent.
50, 51. SJ 16, 26. MA, MO. (2) 1978.

Ascochyta primulae Trail
Leaf spot on *Primula vulgaris*; frequent.
51. Recorded only from Mold (SJ 26) since 1976.

Ascochyta syringae Bres.
Leaf spot of lilac; frequent.
51. Recorded only from Mold (SJ 26) since 1978.

Ascochyta tenerrima Sacc. & Boud.
Leaf spot on *Symphoricarpus*; frequent.
50. SJ 16, 45. MA. (2) 2016.

Peyronellaea curtisii (Berk.) Aveskamp, Gruyter & Verkley
Stagonospora curtisii (Berk.) Sacc.
Leaf scorch of *Narcissus* species; frequent.
51. Recorded from Flintshire, without locality, by Grove (1935).

Phoma complanata (Tode) Desm.
On dead stems of *Heracleum sphondylium*; common.
51. Recorded only from Mold (SJ 26) since 1978.

Phoma mahoniae Thüm.
On dead branches of *Mahonia aquifolium*; rare.
51. Recorded from Flintshire by Grove (1935) without locality.

Phoma tussilaginis Cooke & Massee
Leaf spot on *Tussilago farfara*; common.
50. Recorded only from Maeshafn (SJ 16) in 2019.

Leptosphaerulina myrtillina (Sacc. & Fautr.) Petrak
Leaf spot on *Vaccinium myrtilllus*; common.
51. Recorded only from Hope Mountain (SJ 25) in 1978.

Unassigned anamorphs of Didymellaceae:

Boeremia hedericola (Durieu & Mont.) Aveskamp, Gruyter & Verkley
Phoma hedericola (Dur. & Mont.) Boerema
Causing white spots on ivy leaves; common.
51. Recorded only from Mold (SJ 26) since 1978. (48, 52)

Neosetophoma samararum (Desm.) Gruyter, Aveskamp & Verkley
Phoma samararum Desm.; *Macrophoma fraxini* Delacr.
On fallen ash fruits; common.
50, 51. SH 85; SJ 26, 34. ER, LO. (4) 1924. (49)

Family Leptosphaeriaceae

Leptosphaeria acuta (Hoffm.) P. Karst. *Nettle Rash*
On dead stems of *Urtica dioica*; common.
50, 51. SJ 15, 16, 25, 26, 34, 35, 43. CE, CN, ER, LO, MO, NF, WW. (13) 1910. (48, 49, 52)

Leptosphaeria coniothyrium (Fuckel) Sacc.
In garden leaf litter, especially on fallen *Laburnum* pods; common.
51. Recorded only from Mold (SJ 26) in 1978, as the *Paraconiothyrium* anamorph.

Leptosphaeria doliolum (Pers.) Ces. & de Not.
On dead stems of *Urtica dioica*; common.
50, 51. SJ 26, 34, 35. ER, MO. (3) 1977. (48)

Family Melanommataceae

Herpotrichia macrotricha (Berk. & Br.) Sacc.
On sycamore leaf litter; common.
51. Recorded only from Loggerheads Country Park (SJ 26) in 1983. (49)

Melanomma fuscidulum Sacc.
On dead, fallen branches of sycamore; uncommon.
51. Recorded only from Loggerheads Country Park (SJ 26) in 1983.

Melanomma pulvis-pyrius (Pers.) Fuckel
On sticks and fallen branches of all kinds; common.
50, 51. SJ 15, 25, 26, 34. ER, LO, NF. (4) 1910. (48, 49, 52)

Family Montagulaceae

Paraphaeosphaeria rusci (Wallr.) O. Erikks.
On dead cladodes of *Ruscus aculeatus*; rare.
50, 51. SH 85; SJ 26. (2) 1977. (48)

Family Phaeosphaeriaceae

Ophiobolus acuminatus (Sow.) Duby
On dead herbaceous stems; frequent.
50. Recorded only from Coed Coch (SH 87) in 1880. (48)

Unassigned anamorph of **Phaeosphaeria**:

Stagonospora atriplicis (Westend.) Lind.
Leaf spot on *Atriplex patula*; uncommon.
51. Recorded only from Prestatyn (SJ 08) in 1988.

Family Pleomassariaceae

Splanchnonema pupula (Fr.) O. Kuntze
On sycamore sticks; uncommon.
51. Recorded only from Loggerheads Country Park (SJ 26) in 1983. (52)

Species Accounts

Unassigned anamorph of **Splanchnonema**:

Helminthosporium corticiorum Höhn.
Parasitic on a resupinate fungus; rare.
51. Recorded only from Hawarden (SJ 36) in 2019.

Family Pleosporaceae

Lewia scrophulariae (Desm.) M.E. Barr & E.G. Simmons
Pleospora scrophulariae (Desm.) Höhnel
On herbaceous litter; frequent.
51. Recorded only from Mold (SJ 26) in 1978.

Unassigned anamorph of **Lewia**:

Alternaria alternata (Fr.) Keissler
On dead stems of garden plants and *Urtica dioica*; common.
51. SJ 26, 36. MO. (2) 1977.

Pleospora epilobii E. Müller
On dead stems of *Chamaerion angustifolium*; uncommon.
50. Recorded only from Alyn Waters Country Park (SJ 35) in 2009.

Pleospora herbarum (Pers.) Rabenh.
On dead herbaceous stems; common.
50, 51. SJ 16, 26, 35, 36. GW, LO, MO. (4) 1976.

Pyrenophora tritici-repentis Drechsler
On living and moribund *Elytrigia repens*; common.
51. Recorded only from Loggerheads Country Park (SJ 26) in 1978, as the *Drechslera* anamorph.

Unassigned anamorphs of Pleosporaceae:

Dendryphion comosum Wallr.
On dead stems of *Urtica dioica*; common.
50, 51. SJ 26, 34. ER, MO. (2) 1977. (48, 49, 52)

Epicoccum nigrum Link
On dead herbaceous stems; common.
51. SJ 25, 26. HM, MO. (2) 1977.

Family Sporormiaceae

Sporormiella intermedia (Auersw.) Ahmed & Cain
On rabbit dung; common.
51. Recorded only from Mold (SJ 26) in 1976. (48, 52)

Unassigned anamorphs of Pleosporales:

Anguillospora crassa Ingold
In foam in fast flowing rivers; common.
50, 51. SJ 16, 17, 24–26, 34, 35. AV, DU, LO, MO, NF. (10) In Rivers Alyn, Cegidog, Clywedog, Dee, Ffrith, Terrig and Wheeler. 1981. (48, 49)

Anguillospora longissima (Sacc. & Syd.) Ingold
In foam in fast flowing rivers; common.
50, 51. SJ 16, 24–26, 34, 35. AV, LO, MO, NF. (7) In Rivers Alyn, Clywedog, Dee and Ffrith. 1981. (48, 49)

Clavariopsis aquatica De Wild.
In foam in fast flowing rivers; common.
50, 51. SJ 16, 17, 24–26, 34, 35. AV, DU, LO, MO, NF. (8) In Rivers Alyn, Clywedog, Dee, Ffrith and Wheeler. 1981. (49)

Mycocentrospora acerina (Hartig) Deighton
Causing rots in a variety of garden vegetables and flowers, also frequent in river foam; common.
51. SJ 16, 26. AV, MO. (3) In River Alyn. 1978.

Periconia byssoides Pers. ex Mèrat
On *Urtica* stems; common.
50. Recorded only from Erddig (SJ 36) in 2016.

Periconia cookei Mason & M.B. Ellis
On dead stems of *Urtica dioica*; common.
51. Recorded only from Mold (SJ 26) since 1977. (48, 49)

Order VENTURIALES
Family Venturiaceae

Atopospora betulina (Fr.) Petrak
On dead birch leaves; frequent.
51. Recorded only from Hawarden (SJ 36) in 1977. (48)

Coleroa alchemillae (Grev.) Winter
Leaf spot on *Alchemilla mollis*; uncommon.
50. Recorded only from Wynnstay Park (SJ 34) in 2019.

Coleroa robertiani (Fr.) E. Müll.
Hormotheca robertiani (Fr.) Höhn.
On living leaves of *Geranium robertianum*; common.
50, 51. SH 85; SJ 16, 17, 26, 34. ER, LO, MO. (7) 1924. (48, 49, 52)

Venturia inaequalis (Cooke) Winter
The cause of scab on cultivated apples; common.
50, 51. SJ 26, 34. AV, ER, MO. (3) 1977.

Species Accounts

Venturia maculiformis (Desm.) Winter
Leaf spot of *Chamaerion angustifolium*; common.
51. Recorded only from Nercwys Mountain (SJ 25) in 1986.

Venturia populina (Vuill.) Fabric.
Leaf spot on poplars; frequent.
50, 51. SJ 15, 16, 36. LO. (4) 2013.

Venturia pyrina Aderh.
Leaf spot on pears; frequent.
50. Recorded only from Erddig (SJ 34) in 1934.

Venturia rumicis (Desm.) Winter
Leaf spot on *Rumex nemoreus* and *R. obtusifolius*; common.
50, 51. SH 85; SJ 16, 25, 26, 34. HM, LO, MO. (6) 1924. (48, 49, 52)

Venturia saliciperda Nüesch
Leaf spot on *Salix* sp.; uncommon.
51. Recorded only from Drury (SJ 26) in 2015.

Order BOTRYOSPHAERIALES
Family Botryosphaeriaceae

Amarenomyces ammophilae (Lasch) O. Erikks.
On dead stems of *Ammophila*; frequent.
51. Recorded only from Point of Ayr (SJ 18) in 1985. (48, 52)

Unassigned anamorph of **Botryosphaeria**:

Sphaeropsis sapinea (Fr.) Dyko & Sutton
On needles and twigs of pines; frequent.
50. SH 77; SJ 34. ER. (2) 2016. (52)

Family Phyllostictaceae

Guignardia aesculi (Peck) Stewart
Leaf blotch of horse chestnut; common.
50, 51. SH 97; SJ 06, 17, 26, 34. BO, MO. (5) 1996.

Guignardia punctoidea (Cooke) J. Schröt.
In oak leaf litter; common.
51. Recorded only from Nant-y-Ffrith (SJ 25) in 1910.

Unassigned anamorphs of **Guignardia**:

Phyllosticta eupatorii All.
Leaf spot on *Eupatorium cannabinum*; uncommon.
50. Recorded only from Alyn Waters Country Park (SJ 35) in 2016.

Phyllosticta heucherae Brun
Leaf spot on *Heuchera sanguinea*; uncommon.
51. Recorded only from Hawarden (SJ 36) in 2019.

Phyllosticta hypoglossi (Mont.) Allesch.
On dying cladodes of *Ruscus aculeatus*; rare.
51. Recorded only from Soughton Hall (SJ 26) in 1977.

Phyllosticta impatientis Fautr.
Leaf spot on *Impatiens glandulifera*; uncommon.
50. Recorded only from Alyn Waters Country Park (SJ 35) in 2016.

Phyllosticta japonica Thüm.
Causing leaf spots on *Mahonia japonica*; rare.
51. Recorded from Flintshire by Grove (1935) without locality.

Phyllosticta lonicerae Westend.
Leaf spot on *Lonicera periclymenum*; frequent.
51. SJ 26. (2) MO. 1935.

Phyllosticta mahoniana All.
Leaf spot on *Mahonia aquifolium*; uncommon.
51. Recorded from Flintshire by Grove (1935) without locality.

Phyllosticta sorbicola All.
Leaf spot on *Sorbus aucuparia*; frequent.
50, 51. SJ 16, 35, 36. MA. (3) 2010.

Phyllosticta syringae Westend.
Leaf spot on lilac; uncommon.
51. Recorded only from Mold (SJ 26) since 1976.

Phyllosticta teucrii Sacc. & Speg.
Leaf spot on *Teucrium scorodonia*; common.
51. Recorded only from Loggerheads Country Park (SJ 26) in 2016.

Order HYSTERIALES
Family Hysteriaceae

Gloniopsis praelonga (Schwein.) Zogg
On dead stems of brambles and roses; common.
50, 51. SJ 17, 18, 26, 34. DU, ER, LO, MO, PA. (5) 1977. (48, 49, 52)

Hysterium angustatum Alb. & Schwein.
On the bark of living trees, especially sycamore; common.
50, 51. SH 77; SJ 16, 26, 34, 36. GW, LO, MO. (6) 1977.

Hysterium pulicare Pers.
On the bark of living trees; common.
50, 51. SH 97; SJ 04, 12, 34. BO, ER. (4) 2007.

Species Accounts

Family Thyridariaceae

Thyridaria rubronotata (Berk. & Br.) Sacc.
On fallen, dead branches of sycamore; uncommon.
51. Recorded only from Loggerheads Country Park (SJ 26) in 1983.

Order MICROTHYRIALES
Family Microthyriaceae

Lichenopeltella salicis (J.P. Ellis) P.M. Kirk
Trichothyrina salicis J.P. Ellis
On leaves of *Salix cinerea* ssp. *oleifolia*; frequent.
51. Recorded only from Ddol Uchaf Nature Reserve (SJ 17) in 1985.

Microthyrium ciliatum Gremmen & de Kam var. **hederae** J.P. Ellis
Leaf spot of *Hedera*; uncommon.
51. Recorded only from Loggerheads Country Park (SJ 26) in 1985. (48, 49)

Microthyrium gramineum Bomm., Rouss. & Sacc.
Leaf spot of *Ammophila arenaria*; frequent.
51. Recorded only from Point of Ayr (SJ 18) in 1985. (48, 52)

Microthyrium macrosporum (Sacc.) Höhn.
On dead box leaves; rare.
50. Recorded only from Rhyd-y-Creuau (SH 85) in 1988. (48)

Order PATELLARIALES
Family Arthrorhaphidaceae

Arthrorhaphis citronella (Ach.) Poelt
Parasitic on the lichen *Baeomyces rufus*; uncommon.
50. Recorded only from Llangollen (SJ 24) in 1899.

(Order uncertain)
Family Paranectriellaceae

Unassigned anamorph of **Paranectriella**:

Titea maxilliforme Rostr.
Tricladium maxilliforme (Rostr.) Ingold
In river foam; frequent.
51. Recorded only from the River Alyn in Mold (SJ 26) in 1986.

Family Polystomellaceae

Platychora ulmi (Fr.) Petrak
Leaf spot on *Ulmus glabra*; common.
50, 51. SJ 16, 17, 24. AV. (4) 1977.

(Family uncertain)

Lidophia graminis (Sacc.) Walker & B. Sutton
On stems of living wheat causing 'twist'; frequent.
50. Recorded only from Llanferres (SJ 16) in 1976, as the *Dilophospora* anamorph. (52)

Rhopographus filicinus (Fr.) Fuckel *Bracken Map*
Causing black streaks on dead stems of bracken; common everywhere.
50, 51. SH 85, 97; SJ 02, 05, 13, 15, 16, 25, 26, 35, 36. AV, CE, HM, GW, LO, MA, NF, NM, WW. (30) 1910. (48, 49, 52)

Unassigned anamorph of **Stuartella**:

Bactrodesmium abruptum (Berk. & Br.) Mason & S.J. Hughes
On rotten oak wood; frequent.
51. Recorded only from Bodelwyddan (SH 97), the type locality, in 1864. (49)

Class EUROTIOMYCETES
Sub-class Chaetothyriomycetidae
Order CHAETOTHYRIALES
Family Herpotrichiellaceae

Capronia nigerrima (Currey) Barr
On fallen, dead branches of sycamore; frequent.
51. Recorded only from Loggerheads Country Park (SJ 26) in 1983. (48, 49)

Capronia pulcherrima (Munk) E. Müll.
On fallen, dead branches of sycamore; frequent.
51. Recorded only from Loggerheads Country Park (SJ 26) in 1983.

Order VERRUCARIALES
Family Verrucariaceae

Muelleriella pygmaea (Körber) D. Hawks.
Parasitic on the lichen *Lecanora campestris*; uncommon.
50. Recorded only from Berwyn (SJ 14) in 1925.
The rest of this order are lichenised and are not dealt with further.

Sub-class Eurotiomycetidae
Order EUROTIALES
Family Elaphomycetaceae

Elaphomyces granulatus Fr. *False Truffle*
A false truffle in soil under beech; frequent.
50, 51. SH 76, 85; SJ 16. CN. (4) 1924. (48, 49, 52)

Elaphomyces muricatus Fr.
A false truffle in soil under beech; frequent.
50, 51. SH 85; SJ 16, 26. CN, LO, MF. (4) 1985. (48, 49, 52)

Species Accounts

Family Trichocomaceae

Unassigned anamorphs of **Emericella**:

Aspergillus flavus Link
On rotting bread and leaf litter; frequent.
51. SJ 17, 26. MO. (2) 1978.

Aspergillus fumigatus Fresen.
In compost heap; frequent. This species is the cause of serious lung infections in Man and farm animals.
51. Recorded only from Mold (SJ 26) in 1976.

Aspergillus parasiticus Spear
On mouldy horse chestnuts; frequent. The source of dangerous toxins in nuts, especially peanuts.
51. Recorded only from Mold (SJ 26) in 1979.

Unassigned anamorphs of **Eupenicillium**:

Penicillium aurantiogriseum Dierckx
Penicillium cyclopium Westl.
On decaying agarics; frequent.
51. Recorded only from Loggerheads Country Park (SJ 26) since 1975.

Penicillium brevicompactum Dierckx
On decaying agarics; frequent.
50, 51. SJ 05, 24, 26. CF, LO, WW. (3) 1975.

Penicillium camemberti Thom
In Brie and Camembert cheese; common.
Not native in the area but commonly introduced with popular white-mould cheeses.

Penicillium citrinum Thom
On old bread; common.
51. Recorded only from Mold (SJ 26) since 1976.

Penicillium coremiforme J. Kickx f.
On rotten wood; frequent.
51. Recorded only from Rhydymwyn Nature Reserve (SJ 26) in 2006.

Penicillium digitatum Sacc.
On mouldy oranges; frequent.
51. Recorded only from Mold (SJ 26) since 1975.

Penicillium expansum Link
On mouldy apples; frequent.
51. Recorded only from Mold (SJ 26) since 1976.

Penicillium gladioli McCulloch & Thom
On stored *Gladiolus* corms; frequent.
51. Recorded only from Mold (SJ 26) in 1975.

Penicillium hirsutum Dierckx
Penicillium corymbiferum Westl.
Causing a rot of *Crocus* corms; frequent.
51. Recorded only from Mold (SJ 26) in 1974.

Penicillium italicum Wehm.
On mouldy oranges; common.
51. Recorded only from Mold (SJ 26) since 1976.

Penicillium roquefortii Thom
In blue cheese; common.
Introduced to the area with popular blue-mould cheeses.

Penicillium vulpinum (Cooke & Massee) Siefert & Samson
Penicillium claviforme Bainier
On mixed leaf litter; frequent.
51. Recorded only from Ddol Uchaf Nature Reserve (SJ 17) in 1975.

Talaromyces duclauxii (Delacr.) Samson, N. Yilmora, Frisner & Seifert
On rotten wood; frequent.
51. SJ 07, 26. AV. (2) 1985. (48) as the anamorph *Penicillium clavigerum* Demel.

Order ONYGENALES
Family Arthrodermataceae

Unassigned anamorphs of **Arthroderma**:

Epidermophyton floccosum (Harz) Lange
A cause of 'athlete's foot' in Man; common.
Recorded for many years in the area but not by mycologists; it has definitely been observed in Mold.

Microsporum audouinii Gruby
The cause of head ringworm or 'tinea capitis' in children; once common, now rare.
Has been frequent in the past in most parts of the region but not recorded for mycological reasons.

Microsporum canis Bodin
The cause of ringworm in domestic cats; now uncommon.
51. Recorded only from Mold (SJ 26) in 1978.

Trichophyton equinum Gedoelet
The cause of ringworm in horses; frequent.
51. Recorded only from Mold (SJ 26) in 1978.

Species Accounts

Trichophyton mentagrophytes (C.P. Robin) Sabaud
Microsporum persicolor (Sabour.) Guiart & Grijnaki; *Trichophyton erinacei* (J.M.B. Sm. & Maupas) Quaife, *T. interdigitale* Priestley
The cause of athlete's foot and ringworm in *Homo, Apodemus, Erinaceus* and *Microtus*; common.
51. Recorded only from Mold (SJ 26) since 1976.

Trichophyton rubrum (Cast.) Saboud
The cause of 'athlete's foot' and body ringworm in Man; common, and beginning to develop resistance to fungicidal drugs. Recorded for generations as a major dermatophyte but not recorded for mycological reasons; the resistant strains were seen in Mold (SJ 26) from 2000 to 2006.

Trichophyton verrucosum Bodin
The cause of ringworm in cattle; frequent.
51. Recorded only from Nannerch (SJ 16) in 1977.

Family Onygenaceae

Onygena corvina Alb. & Schwein. *Feather Stalkball*
On feathers and bird pellets; uncommon.
51. Recorded only from Nant-y-Ffrith (SJ 25) in 1978. (49, 52)

Sub-class Mycocaliciomycetidae
Order MYCOCALICIALES
Family Mycocaliciaceae

Stenocybe pullulata (Ach.) R. Stein.
On unattached alder twigs; uncommon.
51. Recorded only from Loggerheads Country Park (SJ 26) in 1973.

Class GEOGLOSSOMYCETES
Order GEOGLOSSALES
Family Geoglossaceae

Geoglossum atropurpureum (Batsch) Pers. *Dark-purple Earthtongue*
Thuemenidium atropurpureum (Batsch) Kunze
In lawns and short turf on acid soil; uncommon.
50, 51. SH 85; SJ 25, 26. (3) 1924. (49)

Geoglossum cookeianum Nannf.
In sandy and calcareous grassland; frequent.
50, 51. SH 77; SJ 17, 18, 24–27, 35. HC, LO, PA. (9) 1910. (48, 49, 52)

Geoglossum fallax Durand
In calcareous turf; uncommon.
50, 51. SH 85, 97; SJ 05, 15, 17, 25, 26, 35. CF, HC, LO, MA, ML. (11) 1979. (48, 49, 52)

Geoglossum umbratile Sacc. *Plain Earthtongue*
Geoglossum nigritum Cooke
In well-grazed grassland; uncommon.
50, 51. SH 97; SJ 08, 15, 17, 23, 25, 26. AV, HC, ML. (7) 1980. (49, 52)

Glutinoglossum glutinosum (Pers.) Hustad, A.N. Mill., Dentiger & P.F. Cannon
Geoglossum glutinosum Pers.
In limestone turf; uncommon.
50, 51. SH 85, 97; SJ 24, 26. LO, ML. (4) 1980. (48, 49, 52)

Trichoglossum hirsutum (Fr.) Boud. *Hairy Earthtongue*
In limestone turf, lawns and sand dune grassland; uncommon.
50, 51. SH 85, 97; SJ 07, 08, 17, 18, 23, 25–27, 35. HC, PA. (12) 1985. (48, 49, 52)

Class LECANOROMYCETES
Sub-class Lecanoromycetidae
Order LECANORALES
Family Dactylosporaceae

Dactylospora parasitica (Flörke) Zopf
Parasitic on lichens such as *Pertusaria* species growing on trees; uncommon.
50, 51. SJ 16. LO. (2) 1973.
The majority of this large class are all lichenised fungi (lichens) which are not treated here.

Sub-class Ostropomycetidae
Order OSTROPALES
Family Stictidaceae

Stictis stellata Wallr.
On fallen branches; frequent.
50, 51. SH 88; SJ 17, 23. DU. (3) 1985. (52)

Class LEOTIOMYCETES
Sub-class Erysiphomycetidae
Order ERYSIPHALES
Family Erysiphaceae

Arthrocladiella mougeotii (Lév.) Vassilkov
Powdery mildew on leaves of *Lycium chinense*; uncommon.
50, 51. SH 97, 98; SJ 08, 17, 25. (5) 1989. (48)

Blumeria graminis (DC.) Speer
Erysiphe graminis DC.
Powdery mildew on leaves of grasses, including *Arrhenatherum elatius*, *Brachypodium sylvaticum*, *Bromus hordeaceus; Cynosurus cristatus*, *Dactylis glomerata*, *Elytrigia repens* and *Poa nemoralis*; common.
50, 51. SH 97; SJ 07, 16–18, 25–27, 36. BO, LO, MF, NF, PA. (12) 1910. (49, 52)

Erysiphe adunca (Wallr.) Fr.
Uncinula adunca (Wallr.) Lév.
Powdery mildew on leaves of *Populus* spp. and *Salix* spp.; common.
50, 51. SH 98; SJ 05, 08, 15, 16, 25–27, 35, 36. AV, CE, LO, MA, MO, NF, PA, WW. (19) 1976. (49, 52)

Species Accounts

Erysiphe alphitoides (Griff. & Maubl.) U. Braun & S. Takam.
Microsphaera alphitoides Griff. & Maubl.
Powdery mildew on leaves of *Quercus petraea, Q. robur* and *Q. x rosacea*; very common.
50, 51. SH 77, 85–88, 96, 97; SJ 05–08, 14–18, 24–27, 34–36, 43, 54. AV, BO, BP, CE, CF, ER, GW, LH, LO, MO, NF, WW. (63) 1974. (48, 49, 52) Forty years ago the ascocarps of this species were rare, only being formed in long hot summers. As a result of climate change they now occur with us every year.

Erysiphe aquilegiae DC. var. **aquilegiae**
Powdery mildew on leaves of *Aquilegia alpina, A. vulgaris* and cultivars, *Caltha palustris, Thalictrum delavayi* and *Clematis vitalba*; common.
50, 51. SH 77, 87, 97; SJ 06, 08, 16, 18, 26, 34–36. BO, ER, LO, MAS, MO, PA. (15) 1974. (49, 52)

Erysiphe aquilegiae var. **ranunculi** (Grev.) Zheng & Chen
Powdery mildew on leaves of *Clematis tangutica* and various *Clematis* cultivars, *Delphinium elatum* and cultivars, *Ranunculus acris, R. auricomus* and *R. repens*; common.
50, 51. SH 87, 94, 97; SJ 06–08, 15–18, 24–27, 34–36. AV, BO, CC, DU, ER, LO, MF, MO, NF, PA. (38) 1973. (48, 49, 52)

Erysiphe arcuata U. Braun, S. Takam. & Heluta
Powdery mildew on leaves of *Carpinus betulus*; increasing. Recorded for some years as the anamorph, *Oidium carpini* U. Braun, but now producing ascocarps regularly in England and Switzerland.
50, 51. SJ 34, 36. ER. (3) 2007.

Erysiphe azaleae (U. Braun) U. Braun & S. Takam.
Microsphaera azaleae U. Braun
Powdery mildew on leaves of *Rhododendron* spp.; increasingly common.
50, 51. SH 77; SJ 14, 25, 26, 36. (7) 2000. (48, 49, 52) The anamorph of this species has been recorded for many years, causing a serious disease of rhododendrons, but the true identity of the fungus was only established in 1998 in Shropshire. The fungus is from North America where it attacks a number of species. The original British infections were on American rhododendron species grown in Scotland.

Erysiphe begoniicola U. Braun & S. Takam.
Microsphaera begoniae Sivanesan
Powdery mildew on leaves of *Begonia rex* and *B. tuberhybrida*; frequent.
51. SJ 08, 26. MO. (2) 1972.

Erysiphe berberidis DC.
Microsphaera berberidis (DC.) Lév.
Powdery mildew on leaves of *Berberis gagnepanii, B. thunbergii, B. vulgaris* and *Mahonia aquifolium*; common.
50, 51. SH 87, 97; SJ 06–08, 15, 26, 35, 36. BO, M). (16) 1976. (49, 52)

Erysiphe betae (Vanha) Weltzien
Powdery mildew on leaves of *Beta maritima*; uncommon.
50, 51. SH 88; SJ 18. PA. (2) 2013. (49)

Erysiphe buhrii U. Braun
Powdery mildew on leaves of *Dianthus* spp., *Silene dioica* and *S. latifolia*; frequent.
50, 51. SH 77, 87, 97; SJ 06–08, 16, 18, 26, 34–36. AV, BO, CE, ER, LO. (18) 1989. (49, 52)

Erysiphe capreae DC. ex Duby
Uncinula adunca var. *regularis* (Zheng & Chen) U. Braun
Powdery mildew on leaves of *Salix caprea*; frequent.
50, 51. SJ 16, 26, 35, 36. AV, LO, MO. (6) 2000.

Erysiphe circaeae Junell
Powdery mildew on leaves of *Circaea lutetiana* and *Fuschsia* cultivars; common.
50, 51. SH 77, 87, 96; SJ 05, 08, 16–18, 24, 26, 34–36. AV, BW, ER, GW, LH, LO, PP, WW. (28) 1974. (48, 49, 52)

Erysiphe convolvuli DC. var. **convolvuli**
Powdery mildew on leaves of *Convolvulus arvensis*; uncommon.
50, 51. SH 87, 98; SJ 07, 08, 16–18, 26, 35. MO, PA. (10) 1989. (49)

Erysiphe convolvuli var. **calystegiae** U. Braun
Powdery mildew on leaves of *Calystegia sepium* and *C. silvatica*; uncommon.
50, 51. SJ 07, 18, 26, 35, 36. MO, PA. (10) 1989.

Erysiphe cotini (U. Braun) U. Braun & S. Takam.
Microsphaera cotini U. Braun
Powdery mildew of leaves of *Cotinus coggyria*; rare.
50, 51. SJ 35, 36. (5) 2003. (49) This species has become more frequent in Britain in recent years, especially on the red-pigmented cultivars.

Erysiphe cruciferarum Junell
Powdery mildew on leaves of *Alliaria petiolata, Brassica napus, B. rapa, Capsella bursa-pastoris, Crambe maritima, Erysimum cheiri, Hesperis matronalis, Lepidium heterophyllum, Lunaria annua, Sisymbrium altissimum, S. officinale, Papaver dubium, P. orientale* and *P. rhoeas*; common.
50, 51. SH 87, 94, 96, 97; SJ 05–08, 15–18, 25–27, 34–37, 54. AV, DU, ER, GW, LO, MA, MO, PA. (47) 1974. (49, 52)

Erysiphe deutziae (Bunkina) U. Braun & S. Takam.
Microsphaera deutziae Bunkina
Powdery mildew on leaves of *Deutzia scabra*; rare.
51. Recorded only from Mold (SJ 26) in 2009. (49) The Mold record was made in August and that from Bangor in September 2009. This species originated in eastern Russia and has recently spread westwards through Europe. It is becoming common in Switzerland but was recorded only in Britain in 2006 and Ireland in 2016.

Erysiphe euonymi DC.
Microsphaera eunonymi (DC.) Sacc.
Powdery mildew on leaves of *Euonymus europaeus*; uncommon.
50, 51. SJ 16, 24, 35. AV. (4) 1990. (49)

Species Accounts

Erysiphe euonymicola U. Braun
Microsphaera euonymi-japonici Viennot-Bourgin
Powdery mildew on leaves of *Euonymus japonicus*; common.
50, 51. SJ 08, 26, 35, 36. MO. (6) 1989. (48, 49, 52)

Erysiphe flexuosa (Peck) U. Braun & S. Takam.
Uncinula flexuosa Peck
Powdery mildew on leaves of *Aesculus hippocastanum*; common and increasing.
50, 51. SH 77, 87, 88, 97; SJ 07, 15, 17, 25-27, 34-36, 44. BO, DU, GW, MO, WW. (28) 2003. This north American species was reported from Europe for the first time in 2000, from two sites in Bern canton, Switzerland. It was found in Surrey in 2001 and in other parts of Switzerland in the same year (Ing & Spooner, 2002). In 2002 it was found in Belgium and Italy and in further Swiss sites. In 2003 it was found in all parts of Switzerland and was spreading in Britain, being found in several sites in the London area and over twenty counties in the Midlands and southeast, and north to Westmorland, including several sites in Cheshire. It is now probably common throughout England and Wales and has been found in Dublin and as far north as Ullapool. Its spread may be associated with climate change. (49, 52)

Erysiphe friesii (Lév.) U. Braun & S. Takam.
Microsphaera friesii Lév.
Powdery mildew on leaves of *Rhamnus cathartica*; uncommon.
51. Recorded only from Loggerheads Country Park (SJ 26) in 2016.

Erysiphe grossulariae (Wallr.) de Bary
Microsphaera grossulariae (Wallr.) Lév.
Powdery mildew on leaves of *Ribes nigrum, R. sanguineum* and *R. uva-crispa*; uncommon.
50, 51. SH 87, 97; SJ 05, 07, 08, 15-17, 26, 34, 35. AV, CN, ER, LO. (16) 1975. This is the European mildew and is more common in semi-natural vegetation on possibly native hosts. (49)

Erysiphe guarinonii (Briosi & Cavara) U. Braun & S. Takam.
Microsphaera guarinonii Briosi & Cavara
Powdery mildew on leaves of *Laburnum x watereri*; rare.
51. SJ 08, 26, 36. MO. (3) 1989.

Erysiphe hedwigii (Lév.) U. Braun & S. Takam.
Microsphaera hedwigii Lév.
Powdery mildew on leaves of *Viburnum lantana*; uncommon.
51. SJ 26. MO, WW. (2) 2000. (49)

Erysiphe heraclei DC.
Powdery mildew on leaves of *Aegopodium podagraria, Angelica sylvestris, Anthriscus sylvestris, Chaerophyllum temulum, Heracleum sphondylium, Pastinaca sativa* and *Smyrnium olusatrum*; common.
50, 51. SH 77, 86-88, 94, 96-98; SJ 05-08, 15-18, 24-27, 34-37, 43, 45, 54. AV, BO, BP, CE, DU, ER, GW, LH, LO, MA, MO, NF, NM, PA, WW. (82) 1972. (48, 49, 52)

Erysiphe howeana U. Braun
Powdery mildew on leaves of *Fuchsia* sp. and *Oenothera glazioviana*; uncommon.
50, 51. SH 77, 98; SJ 08, 18, 26, 35-37. MO, PA. (10) 1985. The first record, from the dunes at Point of Ayr, was the first British record. It is now recorded from many dune systems and also on garden plants.

Erysiphe hyperici (Wallr.) S.Blumer
Microsphaera hypericacearum U. Braun
Powdery mildew on leaves of *Hypericum androsaemum, H. calycinum, H. hirsutum, H. maculatum* and *H. perforatum*; common.
50, 51. SH 97; SJ 06, 08, 15-18, 24-26, 34, 35. AV, BO, BP, ER, LH, LO, MO, NF. (25) 1910. (49)

Erysiphe intermedia (U. Braun) U. Braun
Microsphaera trifolii var. *intermedia* U. Braun
Powdery mildew on leaves of *Lupinus arboreus* and *L. x regalis*; common.
50, 51. SJ 08, 18, 26, 34-36. ER, MO, PA. (8) 1974. (49, 52)

Erysiphe knautiae Duby
Powdery mildew on leaves of *Scabiosa caucasica* and *Succisa pratensis*; frequent.
50, 51. SH 87; SJ 08, 16, 26. AV, LO. (5) 1976.

Erysiphe lonicerae DC.
Microsphaera lonicerae (DC.) Wint.
Powdery mildew on leaves of *Lonicera periclymenum* and *L. xylosteum*; frequent.
50, 51. SH 77, 97; SJ 08, 16, 18, 24, 26, 35, 36. LO, MO. (12) 1973. (48, 49, 52)

Erysiphe lycopsidis Zheng & Chen
Powdery mildew on leaves of *Anchusa arvensis* and *Pentaglottis sempervirens*; uncommon.
50, 51. SH 98; SJ 07, 08, 18, 24, 35. PA. (6) 1989.

Erysiphe lythri L. Junell
Powdery mildew on leaves of *Lythrum salicaria*; rare.
50. Recorded only from Erddig (SJ 34) in 2016.

Erysiphe macleyae R.Y. Zeng & G.Q. Chen
Powdery mildew on *Papaver cambrica*; rare.
51. Recorded only from Mold (SJ 26) in 1990. This is the first record from Wales.

Erysiphe magnifica (U. Braun) U. Braun & S. Takam.
Microsphaera magnifica U. Braun
Powdery mildew on leaves of *Magnolia x soulangeana*; rare.
50, 51. SH 77; SJ 36. BO. 2009. This species is a new arrival from North America; it was first found in Britain in 2007, in Surrey (Kew and Wisley) and is now well established in Cheshire and is likely to spread and may become a problem.

Erysiphe mayorii S. Blumer
Powdery mildew on *Cirsium arvense*; uncommon.
50. Recorded only from Maeshafn (SJ 16) in 2019.

Erysiphe necator Schwein. var. **necator**
Uncinula necator (Schwein.) Burr.
Powdery mildew on leaves of *Vitis vinifera*; frequent.
50, 51. SJ 08, 34. ER. (2) 1989. (49)

Species Accounts

Erysiphe necator var. **ampelopsidis** (Peck) U. Braun & S. Takam.
Uncinula necator var. *ampelopsidis* (Peck) U. Braun
Powdery mildew on *Parthenocissus quinquefolia*; uncommon.
51. Recorded only from Gronant (SJ 08) in 1990.

Erysiphe ornata (U. Braun) U. Braun & S. Takam. var. **europaea** U. Braun
Microsphaera ornata U. Braun var. *europaea* U. Braun
Powdery mildew on leaves of *Betula pendula*, *B. pubescens* and *B. utilis*; uncommon.
50, 51. SH 97; SJ 07, 16-18, 25, 26, 35. AV, BO, BW, MO, NF. (11) 1976. (48, 49)

Erysiphe penicillata (Wallr.) Link
Microsphaera penicillata (Wallr.) Lév.
Powdery mildew on leaves of *Alnus glutinosa*; uncommon.
50, 51. SJ 07, 08, 15, 16, 25, 26, 35. AV, CE, LO, NF. (11) 1991. (49, 52)

Erysiphe pisi DC. var. **pisi**
Powdery mildew on leaves of *Medicago lupulina*, *Pisum sativum* and *Vicia cracca*; common.
50, 51. SH 97, 98; SJ 07, 08, 16-18, 24-27, 35, 36. MO. (16) 1974.

Erysiphe pisi var. **cruchetiana** (S. Blumer) U. Braun
Powdery mildew on leaves of *Ononis repens*; uncommon.
51. SJ 18, 37. PA. (2) 2016.

Erysiphe platani (Howe) U. Braun & S. Takam.
Microsphaera platani Howe
Powdery mildew on leaves of *Platanus x hispanica*; becoming common and widespread.
50, 51. SJ 27, 35, 36. (3) 2007. (48, 49) This species has spread across Europe from the south-east during the last thirty years and has become quite common in the southern half of England in the last fifteen years. It rarely produces ascospores in Britain but these have been found recently in London. In France, Italy and Spain the fungus causes serious damage to street trees.

Erysiphe polygoni DC.
Powdery mildew on leaves of *Fallopia baldschuanica*, *Persicaria maculosa*, *Polygonum aviculare* and *Rumex acetosella*; common.
50, 51. SH 87, 96-98; SJ 06-08, 15-18, 24-27, 34-36, 54. AV, BO, ER, LO, MO, PA, WW. (31) 1974. (48, 49, 52)

Erysiphe prunastri DC.
Uncinula prunastri (DC.) Sacc.
Powdery mildew on leaves of *Prunus spinosa*; frequent.
50, 51. SH 77; SJ 08, 16, 26, 36. LO, MA. (8) 1983.

Erysiphe russellii (Clinton) U. Braun & S. Takam.
Microsphaera russellii Clinton
Powdery mildew on leaves of *Oxalis stricta*; uncommon.
51. Recorded only from Mold (SJ 26) in 2009. (48)

Erysiphe symphoricarpi (Howe) U. Braun & S. Takam.
Microsphaera symphoricarpi Howe
Powdery mildew on leaves of *Symphoricarpus albus*; rare.
50, 51. SH 97; SJ 16, 17, 26, 35. BO, LO, MO. (7) 1977.

Erysiphe syringae Schwein.
Microsphaera syringae (Schwein.) Magnus
Powdery mildew on leaves of *Forsythia* sp., *Ligustrum ovalifolium*, *L. vulgare* and *Syringa vulgaris*; becoming common.
50, 51. SH 97; SJ 06, 08, 16, 18, 26, 35, 36. MO, PA. (10) 1976.

Erysiphe tortilis (Wallr.) Link
Microsphaera tortilis (Wallr.) Speer
Powdery mildew on leaves of *Cornus sanguinea* and *C. alba*; frequent.
50, 51. SH 97; SJ 08, 15, 17, 26, 34–36. BO, BW, CE, ER, MO. (12) 1974. (52)

Erysiphe trifoliorum (Wallr.) U. Braun
Microsphaera trifolii (Grev.) U. Braun
Powdery mildew on leaves of *Cytisus scoparius*, *Lathyrus latifolius*, *L. montanus*,
L. odoratus, *L. pratensis*, *Trifolium pratense*, *T. repens*, *Ulex europaeus* and *Wisteria sinensis*; common.
50, 51. SH 94, 97, 98; SJ 04–08, 16–18, 24–27, 34–37, 43. AV, BO, DU, ER, LH, LO, MO, NF, NM, PA, WW. (39) 1969. (48, 49, 52)

Erysiphe ulmariae Desm.
Powdery mildew on leaves of *Filipendula ulmaria*; uncommon.
50, 51. SJ 15, 16, 26. LO. (4) 2013. (48, 49, 52)

Erysiphe urticae (Wallr.) Blumer
Powdery mildew on leaves of *Urtica dioica*; frequent.
50, 51. SH 88, 97; SJ 06–08, 16–18, 25–27, 34–36. AV, BO, CE, ER, MA, MO. (24) 1974 (49, 52).

Erysiphe vanbruntiana (Gerard) U. Braun & S. Takam.
var. **sambuci-racemosae** (U. Braun) U. Braun & S. Takam.
Microsphaera vanbruntiana Gerard var. *sambuci-racemosae* U. Braun
Powdery mildew on leaves of *Sambucus racemosus*; rare but increasing.
51. Recorded only from Hawarden (SJ 36) in 2002. This species was first found in Britain in 1997 and is spreading, as it has done across Europe in the last twenty years. The Hawarden record is the first for Wales; however, the bush has recently been destroyed.

Erysiphe viburni Duby
Microsphaera sparsa Howe
Powdery mildew on leaves of *Viburnum opulus*; frequent.
50, 51. SH 97; SJ 07, 15–17, 26, 35, 36. LO, MO. (10) 1974. (49, 52)

Unassigned anamorph of **Erysiphe**:

Pseudoidium hortensiae (Jorst.) U. Braun & R.T.A. Cook
Oidium hortensiae Jorst.
Powdery mildew on leaves of *Hydrangea macrophylla*; frequent.
51. Recorded only from Drury (SJ 26) in 2015. (49, 52)

Golovinomyces artemisiae (Grev.) V.P. Gelyuta
Erysiphe artemisiae Grev.
Powdery mildew on leaves of *Artemisia abrotanum*, *A. absinthium* and *A. vulgaris*; common.
50, 51. SJ 08, 16–18, 25–27, 34–37. AV, CN, ER, LO, MO, PA. (16) 1882. (48, 49, 52)

Species Accounts

Golovinomyces asterum (Schwein.) U. Braun var. **asterum**
Erysiphe cichoracearum DC. in part
Powdery mildew on leaves of *Bellis perennis*; frequent.
51. Recorded only from Mold (SJ 26) in 2013.

Golovinomyces asterum var. **moroczkovskii** (Heluta) U. Braun
Erysiphe cichoracearum DC. in part
Powdery mildew on *Symphyotrichum* spp.; common.
50, 51. SH 77, 88, 98; SJ 06–08, 17, 18, 25, 26. MO, PA. (16) 1974. (48, 49, 52)

Golovinomyces asterum var. **solidaginis** U. Braun
Powdery mildew on leaves of *Solidago altissima, S. canadensis* and *S. virgaurea*; common.
50, 51. SH 87; SJ 08, 17, 26, 35. DU, MO. (7) 1974. (48, 49, 52)

Golovinomyces biocellatus (Ehrenb.) V.P. Gelyuta
Erysiphe biocellata Ehrenb.
Powdery mildew on leaves of *Ajuga reptans, Melissa officinalis, Mentha aquatica, M. rotundifolia* and *Salvia officinalis*; common.
50, 51. SH 97; SJ 06, 08, 16, 17, 26, 34–36. AV, BO, ER, LH, MO. (13) 1981. (48, 49)

Golovinomyces chrysanthemi (Rabenh.) M. Bradshaw, U. Braun, Meebon & S. Tamak.
Euoidium chrysanthemi (Rabenh.) U. Braun & R.T.A. Cook; *Oidium chrysanthemi* Rabenh.
Powdery mildew on leaves of Chrysanthemum x grandiflorum; frequent.
51. Recorded only from Shotton (SJ 26) in 1976.

Golovinomyces cichoracearum (DC.) V.P. Gelyuta
Erysiphe cichoracearum DC.
Powdery mildew on leaves of *Hieracium* spp., *Hypochaeris* spp.; *Lactuca serriola, Mycelis muralis* and *Tragopogon pratense*; common.
50, 51. SH 85; SJ 08, 16–18, 26, 36. LO, MO, PA. (9) 1974. (48, 49, 52)

Golovinomyces circumfusus (Schltdl.) U. Braun
Erysiphe cichoracearum DC. in part
Powdery mildew on leaves of *Eupatorium cannabinum*; uncommon.
50, 51. SJ 08, 16–18, 35. BP, LO, PA. (6) 1974.

Golovinomyces cucurbitacearum (Zheng & Chen) U. Braun & S. Takam.
Erysiphe cucurbitacearum Zheng & Chen
Powdery mildew on leaves of cucumber, courgette and marrow; only recently separated from other taxa and often incorrectly identified in the past; probably common.
50, 51. SJ 06, 08, 17, 26, 36. MO. (7) 1976.

Golovinomyces cynoglossi (Wallr.) V.P. Gelyuta
Erysiphe cynoglossi (Wallr.) U. Braun
Powdery mildew on leaves of *Borago officinalis, Myosotis arvensis, M. sylvatica* and cultivars, *Pulmonaria officinalis, Symphytum officinale* and *S. x uplandicum*; common.
50, 51. SH 87, 96–98; SJ 05–08, 15–18, 24, 26, 34–36, 43. AV, CC, ER, LO, MO. (32) 1880. (48, 49, 52)

Golovinomyces depressus (Wallr.) V.P. Gelyuta
Erysiphe depressa (Wallr.) Schlecht.
Powdery mildew on leaves of *Arctium lappa, A. minus* and *Centaurea montana*; common.
50, 51. SH 87, 97; SJ 05–08, 15–18, 24–27, 34–36, 54. BO, ER, LH, LO, MO. (26) 1882. (48, 49, 52)

Golovinomyces echinopis (U. Braun) Heluta
Powdery mildew on leaves and stems of *Echinops* spp.; uncommon.
50, 51. SH 77; SJ 36. (2) 2013.

Golovinomyces fischeri (S. Blumer) U. Braun & R.T.A. Cook
Erysiphe cichoracearum var. *fischeri* (Blumer) U. Braun
Powdery mildew on leaves of *Senecio vulgaris*; common.
50, 51. SH 87, 97; SJ 07, 08, 16–18, 25–27, 35–37. DU, LO, MO. (21) 1973. (52)

Golovinomyces longipes (Noordel.) L. Kiss
Euoidium longipes (Noordel. & Loer.) U. Braun & R.T.A. Cook;
Oidium longipes Noordel. & Loer.
Powdery mildew on leaves of *Petunia x hybrida*; frequent.
50, 51. SJ 07, 08, 26, 35, 36. MO. (6) 1990.

Golovinomyces macrocarpus (Speer) U. Braun
Erysiphe cichoracearum DC. in part
Powdery mildew on leaves of *Achillea millefolium, Matricaria discoidea, Tanacetum parthenium, T. vulgare* and *Tripleurospermum maritimum*; frequent.
50, 51. SJ 16–18, 26, 34–36. ER, LO, MO, PA. (10) 1974. (49)

Golovinomyces magnicellulatus (U. Braun) V.P. Gelyuta
Erysiphe magnicellulata U. Braun
Powdery mildew on leaves of *Phlox drummondii* and *P. paniculata*; common.
50, 51. SH 77, 87; SJ 06, 08, 18, 26. MO. (7) 1989.

Golovinomyces montagnei U. Braun
Erysiphe cichoracearum DC. in part
Powdery mildew on leaves of *Carduus tenuiflorus, Centaurea cyanus, C. nigra, C. scabiosa, Cirsium arvense* and *C. vulgare*; common.
50, 51. SH 85, 94; SJ 06–08, 16–18, 25–27, 35, 36, 54. CG, LH, LO, MA, PA. (21) 1924. (48, 52)

Golovinomyces orontii (Cast.) V.P. Gelyuta
Erysiphe orontii Cast.
Powdery mildew on leaves of *Antirrhinum majus, Campanula persicifolia, Chrysanthemum rubellum, Dahlia variabilis, Hydrangea macrophylla, Penstemon* sp., and *Streptocarpus x hybridus*; common.
50, 51. SH 97; SJ 07, 08, 17, 34, 26, 36. ER, MO. (11) 1989. (48, 49)

Golovinomyces senecionis U. Braun
Erysiphe cichoracearum DC. in part
Powdery mildew on leaves of *Senecio* spp. and *Tussilago farfara*; frequent.
50, 51. SH 98; SJ 07, 08, 16–18, 25–27, 35–37. LO, MA, MO, PA. (18) 1974.

Golovinomyces sonchicola U. Braun & R.T.A. Cook
Golovinomyces cichoracearum in part
Powdery mildew on leaves of *Sonchus arvensis*, *S. asper* and *S. oleraceus*; common.
50, 51. SH 87, 88, 94, 96–98; SJ 05–08, 16–18, 24–27, 34–36, 54. BO, ER, LO, MA, MO, PA. (39) 1973. (48, 49, 52)

Golovinomyces sordidus (L. Junell) V.P. Gelyuta
Erysiphe sordida Junell
Powdery mildew on leaves of *Plantago major* and *P. maritima*; common.
50, 51. SH 77, 87, 88, 94, 97, 98; SJ 06–08, 15–18, 24–27, 34–37, 43. AV, BO, ER, GW, LH, LO, MA, MO, NF, NM, PA, WW. (52) 1973. (48, 49, 52)

Golovinomyces valerianae (Jacz.) V.P. Gelyuta
Erysiphe valerianae (Jacz.) S. Blumer
Powdery mildew on leaves of *Patrinia scabiosella* and *Valeriana officinalis*; uncommon.
50, 51. SH 77; SJ 36. MO. (2) 2009. (48)

Golovinomyces verbasci (Jacz.) V.P. Gelyuta
Erysiphe verbasci (Jacz.) Blumer
Powdery mildew on leaves of *Verbascum thapsus*; frequent.
50, 51. SH 77; SJ 06–08, 16, 17, 24–27. AV, LO, MO. (13) 1974. (48, 49, 52)

Golovinomyces vincae U. Braun & S. Takam.
Erysiphe orontii Cast. in part
Powdery mildew on *Vinca major*; uncommon.
50. Recorded from Kinmel Park (SH 98) in 2000.

Leveillula contractirostris Heluta & Simonyan
Powdery mildew on leaves of *Sidalcea malviflora*; rare.
50. Recorded only from Tyddyn Bach (SH 97) in 1989.

Leveillula taurica (Lév.) G. Arnaud *sensu stricto*
Powdery mildew on leaves of *Echium vulgare*; rare.
51. Recorded only from Point of Ayr (SJ 18) in 2017. This is the first record of this restricted taxon in the UK.

Unassigned anamorph of **Leveillula**:

Oidiopsis cisti (Jaap) Golovin
Leveillula taurica (Lév.) Arnaud in part
Powdery mildew on leaves of *Cistus* sp.; uncommon.
51. SJ 08, 16. (2) 1989.

Neoerysiphe galeopsidis (DC.) U. Braun
Erysiphe galeopsidis DC.
Powdery mildew on leaves of *Acanthus* spp., *Ballota nigra*, *Lamium album*, *L. purpureum*, *Monarda didyma* and *Stachys sylvatica*; common.
50, 51. SH 77, 87, 97; SJ 05–08, 15–18, 24–27, 34–36. AV, BO, ER, LO, MA, MF, MO, PA, WW. (33) 1974. (49, 52)

Neoerysiphe galii (S. Blumer) U. Braun
Erysiphe galii Blumer
Powdery mildew on leaves of *Cruciata laevipes* and *Galium aparine*; frequent.
50, 51. SJ 07, 08, 15-18, 25, 26. AV, LH, LO, MO. (11) 1974. (49, 52)

Neoerysiphe nevoi Heluta & S. Takam.
Powdery mildew on leaves of *Lapsana communis*; rare.
51. Recorded only from Ewloe (SJ 26) in 2016.

Phyllactinia betulae (DC.) Fuss.
Phyllactinia guttata (Wallr.) Lév. in part
Powdery mildew on leaves of *Betula* spp. esp. *B. utilis*; frequent.
50, 51. SJ 25, 26, 35, 36. MO, NF. (8) 1976. (49)

Phyllactinia carpini (Rabenh.) Fuss
Phyllactinia guttata (Wallr.) Lév. in part
Powdery mildew on leaves of *Carpinus betulus*; uncommon.
51. Recorded only from Ewloe (SJ 26) in 2016. First record for Wales.

Phyllactinia fraxini (DC.) Fuss.
Powdery mildew on leaves of *Fraxinus excelsior*; uncommon but increasing.
50, 51. SH 95, 97; SJ 08, 15, 16, 24-26, 34-36. AV, ER, GW, LO, MA, MO. (16) 1989. (48, 52)

Phyllactinia guttata (Wallr.) Lév.
Powdery mildew on leaves of *Corylus avellana* and *C. maxima*; frequent.
50, 51. SH 77, 97; SJ 02, 05-08, 15-18, 24, 26, 35. AV, BO, BP, CF, CN, DU, LO, MA, MO. (33) 1974. (48, 49, 52)

Phyllactinia marissallii (Westend.) U. Braun
Powdery mildew on leaves of *Acer pseudoplatanus*; rare.
50. Recorded only from Loggerheads Country Park (SJ 16) in 2013. This is the first record from Wales.

Phyllactinia orbicularis (Ehrenb.) U. Braun
Phyllactinia guttata (Wallr.) Lév. in part
Powdery mildew on leaves of *Fagus sylvatica*, often noticed only after they have fallen; frequent.
50, 51. SH 97; SJ 02, 16, 26, 36. BO, CN, GW, LO. (8) 1996.

Podosphaera amelanchieris Maurizio
Powdery mildew on leaves of *Amelanchier lamarckii*; uncommon.
50, 51. SJ 24, 26, 36. MO. (4) 1990. (49)

Podosphaera aphanis (Wallr.) U. Braun & S. Takam.
Sphaerotheca aphanis (Wallr.) U. Braun
Powdery mildew on leaves of *Alchemilla glabra, Geum urbanum, Potentilla anglica, P. erecta, P. reptans, Rubus fruticosus* agg. and *R. idaeus*; common.
50, 51. SH 87, 94, 97; SJ 05, 07, 08, 15-18, 24-27, 34, 35. AV, BO, BO, DU, ER, LO, MA, MF, MO, WW. (31) 1974. (48, 49, 52)

Species Accounts

Podosphaera aucupariae Erikks.
Podosphaera clandestina var. *aucupariae* (Erikss.) U. Braun
Powdery mildew on leaves of *Sorbus aria* and *S. aucuparia*; frequent.
50, 51. SJ 08, 16, 25, 26, 35, 36. LO, NF. (6) 1976. (48, 49)

Podosphaera clandestina (Wallr.) Lév.
Powdery mildew on leaves of *Crataegus monogyna*; common.
50, 51. SH 86-88, 94, 97; SJ 05-08, 14-18, 24-27, 34-37. AV, BO, BP, ER, LO, MA, MO, NM. (46) 1974. (48, 49, 52)

Podosphaera delphinii (P. Karst.) U. Braun & S. Takam.
Sphaerotheca delphinii (P. Karst.) S. Blumer
Powdery mildew on leaves of *Delphinium* cv.; uncommon.
51. Recorded only from Hawarden (SJ 36) in 2016. This is the first record for Wales.

Podosphaera dipsacearum (Tul. & C. Tul.) U. Braun & S. Takam.
Sphaerotheca dipsacearum (Tul. & C. Tul.) L. Junell
Powdery mildew on leaves of *Dipsacus fullonum*; frequent.
50, 51. SJ 08, 16, 26, 35. AV, MO. (4) 1982.

Podosphaera epilobii (Wallr.) U. Braun & S. Takam.
Sphaerotheca epilobii (Wallr.) Sacc.
Powdery mildew on leaves of *Chamerion angustifolium, Epilobium hirsutum,*
E. montanum, E. parviflorum and *E. tetragonum*; common.
50, 51. SH 85, 87, 94, 95, 97, 98; SJ 06-08, 16-18, 24-26, 34-36, 54. AV, BO, CE, ER, MA, MO, NF. (28) 1974. (48, 49, 52)

Podosphaera erigerontis-canadensis (Lév.) U. Braun & T.Z. Liu
Sphaerotheca fusca (Fr.) Blumer in part
Powdery mildew on leaves of *Crepis vesicaria, Lapsana communis, Matricaria discoidea* and *Taraxacum* spp.; common.
50, 51. SH 77, 85, 87, 88, 94, 96-98; SJ 05-08, 15-18, 24-26, 34-36, 54. AV, DU, LO, MO, NF, PA, WW. (48) 1973. (48, 49, 52)

Podosphaera euphorbiae (Cast.) U. Braun & S. Takam.
Sphaerotheca euphorbiae (Cast.) Salmon
Powdery mildew on leaves of *Euphorbia aeruginosa, E. amygdaloides, E. balsamifera, E. caput-medusae, E. cylindrica, E. decepta, E. donii, E. dulcis, E. epithymoides, E. esculenta, E. ferox, E. globosa, E. griffithii, E. hamata, E. helioscopia, E. horrida, E. inconstantia, E. inermis, E. lathyris, E. mellifera, E. meloformis, E. moratii, E. neriifolia, E. obesa, E. opuntioides, E. palustris, E. peplus, E. polygona, E. pulvinata, E. resinifera, E. schillingii, E. serrrulata, E. submammilaris, E. susannae, E. uhligiana, E. villosa* and *E. vigueri*; frequent.
50, 51. SH 77, 87; SJ 07, 18, 26, 35, 36. MO. (13) 1974. As can be seen from the host list this common mildew of hardy spurges can also infect greenhouse succulent species.

Podosphaera ferruginea (Schlecht.) U. Braun & S. Takam.
Sphaerotheca ferruginea (Schlecht.) L. Junell
Powdery mildew on leaves of *Sanguisorba minor*; uncommon.
50. Recorded only from Bryn Euryn (SH 87) in 1989. (52)

Podosphaera filipendulae (Z.Y. Zhao) T.Z. Liu & U. Braun
Sphaerotheca spiraeae Sawada in part
Powdery mildew on leaves and stems of *Filipendula ulmaria*; common.
50, 51. SH 87; SJ 08, 16, 17, 26, 34, 36. AV, BW, CC, ER, LO. (9) (49, 52)

Podosphaera fugax (Penz. & Sacc.) U. Braun & S. Takam.
Sphaerotheca fugax Penz. & Sacc.
Powdery mildew on leaves of *Geranium dissectum, G. molle, G. phaeum, G. pratense, G. pyrenaicum, G. robertianum, G. sanguineum* and *G. subcaulescens*; common.
50, 51. SH 77, 87, 98; SJ 06–08, 16, 26, 27, 34, 36, 37. AV, ER, LO, MA, MO. (16) 1974. (48)

Podosphaera fuliginea (Schlecht.) U. Braun & S. Takam.
Sphaerotheca fuliginea (Schlecht.) Poll.
Powdery mildew on leaves of *Veronica montana* and *V. spicata* cultivar; uncommon.
50, 51. SJ 08, 16, 18. MA, MF. (5) 1978. (49)

Podosphaera helianthemi (L. Junell) U. Braun & S. Takam.
Sphaerotheca helianthemi L. Junell
Powdery mildew on leaves of *Helianthemum nummularium* cultivars; rare.
51. Recorded only from Mold (SJ 26) in 1976.

Podosphaera leucotricha (Ell. & Ev.) Salmon
Powdery mildew on leaves of *Malus domestica, M. sylvestris* and *Pyrus communis*; common.
50, 51. SH 97; SJ 07, 08, 16–18, 26, 27, 34–36. AV, DU, ER, LO, MO, PA. (16) 1974. The apple mildew is presumably far commoner than these records suggest. (53)

Podosphaera macrospora (U. Braun) U. Braun & V. Kummer
Powdery mildew on leaves of *Heuchera sanguinea*; uncommon.
51. Recorded only from Hawarden (SJ 36) in 2015. Another recent arrival in the UK.

Podosphaera macularis (Wallr.) U. Braun & S. Takam.
Sphaerotheca macularis (Wallr.) Lind
Powdery mildew on leaves of *Humulus lupulus*; uncommon.
51. Recorded only from Mold (SJ 26) since 1974.

Podosphaera mors-uvae (Schwein.) U. Braun & S. Takam.
Sphaerotheca mors-uvae (Schwein.) Berk. & M.A. Curt
Powdery mildew on leaves of *Ribes sanguineum* and *R. uva-crispa*; common.
50, 51. SJ 07, 08, 16–18, 25, 26, 34, 36. DU, ER, MO. (13) 1976. This is the familiar American mildew of gooseberries and currants, which is rarely found in semi-natural vegetation.

Podosphaera myrtillina (Schub.) Kunze
Powdery mildew on leaves of *Vaccinium myrtillus*; frequent.
50, 51. SH 85; SJ 17, 25. HM, MF, NM. (5) 1924. (48)

Podosphaera pannosa (Wallr.) de Bary
Sphaerotheca pannosa (Wallr.) Lév.
Powdery mildew on leaves of *Prunus laurocerasus, P. lusitanica, Rosa arvensis, R. canina, R. pimpinellifolia, R. villosa* and cultivated roses; common.
50, 51. SH 77, 87, 97; SJ 06–08, 14–18, 24–27, 34–36. AV, BO, BP, BW, DU, ER, HM, LH, LO, MO, WW. (48) 1910. This is the well-recorded rose mildew, especially prevalent on climbing varieties; it also causes considerable damage to hedges of cherry laurel. (48, 49, 52)

Podosphaera phtheirospermi (Henn. & Shirai) U. Braun & T.Z. Liu
Sphaerotherca fusca (Fr.) Blumer in part
Powdery mildew on leaves of *Odontites verna*; uncommon.
50, 51. SH 97; SJ 15, 26. BO, CG, MO. (3) 1974.

Podosphaera plantaginis (Cast.) U. Braun & S. Takam.
Sphaerotheca plantaginis (Cast.) L. Junell
Powdery mildew on leaves of *Plantago lanceolata*; common.
50, 51. SH 77, 87, 98; SJ 06–08, 17, 18, 26, 27, 35–37, 43, 54. BP, GW, MO, PA. (19) 1974. (52)

Podosphaera senecionis U. Braun
Sphaerotheca fusca (Fr.) Blumer in part
Powdery mildew on leaves of *Senecio ambigua, S. fluviatilis, Jacobaea vulgaris* and *J. squalidus*; frequent.
50, 51. SH 97; SJ 16, 18, 26, 27, 34, 36, 45. BO, LO, MA, MO. (14) 1914.

Podosphaera spiraeae (Sawada) U. Braun & S. Takam.
Sphaerotheca spiraeae Sawada in part
Powdery mildew on leaves of *Spiraea* spp.; frequent.
50, 51. SJ 26, 34–36. MO. (5) 1975. (49, 52)

Podosphaera thalictri (Junell) U. Braun & S. Takam.
Sphaerotheca thalictri Junell
Powdery mildew on leaves of *Thalictrum aquilegifolium* and *T. sphaerostachyum*; uncommon.
50, 51. SH 77; SJ 26. BO, MO. (2) since 2009. Probably new to Wales.

Podosphaera tridactyla (Wallr.) de Bary
Powdery mildew on leaves of *Prunus cerasifera* and *P. spinosa*; frequent.
50, 51. SH 94, 97; SJ 08, 16, 26, 35, 36. BO, BP, LO, MO. (9) 1949. (49, 52)

Podosphaera xanthii (Cast.) U. Braun & S. Takam.
Sphaerotheca fusca (Fr.) Blumer in part
Powdery mildew on leaves of *Calendula officinalis* and *Impatiens glandulifera*; uncommon.
50, 51. SJ 07, 08, 18, 26, 35, 36. MO, PA. (10) 1985. (48, 49, 52)

Unassigned anamorphs of **Podosphaera**:

Fibroidium balsaminae (Rajd.) U. Braun & R.T.A. Cook
Powdery mildew on *Impatiens glandulifera*; uncommon.
51. Recorded only from Talacre (SJ 18) in 1989.

Fibroidium primulae-obconicae (Ciocan & Calnegro) U. Braun & R.T.A. Cook
Oidium primulae-obconicae Ciocan & Calnegro
Powdery mildew on leaves of *Primula malacoides*; frequent.
51. Recorded from Flintshire without locality, in 1956.

Sawadaea bicornis (Wallr.) Homma
Uncinula bicornis (Wallr.) Lév.
Powdery mildew on leaves of *Acer campestre* and *A. pseudoplatanus*; very common.
50, 51. SH 75, 77, 85–88, 94, 96–98; SJ 05–08, 14–18, 24–27, 34–36, 54. AV, BO, BP, BW, CC, DU, ER, GW, LO, MA, MF, ML, MO, NF, WW. (70) 1882. (49, 52)

Sawadaea tulasnei (Fuckel) Homma
Powdery mildew on leaves of *Acer platanoides* and *A. palmatum*; increasing.
50, 51. SH 88, 97; SJ 07, 17, 25–27, 35, 36. BO, MO, WW. (13) 1989. The species is now spreading rapidly from its headquarters in eastern and Central Europe. It is especially noticeable on purple-leaved cultivars of *A. platanoides*

Sub-class Leotiomycetidae
Order HELOTIALES
Family Amorphothecaceae

Amorphotheca resinae Parberry
On paraffin-soaked table mat; uncommon.
50. Recorded only from Llanferres (SJ 16) in 1976, as the *Hormoconis* anamorph. (49)

Polydesmia pruinosa (Berk. & Br.) Boud.
On old perithecia on dead wood; common.
50, 51. SH 85, 87; SJ 05, 15, 16, 26, 27, 34–36. CC, CF, ER, GW, LO. (11) 1880. (48, 49, 52)

Family Arachnopezizaceae

Arachnopeziza eriobasis (Berk.) Korf
On dead herbaceous stems; frequent.
51. Recorded only from Loggerheads Country Park (SJ 26) in 1972.

Eriopezia caesia (Pers.) Rehm
On rotten oak wood; uncommon.
50. Recorded only from Hafod Wood, near Betws-y-Coed (SH 85) in 1924. (48, 49)

Family Bulgariaceae

Bulgaria inquinans (Pers.) Fr. *Black Bulgar*
On the bark of newly-fallen oaks; common.
50, 51. SH 77, 85, 97; SJ 06–08, 15–18, 23, 26, 27, 34–36. AV, BW, ER, GW, HC, LH, LO, ML, MO, WW. (32) 1910. (48, 49, 52)

Species Accounts

Family Calloriaceae

Calloria neglecta (Lib.) Hein
On dead nettle stems; common.
50, 51. SJ 15–17, 25, 26, 35. CE, CN, MO. (13) 1976. Often only as the *Cylindrocolla* anamorph. (48, 49, 52)

Cistella acuum (Alb. & Schwein.) Raitv.
Dasyscyphus acuus (Alb. & Schwein.) Sacc.
On conifer litter; frequent.
51. Recorded only from Loggerheads Country Park (SJ 26) in 1985.

Cistella fugiens (Bucknall) Mathies
Dasyscyphus fugiens (Bucknall) Massee
On dead rush stems; frequent.
51. Recorded only from Hope Mountain (SJ 25) in 1978.

Cistella grevillei (Berk.) Raschle
Discocistella grevillei (Berk.) Svrček; *Dasyscyphus grevillei* (Berk.) Massee
On dead *Heracleum* stems; uncommon.
51. Recorded only from Hawarden Woods (SJ 36) in 1972.

Cistella stereicola (Cooke) Dennis
On old *Stereum* fruit bodies; rare.
50. Recorded only from Coed Coch (SH 87) in 1880.

Diplonaevia exigua (Desm.) B. Hein
On dead stems of *Juncus* species; uncommon.
51. Recorded only from Hope Mountain (SJ 25) in 1978.

Ploettnera exigua (Niessl.) Höhn.
On dead bramble stems; uncommon.
50. Recorded only from Erddig (SJ 34) in 1985. (49)

Unassigned anamorphs of Calloriaceae:

Tetracladium marchalianum De Wild.
In foam rivers and lakes; common.
50, 51. SJ 17, 24–26, 34, 35. AV, DU, HM, LO, MO, NF. (11) In Rivers Alyn, Cegidog, Clywedog, Dee, Ffrith, Terrig and Wheeler. 1978. (48, 49)

Tetracladium setigerum (Grove) Ingold
In foam from Rivers Alyn and Wheeler; frequent.
50, 51. SJ 16, 17, 26, 35. AV, DU, LO, MO. (5) 1981. (48, 49)

Family Chlorociboriaceae

Chlorociboria aeruginascens (Nyl.) Ramam., Korf & Batta *Green Elfcup*
Chlorosplenium aeruginascens (Nyl.) P. Karst.
On dead wood of all kinds, colouring it a bright verdigris-green; very common.
50, 51. SH 77, 85–87, 97; SJ 05, 06, 15–18, 25–27, 34. AV, CC, CE, DU, ER, HC, LH, LO, MA, ML, NF, WW. (41) 1880. (48, 49, 52)

Family Dermateaceae

Catinella olivacea (Batsch) Boud.
On damp sticks and rotten branches; uncommon.
50, 51. SJ 34, 36. ER, GW. (3) 1972. (48, 49)

Pezicula eucrita (P. Karst.) P. Karst.
On pine bark; uncommon.
50. Recorded only from Hafod Wood (SH 85) in 1924.

Pezicula livida (Berk. & Br.) Rehm
On fallen pine cones; common.
50. SH 85; SJ 16. MF. (2) 1924. (48, 49)

Pezicula rubi (Lib.) Niessl.
On dead bramble stems; frequent.
50. Recorded only from the Alyn Waters Country Park (SJ 35) in 1999. (48, 49)

Family Discinellaceae

Unassigned anamorph of Discinellaceae:

Tetrachaetum elegans Ingold
In foam in fast flowing rivers; common.
50, 51. SJ 16, 24–26, 34, 35. AV, LO, MO, NF. (9) In Rivers Alyn, Cegidog, Clywedog, Dee, Ffrith and Terrig. 1981. (49)

Family Drepanopezizaceae

Diplocarpon earlianum (Ellis & Ev.) Wolf
Black spot on leaves of *Fragaria vesca* and *Potentilla reptans*; frequent.
51. SJ 17, 26. DU, LH, MO. (3) 1978. Only as the *Marssonina* anamorph. (49)

Diplocarpon mespili (Sorauer) B. Sutton
Leaf spot on *Crataegus monogyna*; frequent.
50. SJ 35, 45. (3) 2016.

Diplocarpon rosae Wolf
Black spot on leaves of wild and cultivated roses; common.
50, 51. SH 87, 96, 97; SJ 06–08, 16–18, 24, 26, 36, 54. AV, BO, LO, MO, PA. (22) 1976. (49)

Species Accounts

Unassigned anamorph of **Diplocarpon**:

Marssonina tiliae Oud.
Leaf spot on *Tilia x europaea*; uncommon.
50. Recorded only from Alyn Waters Country Park (SJ 35) in 2016.

Drepanopeziza ribis (Kleb.) Höhn.
On stems of *Ribes alpinum*; uncommon.
50. Recorded only from Aberduna Nature Reserve, Maeshafn (SJ 26) in 2001. (49)

Drepanopeziza sphaerioides (Pers.) Hohnel
On stems of willows, causing anthracnose; uncommon.
51. Recorded only from Mold (SJ 26) in 1978, as the *Marssonina* anamorph.

Leptotrochila medicaginis (Fuckel) Schüepp
On living leaves of *Medicago lupulina*; common.
51. Recorded only from Ddol Uchaf Nature Reserve (SJ 17) in 1985.

Leptotrochila ranunculi (Fr.) Schüepp
On living leaves of *Ranunculus repens*; common.
50, 51. SH 85; SJ 05, 17, 26, 34. CF, DU, ER, MO. (5) 1977. (49)

Leptotrochila verrucosa (Wallr.) Schüepp
On living leaves of *Galium saxatile*; frequent.
50. Recorded only from Coed Clwyd on Moel Famau (SJ 16) in 1985. (49)

Spilopodia melanogramma Boud.
On dead stems of *Mercurialis perennis*; uncommon.
50. Recorded only from Loggerheads Country Park (SJ 16) in 1988.

Family Gelatinodiscaceae

Ascocoryne cylichnium (Tul.) Korf
On dead wood; uncommon.
50, 51. SJ 05, 16, 26, 34, 35. AV, CF, ER, MF. (8) 1976. (48, 49, 52)

Ascocoryne sarcoides (Jacq.) Groves & Wilson *Purple Jellydisc*
On dead wood; very common.
50, 51. SH 77, 85, 87, 95, 97; SJ 05-07, 15-18, 24-26, 34-36. AV, BW, CC, CE, DU, ER, GW, LH, LO, MA, MF, MO, NF, PP, WW. (48) 1880. (48, 49, 52)

Unassigned anamorph of **Ascocoryne**:

Coryne atrovirens (Pers.) Sacc.
On fallen log; presumably common.
50. Recorded only from Coed Coch (SH 87) in 1880.

Neobulgaria pura (Fr.) Petrak
On fallen trunks; frequent.
50, 51. SJ 15, 17, 24, 26, 34. CE, ER, LH, LO, MA, PP. (11) 1978. (48, 49, 52)

Family Hamatocanthoscyphaceae

Ciliolarina laricina (Raitv.) Svrček
On fallen larch cones; uncommon.
50. Recorded only from Coed Clwyd, Moel Famau (SJ 16) in 1985.

Microscypha grisella (Rehm) H. Sydow & Sydow
On dead bracken fronds; common.
50, 51. SH 85; SJ 26. LO. (2) 1972. (48, 49)

Family Helotiaceae

Cyathicula culmicola (Desm.) de Not.
Crocicreas culmicola (Desm.) S. Carp.
On dead grass and rush stems; frequent.
51. SJ 18, 25. HM, PA. (2) 1978.

Cyathicula cyathoidea (Bull.)
Crocicreas cyathoideum (Bull.) S. Carp.
On dead herbaceous stems, esp. bracken and nettle; very common.
50, 51. SH 87; SJ 16, 25, 26, 34, 35. AV, CC, ER, LO, WW. (6) 1880. (48, 49, 52)

Hymenoscyphus albidus (Desm.) Phill.
On fallen leaf stalks of ash; frequent.
50, 51. SJ 16, 17, 26, 34. AV, DU, ER, LO. (6) 1979. (52)

Hymenoscyphus calyculus (Sow.) Phill.
On small, fallen branches, especially of beech; common.
50, 51. SH 85, 87; SJ 08, 17, 34, 35. CC, ER, LH. (8) 1910. (48, 49, 52)

Hymenoscyphus caudatus (P. Karst.) Dennis
On rotting leaves in litter; common.
50, 51. SH 85; SJ 17, 25, 26, 34. DU, ER, LO, NF. (6) 1976. (48, 49)

Hymenoscyphus fructigenus (Bull.) Gray *Nut Disco*
On fallen hazel nuts and beech mast; common.
50, 51. SH 85, 97; SJ 15, 16, 26, 34, 36. AV, BO, CE, ER, GW, LO, MA (11) 1924. (48, 49, 52)

Hymenoscyphus herbarum (Pers.) Dennis
On dead stems of stinging nettle; common.
50, 51. SH 85; SJ 05–08, 16, 17, 24–26, 34, 35, 43. CF, DU, ER, LO, MA, MF, MO, NF, WW. (20) 1910. (48, 49)

Hymenoscyphus imberbis (Bull.) Dennis
On dead twigs in litter; frequent.
50, 51. SJ 26, 34. ER, LO. (2) 1985.

Hymenoscyphus infarciens (Ces.) Dennis
On rotten *Laburnum* wood; uncommon.
50. Recorded only from Hafod Wood, near Betws-y-Coed (SH 85) in 1924.

Hymenoscyphus repandus (Phill.) Dennis
On dead stems of Rosebay Willowherb; common.
50, 51. SJ 08, 26, 35. MO. (3) 1977. (48, 49, 52)

Hymenoscyphus rokebyensis (Srvček) Mathies
On fallen beech cupules; uncommon.
50. Recorded only from Erddig (SJ 34) in 1985. (49, 52)

Hymenoscyphus salicellus (Fr.) Dennis
On fallen willow twigs; frequent.
50. Recorded only from Loggerheads Country Park (SJ 16) in 1972.

Hymenoscyphus scutula (Pers.) Phill.
On dead herbaceous stems; common.
50, 51. SH 85; SJ 08, 16, 17, 24–26, 34–36. AV, ER, GW, LH, LO, MF, MO, NF. (15) 1930. (48, 49, 52)

Hymenoscyphus splendens Abdullah et al.
In foam in fast flowing rivers; common.
50, 51. SJ 16, 24–26, 34, 35. AV, ER, LO, MO, NF. (11) In Rivers Alyn, Cegidog, Clywedog, Dee, Ffrith and Terrig. 1981. (48, 49) Collected as the anamorph – *Variocladium splendens* (Ingold) Descals & Marvonova.

Hymenoscyphus tetracladia Abdullah et al.
In foam in fast flowing rivers; common.
50, 51. SJ 16, 24, 26, 34, 35. AV, ER, LO, MO. (7) In Rivers Alyn, Clywedog and Dee. 1981. (48, 49) Collected as the anamorph – *Articulospora tetracladia* Ingold.

Hymenoscyphus vitellinus (Rehm) O. Kuntze
On dead herbaceous stems; frequent.
50, 51. SJ 35, 36. GW. (3) 2015. (49, 52)

Phaeohelotium carneum (Fr.) Hengstm.
On small, decorticated branches on damp ground; uncommon.
51. Recorded only from Ddol Uchaf Nature Reserve (SJ 17) in 1978.

Phaeohelotium epiphyllum (Pers.) Hengstm.
Hymenoscyphus epiphyllus (Pers.) Kauffman
On leaf litter; common.
50, 51. SH 85; SJ 25, 35. NF. (3) 1924. (48, 49, 52)

Phaeohelotium fagineum (Pers.) Hengstm.
Hymenoscyphus fagineus (Pers.) Dennis
On fallen beech cupules; frequent.
50. Recorded only from Erddig (SJ 34) since 1910. (48, 49, 52)

Phaeohelotium geogenum (Cooke) Svrček & Mathies
On moss on fallen bark; rare.
50. Recorded only from Clocaenog Forest (SJ 05) in 1988.

Phaeohelotium vernum (Boud.) Declercq
Hymenoscyphus vernus (Boud.) Dennis
On alder sticks on wet soil, in spring; rare.
51. Recorded only from the Rhydymwyn Nature Reserve (SJ 26) in 2007. (49)

Family Hemiphacidiaceae

Trochila craterium (DC.) Fr.
On dead ivy leaves in litter; common.
50, 51. SH 85; SJ 26, 27, 34. ER, LO. (4) 1972. (48, 49)

Trochila ilicina (Nees) Greenhalgh & Morgan-Jones *Holly Speckle*
On dead holly leaves in litter; common.
50, 51. SH 77, 85; SJ 05–07, 15–17, 25–27, 34, 36. BW, CE, CF, DU, ER, GW, LO, MA, MF, MO, NM, WW. (27) 1910. (48, 49, 52)

Family Heterosphaeriaceae

Heterosphaeria patella (Tode) Grev.
On dead herbaceous stems; common.
51. SJ 17, 36. DU, GW. (2) 1972. (48, 52)

Family Hyaloscyphaceae

Hyaloscypha aureliella (Nyl.) Huhtinen
Hyaloscypha velenovskyi Graddon
On rotten conifer wood; uncommon.
50. SH 85; SJ 16. MF. (2) 1985. (48, 49)

Hyaloscypha hyalina (Pers.) Boud.
On rotten branches, esp. of oak; common.
50, 51. SJ 26, 34. ER, LO. (2) 1985. (48, 49)

Family Hydrocinaceae

Unassigned anamorph of Hydrocinaceae:

Varicosporium elodeae Kegel
In foam at lake margins and from rivers; frequent.
50, 51. SJ 25, 26. LO, MO, NF. (4) In Rivers Alyn and Ffrith. 1978. (48, 49)

Family Hyphodiscaceae

Fuscolachnum pteridis (Alb. & Schwein.) J.H. Haines
Dasyscyphus pteridis (Alb. & Schwein.) Massee
On dead bracken fronds; uncommon.
51. Recorded only from Hope Mountain (SJ 25) in 1978. (49)

Species Accounts

Family Lachnaceae

Albotricha acutipila (P. Karst.) Raitv.
Dasyscyphus acutipilus (P. Karst.) Sacc.
On grass litter; frequent.
50, 51. SJ 15, 18, 34. ER, PA. (3) 1910. (48, 49, 52)

Belonidium mollissimum (Lasch) Raitv.
Dasyscyphus mollissimus (Lasch) Dennis; *Lachnum mollissimum* (Lasch) Karst.; *Trichopezizella mollissima* (Lasch) Fuckel
On dead herbaceous stems; common.
51. Recorded only from Mold (SJ 26) in 1977. (48, 49, 52)

Belonidium sulphureum (Pers.) Raitv.
Dasyscyphus sulphureus (Pers.) Massee
On dead stems of *Urtica*; frequent.
50. SH 85; SJ 05, 34. CF, ER. (3) 1988. (49, 52)

Brunnipila clandestina (Bull.) Baral
Dasyscyphus clandestinus (Bull.) Fuckel
On dead herbaceous stems; common.
50, 51. SJ 17, 26, 35. DU, MO. (3) 1975. (48, 49, 52)

Brunnipila dumorum (Desm.) Baral, Hellema & Honraid
Dasyscyphus dumorum (Desm.) Massee; *Lachnum dumorum* (Desm.) Huht.
On bramble leaves in litter; common.
50. SH 85; SJ 16, 34. ER, MF. (3) 1985. (48, 49, 52)

Brunnipila palearum (Desm.) Baral
Dasyscyphus palearum (Desm.) Massee; *Lachnum palearum* (Desm.) Massee
On dead marram stems; frequent.
51. Recorded only from Point of Ayr (SJ 18) in 1978.

Dasyscyphella nivea (Hedw.) Raitv.
Dasyscyphus niveus (Hedw.) Sacc.
On rotten wood; common.
50, 51. SH 85; SJ 15, 16, 25, 26. LO, MO, NF. (8) 1910. (48, 49)

Incrustipilum ciliare (Schrad.) Baral
Dasyscyphus ciliaris (Schrad.) Sacc.; *Lachnum ciliare* (Schrad.) Rehm
In mixed litter; frequent.
50, 51. SH 85; SJ 16, 26. LO, MF. (4) 1985. (48, 49)

Lachnellula occidentalis (Hahn & Ayers) Dharne *Larch Disco*
On fallen larch twigs; common.
50, 51. SH 87; SJ 15–17, 26. CC, LO, MF, WW. (7) 1880. (48, 49)

Lachnellula subtilissima (Cooke) Dennis *Conifer Disco*
On fallen pine twigs; common.
50, 51. SJ 16, 25, 26. AV, MF, NF. (5) 1910. (48, 49)

Lachnum apalum (Berk. & Br.) Nannf. *Rush Disco*
Dasyscyphus apalus (Berk. & Br.) Dennis
On decaying rush stems; common.
50, 51. SH 85; SJ 26, 43. MO. (3) 1977. (48, 49)

Lachnum brevipilosum Baral
On dead wood; frequent.
51. Recorded only from Rhydymwyn Nature Reserve (SJ 26) in 2007. (48, 52)

Lachnum carneolum (Sacc.) Rehm
Dasyscyphus carneolus (Sacc.) Sacc.
On dead grasses; common.
50. SJ 16, 34. ER, MF. (2) 1985. (48, 49, 52)

Lachnum diminutum Desm.
Dasyscyphus diminutus (Desm.) Sacc.
On dead rush stems; frequent.
51. SJ 25, 36. GW, HM. (2) 1972. (48, 49, 52)

Lachnum nudipes (Fuckel) Nannf.
Dasyscyphus nudipes (Fuckel) Sacc.
On dead meadow-sweet stems; frequent.
51. Recorded only from Mold (SJ 26) in 1977. (48, 52)

Lachnum rhodoleucum (Sacc.) Rehm
Dasyscyphus rhodoleucus (Sacc.) Sacc.
On dead marram grass; uncommon.
51. Recorded only from Point of Ayr (SJ 18) in 1985.

Lachnum rubi (Bres.) Raitv.
Capitotrichum rubi (Bres.) Baral; *Dasyscyphus bicolor* (Bull.) Fuckel
On dead raspberry canes; common.
50, 51. SJ 16, 25, 26, 34, 35. ER, LO, MA, MO, NF. (6) 1910.

Lachnum tenuissimum (Quélet) Korf & Zhuang
Dasyscyphus tenuissimus (Quélet) Dennis
On dead grasses; frequent.
50, 51. SJ 15, 26. LO. (2) 1985.

Lachnum virgineum (Batsch) P. Karsten *Snowy Disco*
Dasyscyphus virgineus (Batsch) Gray
On dead wood; very common.
50, 51. SH 85, 87; SJ 07, 15–18, 25, 26, 34, 36. AV, CC, CN, GW, LO, MO. (17) 1880. (48, 49, 52)

Lasiobelonium corticale (Pers.) Nannf.
Lachnum corticale (Pers.) Nannf.
On the bark of living trees and shrubs; uncommon.
50, 51. SJ 12, 26. MO. (2) 2008.

Species Accounts

Lasiobelonium nidulum (Schm. & Kunze) Spooner
Dasyscyphus nidulus (Schm. & Kunze) Massee
On dead herbaceous stems; frequent.
50, 51. SJ 16, 17, 26, 35. DU, LO, MO. (5) 1972.

Neodasyscypha cerina (Pers.) Spooner
Dasyscyphus cerinus (Pers.) Fuckel; *Lachnum cerinum* (Pers.) Nannf.
On dead gorse stems; frequent.
51. Recorded only from Coed Bach-y-Graig (SJ 07) in 1992. (49)

Proliferodiscus pulveraceus (Alb. & Schwein.) Baral
Dasyscyphus pulveraceus (Alb. & Schwein.) Höhn.
On fallen branches; frequent.
50. Recorded only from Hafod Wood, near Betws-y-Coed (SH 85) in 1988. (48, 49, 52)

Family Leotiaceae

Leotia lubrica (Scop.) Pers. *Jellybaby*
On soil in damp woods; frequent, especially in western Britain.
50, 51. SH 85, 87, 97; SJ 05, 15–17, 24, 26, 34, 36. AV, CC, CE, CF, ER, GW, LH, LO, MA, MF, ML, PP, WW. (23) 1880. (48, 49, 52)

Microglossum olivaceum (Pers.) Gillet *Olive Earthtongue*
In limestone turf; rare.
50, 51. SH 97; SJ 08, 15, 17, 25, 26. AV, HC, ML. (7) 1980. (48, 49, 52)

Microglossum viride (Pers.) Gillet *'Green Earthtongue'*
Among mosses in damp woodland; uncommon.
50. SH 85, 87; SJ 05. CC, CF. (4) 1880. (48, 49, 52)

Family Loramycetaceae

Loramyces juncicola Weston
On rush stems under water; uncommon.
51. Recorded only from Hope Mountain (SJ 25) in 1978.

Family Mollisiaceae

Mollisia amenticola (Sacc.) Rehm
On fallen female catkins of *Alnus glutinosa*; uncommon.
50, 51. SJ 26, 34. CE, ER. (2) 2003. (49, 52)

Mollisia cinerea *sensu lato*
On dead wood of all kinds; common.
50, 51. SH 85, 87; SJ 05, 15–17, 24–26, 34–36. AV, CC, CN, DU, ER, LH, LO, MF, MO, NF, WW. (27) 1880. No attempt has been made to separate the closely similar members of this species-group. (48, 49, 52)

Mollisia discolor (Mont.) Phill.
On fallen oak twigs; common.
51. Recorded only from Ysceifiog (SJ 17) in 1978. (48)

Mollisia escharodes (Berk. & Br.) Gremmon
Pyrenopeziza escharodes (Berk. & Br.) Rehm
On raspberry canes; frequent.
50. Recorded only from Rhyd-y-Creuau (SH 85) in 1988.

Mollisia ligni (Desm.) P. Karsten
On damp, fallen branches and tree fruits; frequent.
50, 51. SJ 05, 17, 34, 36. CF, ER, LH. (5) 1977. (48, 49)

Mollisia melaleuca (Fr.) Sacc.
On damp, fallen branches; frequent.
50, 51. SJ 17, 26, 34. DU, ER, LO. (4) 1910. (48, 49)

Mollisia olivascens (Fettzen) Le Gal & F. Mangenot
Haglundia perelegans Nannf.
On dead gorse stems; common.
51. Recorded only from Hope Mountain (SJ 25) in 1978. (48)

Mollisia palustris (Desm.) P. Karst.
On rotten *Juncus* stems; common.
51. SJ 18, 25, 26. HM, MO, PA. (3) 1977. (49, 52)

Mollisia ramealis (P. Karst.) P. Karst.
On fallen birch twigs; frequent.
50. Recorded only from Coed Hafod, near Betws-y-Coed (SH 85) in 1988. (49)

Pyrenopeziza chailletii Fuckel
On dead stems of *Heracleum*; frequent.
50. Recorded only from Lady Bagot's Drive, Rhewl (SJ 15) in 2001. (48, 49, 52)

Pyrenopeziza fuckelii Nannf.
On rotten willow leaves; uncommon.
51. Recorded only from Mold (SJ 26) in 1977.

Tapesia fusca (Pers.) Fuckel
Mollisia fusca (Pers.) P. Karst.
On damp, rotten wood; common.
50, 51. SJ 05, 08, 36. GW. (3) 1998. This species is surely more widespread. (48, 49)

Family Pezizellaceae

Allophylaria campanuliformis (Fuckel) Svrček
Pezizella campanuliformis (Fuckel) Dennis
On decaying fern fronds; uncommon.
50. SH 85; SJ 34. ER. (2) 1910.

Allophylaria clavuliformis (P. Karst.) P. Karst.
On dead *Urtica* stems; uncommon.
50. SH 85; SJ 05. CF. (2) 1988.

Allophylaria macrospora (Jaap) Nannf.
On dead *Urtica* stems; uncommon.
50. Recorded only from Rhyd-y-Creuau (SH 85) in 1988.

Bisporella citrina (Batsch) Korf & Carpenter *Lemon Disco*
On rotten wood; very common.
50, 51. SH 85, 97; SJ 05, 07, 08, 13, 15–17, 23–26, 34–36. AV, BO, CE, DU, ER, GW, LH, LO, MA, ML, NF, PP, WW. (43) 1910. (48, 49, 52)

Bisporella sulfurina (Quél.) Carpenter
On old pyrenomycetes on dead wood; frequent.
50, 51. SJ 15–17, 25, 26, 34. AV, DU, ER, LO, NF. (7) 1912. (48, 49, 52)

Calycellina fagina (Pers.) A.F.W. Schmidt, Arendh. & Baral
Scutoscypha fagi Graddon
On beech leaves in litter; uncommon.
50, 51. SH 85; SJ 26, 34. ER, LO. (4) 1985.

Calycellina indumenticola Graddon
On fallen sallow leaves in litter; uncommon.
51. Recorded only from Ddol Uchaf Nature Reserve (SJ 17) in 1985.

Calycellina punctata (Fr.) Lowen & Dumont
On fallen oak leaves in litter; common.
50, 51. SJ 16, 26. LO, MF. (2) 1972. (49, 52)

Calycellina rubescens (Mouton) van Vooren
Pezizella rubescens Mouton
On dead leaves in litter; uncommon.
50. Recorded only from Coed Clwyd, Moel Famau (SJ 16) in 1985.

Calycina marina (Boyd) Rämä & Baral
Laetinaevia marina (Boyd) Spooner
On cast-up seaweed on the strand-line; probably common but rarely recorded.
51. Recorded only from Point of Ayr (SJ 18) in 1985.

Mollisina rubi (Rehm) Höhn.
On bramble litter; frequent.
50. SH 85. (2) 1988. (48)

Pezizella muscicola Graddon
On *Polytrichum* mosses; uncommon.
51. Recorded only from Loggerheads Country Park (SJ 26) in 1985.

Pezizella punctoidea (P. Karst.) Rehm
On dead willow-herb stems; common.
51. Recorded only from Mold (SJ 26) in 1977.

Pezizella vulgaris (Fr.) Sacc.
On fallen twigs; uncommon.
50. Recorded only from Coed Coch (SH 87) in 1880.

Phialina flaveola (Cooke) Raitv.
Hyaloscypha flaveola (Cooke) Nannf.
On dead bracken fronds; common.
50, 51. SH 85; SJ 16, 36. GW, MF. (3) 1972.

Phialina lachnobrachya (Desm.) Raitv.
Hyaloscypha lachnobrachya (Desm.) Nannf.
On sycamore litter; rare.
50. Recorded only from Rhyd-y-Creuau (SH 85) in 1988.

Phialina pseudopuberula (Graddon) Raitv.
Hyaloscypha pseudopuberula Graddon
On oak leaves in litter; uncommon.
51. Recorded only from Loggerheads Country Park (SJ 26) in 1985.

Psilachnum chrysostigma (Fr.) Raitv.
Pezizella chrysostigma (Fr.) Sacc.
On dead fern stems; common.
50, 51. SH 85; SJ 26, 34, 36. ER, GW, LO. (5) 1910.

Psilachnum inquilinum (P. Karsten) Dennis
On dead *Equisetum* stems; uncommon.
51. Recorded only from Point of Ayr (SJ 18) in 1985. (52)

Family Phacidiaceae

Phacidium lauri (Sowerby) Crous & D. Hawksw.
Phacidium multivalve (DC.) Schm.
On dead holly leaves in litter; common.
50, 51. SH 85; SJ 15, 17, 26, 34, 36. GW, MA. (7) 1910. (48, 49, 52)

Unassigned anamorph of **Phacidium**:

Ceuthospora mahoniae Grove
Leaf spot on *Mahonia aquifolium*; common.
51. Recorded from Flintshire, without locality, by Grove (1935).

Family Ploettnerulaceae

Pseudopeziza medicaginis (Lib.) Sacc.
On living leaves of *Medicago lupulina*; common.
50. Recorded only from Llanferres (SJ 26) in 1976.

Pseudopeziza trifolii (Biv.-Bern.) Fuckel
On living leaves of *Trifolium* species; common.
50, 51. SJ 16, 17, 24, 26, 27, 34, 45. DU, ER, LO, MF, MO. (12) 1972. (48)

Species Accounts

Unassigned anamorph of Ploettnerulaceae:

Mastigosporium album Riess
Leaf spots on *Alopecurus pratensis*; frequent.
51. Recorded only from Mold (SJ 26) in 1978. (49)

Family Rutstroemiaceae

Lanzia luteovirescens (Desm.) Korf & Dumont
On sycamore petioles in litter; frequent.
50, 51. SJ 16, 17, 25, 26, 34, 35. BW, CE, CN, DU, ER, LO, NF, WW. (10) 1978. (48, 49, 52)

Rutstroemia echinophila (Bull.) Höhn.
On decaying *Castanea* cupules; uncommon.
51. Recorded only from Gwernymynydd (SJ 26) in 2015. (49, 52)

Rutstroemia firma (Pers.) P. Karst. *Brown Cup*
Poculum firmum (Pers.) Dumont
On oak twigs; frequent.
50, 51. SH 77, 85, 87; SJ 06, 07, 15, 17, 24, 26, 36. AV, CC, CE, GW, LH, PP, WW. (17) 1880. (48, 49, 52)

Rutstroemia petiolorum (Desm.) Gillet
Poculum petiolorum (Desm.) Dumont & Korf
On beech leaves in litter; frequent.
50, 51. SJ 16, 26, 35. CC, ER, LO, WW. (4) 1880. (49)

Rutstroemia sydowiana (Rehm) White *Oakleaf Cup*
Poculum sydowianum (Rehm) Dumont
On oak leaves in litter; frequent.
50, 51. SH 85; SJ 16, 26, 34. CE, ER, LO, MF. (5) 1978. (49, 52)

Family Sclerotiniaceae

Botryotinia fuckeliana (de Bary) Whetzel
Found as 'grey mould' on a wide range of wild and cultivated plants; very common.
50, 51. SH 87; SJ 16, 17, 25, 26, 27. CC, DU, LO, MO, NF. (8) 1880. Only found as the *Botrytis* anamorph. (48, 49)

Botryotinia globosa Buchwald
On living leaves of *Allium ursinum*; frequent.
50, 51. SJ 16, 17, 26, 34. AV, ER, LO, WW. (6) 1974. Only found as the *Botrytis* anamorph. (48, 49, 52)

Ciboria amentacea (Balb.) Fuckel
On fallen hazel catkins; uncommon.
51. SJ 17, 26. DU. (2) 1986.

Ciboria americana Durand
Rutstroemia americana (Durand) White
On fallen sweet chestnut burrs; uncommon.
50, 51. SJ 15, 34, 36. ER, GW. (3) 1964.

Ciboria batschiana (Zopf) Buchwald
On fallen acorns in leaf litter; uncommon.
50, 51. SJ 15, 26, 36. CE, GW. (3) 1977. (49)

Ciboria viridifusca (Fuckel) Höhnel
On fallen alder catkins; uncommon.
51. Recorded only from Ysceifiog (SJ 17) in 1978.

Ciborinia candolleana (Lév.) Whetzel
On oak leaves in litter; frequent.
51. SJ 17, 26. MO. (2) 1977.

Dumontinia tuberosa (Hedw.) Kohn *Anemone Cup*
On soil, arising from rhizomes of *Anemone nemorosa*; uncommon.
50, 51. SJ 15–17, 25. BW, CN, NF. (10) 1959.

Encoelia furfuracea (Roth) P. Karst. *Spring Hazelcup*
On hazel twigs still attached to the bush; uncommon.
50, 51. SJ 15–17, 25, 26, 35. BW, CN, NF, WW. (7) 1910. (48, 49, 52)

Mitrula paludosa Fr. *Bog Beacon*
On rotting vegetation at edge of water; frequent in spring.
50, 51. SH 95–97; SJ 17. LH, ML. (4) 1978. (48, 49)

Monilinia fructigena Whetzel
On apples, causing 'brown rot'; common.
50, 51. SJ 26, 34. ER, MO. (2) 1910, as the *Monilia* anamorph, but surely more widespread.

Monilinia johnsonii (Ellis & Ev.) Honey
On mummified fruits of hawthorn; uncommon.
51. SJ 17, 25. LH. (3) 1977. (49, 52)

Monilinia laxa (Alderh. & Ruhland) Honey
On leaves and fruit of *Prunus*; uncommon.
50. Recorded only from Erddig (SJ 34) in 2013. (49)

Myriosclerotinia curreyana (Currey) Buchw.
On living and moribund stems of rushes; uncommon.
50, 51. SH 85, 95, 97; SJ 15, 25, 26. CE, NF. (10) 1910. (48, 49)

Myriosclerotinia scirpicola (Rehm) Buchw.
On stems of *Schoenoplectus tabernaemontani*; uncommon.
51. Recorded only from Point of Ayr (SJ 18) in 1975.

Species Accounts

Myrioscletorinia sulcatula T. Schumach. & L.M. Kohn
Myriosclerotinia sulcata (Whetzel) Buchw.
On stems of *Carex* species; uncommon.
51. Recorded only from Point of Ayr (SJ 18) in 1975. (49)

Sclerotinia sclerotiorum (Lib.) de Bary
On dead herbaceous stems, especially *Heracleum*; frequent.
51. Recorded only from Mold (SJ 26) in 1977. (48)

Stromatinia cepivora (Berk.) Whetzel
On onion and garlic bulbs; uncommon.
51. Recorded from Mold (SJ 26) in 1978, as the *Sclerotium* anamorph.

Zoellneria rosarum Velen.
On fallen rose leaves; uncommon.
51. SJ 17, 25. LH, LO. (2) 1977.

Unassigned anamorphs of Sclerotiniaceae:

Cristulariella depraedans (Cooke) Höhnel
White leaf spots on sycamore; common.
50, 51. SJ 05, 24, 26. CF, LO. PP, WW. (4) 1985. (49)

Sclerotium tuliparum Kleb.
Black spot on tulip bulbs; frequent.
51. Recorded only from Mold (SJ 26) in 1977.

Family Tricladiaceae

Cudoniella acicularis (Bull.) Schröt. *Oak Pin*
On dead wood, esp. stumps; common.
50, 51. SH 85, 87; SJ 15-17, 24-26, 34. CC, LO, MA, MO, NF, WW. (13) 1978. (48, 49)

Cudoniella clavus (Alb. & Schwein.) Dennis *Spring Pin*
On dead wood in damp places; common.
50, 51. SJ 17, 24, 25, 34. ER, LH, NF. (5) 1977. (48, 49, 52)

Graddonia coracina (Bres.) Dennis
On submerged wood in streams; uncommon.
51. Recorded only from Ddol Uchaf Nature Reserve (SJ 17) in 1985.

Unassigned anamorphs of Tricladiaceae:

Tricladium angulatum Ingold
In foam in fast flowing rivers; common.
50, 51. SJ 16, 17, 24-26, 34, 35. AV, DU, LO, MO, NF. (10) In Rivers Alyn, Cegidog, Clywedog, Dee, Ffrith, Terrig and Wheeler. 1981. (49)

Tricladium chaetocladium Ingold
In foam in fast flowing rivers; frequent.
51. Recorded only from the River Alyn near Cilcain (SJ 16) in 1981.

Varicladium giganteum (S.H. Iqbal) Descals & Marvanová
Tricladium giganteum S.H. Iqbal
In foam in fast flowing rivers; frequent.
51. Recorded only from the River Alyn near Cilcain (SJ 16) in 1981. (48)

(Family uncertain)

Crocicreas coronatum (Bull.) S. Carp.
On dead herbaceous stems; common.
50, 51. SJ 08, 15, 16, 26. LO, MA, MF. (5) 1930. (49, 52)

Crocicreas dolosellum (P. Karst.) S. Carp.
On rotting ash petioles; frequent.
50. Recorded only from Rhyd-y-Creuau (SH 85) in 1988. (49)

Crocicreas subhyalinum (Rehm) S. Carp.
On fallen sycamore leaf stalks; frequent.
51. Recorded only from Mold (SJ 26) in 1977.

Mollisiopsis dennisii Graddon
On dead gorse twigs; uncommon.
50. Recorded only from Coed Clwyd, Moel Famau (SJ 16) in 1985. (49)

Unassigned anamorphs of Helotiales:

Dactylaria parvispora (Preuss) M. Hoog & Arx
Pleurophragmium parvisporum (Preuss) Holubova-Jechova
On dead stems of *Urtica dioica*; frequent.
51. Recorded only from Mold (SJ 26) in 1977. (48, 49, 52)

Lemmoniera aquatica De Wild.
In foam in fast flowing rivers; common.
50, 51. SJ 16, 17, 24–26, 34, 35. AV, DU, LO, MO, NF. (11) In Rivers Alyn, Cegidog, Clywedog, Dee, Ffrith, Terrig and Wheeler. 1981. (49)

Lemmoniera terrestris Tubaki
In foam in fast flowing rivers; common.
50, 51. SJ 24, 26, 35. LO, MO. (4) In Rivers Alyn and Dee. 1985. (48, 49)

Margaritispora aquatica Ingold
In foam in fast flowing rivers; uncommon.
50, 51. SJ 16, 26. AV, LO, MO. (3) In River Alyn. 1985. (49)

Rhynchosporium orthosporum Caldwell
Leaf blotch on *Dactylis glomerata*; frequent.
51. Recorded only from Mold (SJ 26) since 1978.

Rhynchosporium secalis (Oud.) Davis
Leaf blotch on *Elytrigia repens*; frequent.
51. Recorded only from Mold (SJ 26) since 1978.

Species Accounts

Order RHYTISMATALES
Family Ascodichaenaceae

Ascodichaena rugosa Butin
On standing trunks of beech trees, looking like sheets of soot; frequent.
50, 51. SH 85; SJ 26. AV. (2) 1924. (48, 49, 52)

Family Cudoniaceae

Spathularia flavida Pers. *Yellow Fan*
In conifer litter; rare, except in Scotland.
50, 51. SJ 05, 17, 26, 35. LO. (4) 1985. (48, 49, 52)

Family Marthamycetaceae

Propolis versicolor (Fr.) Fr.
On dead, fallen branches of broad-leaved trees; common.
50, 51. SJ 05, 15–17, 26, 35. AV, CF, CN, DU, MO. (9) 1976. (48, 49, 52)

Family Rhytismataceae

Coccomyces dentatus (Kunze & Schm.) Sacc.
On fallen oak leaves; common.
50, 51. SH 85; SJ 16, 17. CN, MF. (4) 1985. (48, 49, 52)

Hypoderma commune (Fr.) Duby
On dead stems of *Sedum spectabile*; frequent.
51. Recorded only from Mold (SJ 26) in 2005. (49)

Hypoderma hederae (Martius) de Not.
On fallen ivy leaves; frequent.
51. Recorded only from Craig Fawr (SJ 08) in 1983. (48, 49, 52)

Hypoderma rubi (Pers.) Chev.
On dead bramble stems; common.
50, 51. SH 85; SJ 05, 16, 34, 36. CF, ER, GW, MF. (7) 1972. (48, 49, 52)

Lophodermium arundinaceum (Schrader) Chev.
On dead marram grass stems; frequent.
51. Recorded only from Point of Ayr (SJ 18) since 1978. (48, 52)

Lophodermium pinastri (Schrad.) Chev.
On living pine needles; common.
50. Recorded only from Coed Clwyd (SJ 16) in 1975. (48, 49, 52)

Lophodermium spiraeae Hazl.
On dead stems of *Filipendula ulmaria*; rare.
51. Reported from Flintshire, without locality, by Grove (1935), as the Leptostroma anamorph.

Rhytisma acerinum (Pers.) Fr. *Sycamore Tarspot*
The cause of 'tar spot' on the leaves of maples, especially sycamore; very common, but decreasing with the reduction in sulphur dioxide levels in the atmosphere. The species is very sensitive to this pollutant and its abundance provides a simple measure of the peak concentration.
50, 51. SH 77, 85–88, 97; SJ 02, 04–08, 13–18, 23–27, 34–36, 43. AV, BO, BP, BW, CC, CE, CF, CN, ER, GW, HC, LH, LO, MA, MF, ML, MO, NF, PA, PP, WW. (125) 1880. (48, 49, 52)

Rhytisma salicinum (Pers.) Fr. *'Willow Tarspot'*
The cause of 'tar spot' on leaves of *Salix* species; frequent.
50, 51. SJ 08, 25, 26, 35. AV, BW. (5) 1994. (48, 49, 52)

Order THELEBOLALES
Family Thelebolaceae

Ascozonus woolhopensis (Renny) Boud.
On rabbit dung; frequent.
51. Recorded only from Mold (SJ 26) in 1976.

Unassigned anamorph of Thelebolales:

Alatospora acuminata Ingold
In foam in fast flowing rivers; common.
50, 51. SJ 16, 17, 24–26, 34, 35. AV, DU, LO, MO, NF. (8) In Rivers Alyn, Cegidog, Clywedog, Dee, Ffrith and Wheeler. 1981. (48, 49)

Class ORBILIOMYCETES
Order ORBILIALES
Family Orbiliaceae

Orbilia auricolor (Berk.) Sacc.
Orbilia curvatispora Boud.
On damp, fallen wood; frequent.
50, 51. SH 85; SJ 08, 26. AV. (3) 1924. (48, 49, 52)

Orbilia delicatula (P. Karst.) P. Karst. *Common Glasscup*
Orbilia xanthostigma auct.
On damp, fallen branches and trunks; common.
50, 51. SH 85; SJ 15, 25, 26, 34–36. CE, ER, GW, LO, NF. (12) 1910. (48, 49, 52)

Orbilia leucostigma (Fr.) Fr.
On damp, fallen wood; uncommon.
51. Recorded only from Bodelwyddan (SH 97) in 1988. (48, 49)

Orbilia sarraziniana Boud.
On damp, fallen wood; frequent.
51. SJ 17, 26. DU, MO. (2) 1978. (48, 49, 52)

Orbilia vinosa (Alb. & Schwein.) P. Karst.
On damp, fallen wood; uncommon.
50. Recorded only from Coed Coch (SH 87) in 1880.

Species Accounts

Unassigned anamorph of **Orbilia**:

Dactylella submersa (Ingold) S. Nilsson
In foam from River Alyn; uncommon.
50, 51. SJ 16, 26. AV, LO, MO. (3) 1981.

Class PEZIZOMYCETES
Order PEZIZALES
Family Ascobolaceae

Ascobolus albidus Crouan
On rabbit dung; frequent.
51. Recorded only from Mold (SJ 26) in 1976. (48, 49, 52)

Ascobolus denudatus Fr.
On decaying sawdust; uncommon.
50. Recorded only from Fairy Glen (SH 85) in 1924. (49)

Ascobolus furfuraceus Pers.
On cow dung; common.
51. SJ 26, 36. GW, MO, WW. (3) 1972. (48, 49, 52)

Ascobolus immersus Pers.
On rabbit dung; uncommon.
51. Recorded only from Mold (SJ 26) in 1976. (48, 49)

Saccobolus glaber (Pers.) Lambotte
On rabbit dung; frequent.
51. Recorded only from Mold (SJ 26) in 1976.

Family Ascodesmidaceae

Lasiobolus papillatus (Pers.) Sacc.
Ascobolus equinus (O.F. Müll.) P. Karst.
On rabbit dung; common.
51. Recorded only from Mold (SJ 26) in 1976. (48, 49, 52)

Family Discinaceae

Gyromitra esculenta (Pers.) Fr. *False morel*
On soil, mycorrhizal with pine but occasionally with other trees; a northern species which has only recently been recorded in Wales. Best treated as **poisonous**.
50, 51. SH 87; SJ 15–17. CC, CN. (5) 1880. (48, 52)

Gyromitra gigas (Krombh.) Cooke
On soil under pines; possibly extinct in Britain, the last collection was made in 1916 from Inverness-shire.
50. Recorded only from Coed Coch (SH 87) in 1880. **RDL/EX**

Hydnotrya tulasnei (Berk. & Br.) Berk. & Br.
A truffle, in soil under beech; uncommon.
51. Recorded only from Coed y Felin Nature Reserve (SJ 16) in 1993.

Family Helvellaceae

Helvella acetabulum (L.) Quélet *Vinegar Cup*
Paxina acetabulum (L.) O. Kuntze
On base-rich soil in woodland; frequent.
50, 51. SH 97; SJ 35. BO. (2) 1979.

Helvella corium (Weberb.) Massee
Cyathipodia corium (Weberb.) Boud.
On woodland soil; uncommon.
50, 51. SJ 16, 17, 35. BW. (4) 1986. (48, 52)

Helvella crispa (Scop.) Fr. *White saddle*
On soil in broad-leaved woods; common. **Poisonous.**
50, 51. SH 85, 97; SJ 05–08, 15–18, 23–26, 34–36. BO, ER, GW, LO, MA, MF, ML. (25) 1910. (48, 49, 52)

Helvella elastica Bull. *Elastic Saddle*
Leptopodia elastica (Bull.) Boud.
In rich woodland soil; uncommon.
50, 51. SH 85, 87, 97; SJ 15, 16, 26, 34. CC, ER, ML, MO. (8) 1880. (48, 49, 52)

Helvella lacunosa Afz. *Elfin saddle*
On soil in woodland; common. **Poisonous.**
50, 51. SH 77, 85, 87, 97; SJ 16, 24–26, 34–36. AV, ER, GW, LO, MA, MF, NM, PP (17) 1974. (48, 49, 52)

Helvella leucomelaena (Pers.) Nannf.
Paxina leucomelas (Pers.) O. Küntze
On soil in open vegetation; uncommon.
50, 51. SJ 16, 35. CN. (2) 2000. (52)

Helvella macropus (Pers.) P. Karst. *Felt Saddle*
Cyathipodia macropus (Pers.) Dennis
On woodland soil; frequent.
50, 51. SH 85; SJ 17, 26, 36. GW, LH, MA. (8) 1924. (48, 49, 52)

Helvella pezizoides Afz.
Leptopodia pezizoides (Afz.) Boud.
On base-rich woodland soil; uncommon.
51. Recorded only from Nant Alyn (SJ 16) in 1977. (49)

Species Accounts

Family Morchellaceae

Disciotis venosa (Pers.) Boud. *Bleach Cup*
On limestone soil in woods, in spring; uncommon.
50, 51. SH 97; SJ 08, 15–17, 24–26, 35. AV, CE, CN, DU, ML, MO, NF, PP, WW. (20) 1901. (49, 52) This species is typical of beechwood so its appearance on a roadside at the edge of Mold in April 2008 was surprising.

Mitrophora semilibera (DC.) Lév. *Semifree Morel*
On mine spoil heaps and in woods, usually over limestone, in spring; frequent.
50, 51. SH 97; SJ 07, 15–18, 25, 26, 34, 35. AV, DU, ER. (20) 1910. (52)

Morchella crassipes (Ventenat) Pers.
In deep litter in limestone woodland; uncommon.
50. Recorded only from Cilygroeslwyd Wood (SJ 15) in 1985. This is a segregate from the *esculenta* group not recognised by all workers.

Morchella elata Fr.
Morchella conica Pers.
On forest bark used as mulch in supermarket car park shrubberies, in spring; frequent. A northern species which appeared in many parts of England and Wales on imported bark. This is probably the species now known as *M. importuna*.
50, 51. SJ 17, 34, 36. MO. (3) 1902. (49)

Morchella esculenta (L.) Pers. *Morel*
On bare soil at the edge of woodland rides, in gardens, usually on base-rich soil; frequent in spring, but some years are more productive than others.
50, 51. SH 97; SJ 07, 16, 17, 25, 35. AV, BO, DU, HM. (9) 1959.

Verpa conica (O. Müll.) Swartz *Thimble Morel*
On base-rich soil under shrubs, in spring; uncommon, but appears in great numbers after hot summers and seems to be increasing in frequency as the climate warms.
50, 51. SH 87; SJ 15, 17, 18, 25, 34, 35. CC, DU, ER. (13) 1880. (52)

Family Pezizaceae

Adelphella babingtonii (Berk.) Pfister, Matočec & I. Kušan
Pachyella babingtonii (Berk.) Boud.
On sticks and branches in streams; uncommon.
50, 51. SJ 17, 34. DU, ER. (3) 1985. (48, 49, 52)

Iodophanus carneus (Pers.) Korf
On rabbit dung; frequent.
51. Recorded only from Mold (SJ 26) in 1976. (48, 49, 52)

Peziza ammophila Durieu & Mont. *Dune Cup*
In sand in marram dunes; frequent.
51. Recorded only from Point of Ayr (SJ 18) in 1985. (48, 52)

Peziza ampliata Pers.
On soil in flower bed; uncommon.
51. Recorded only from Mold (SJ 26) in 1978. (49)

Peziza badia Pers. *Bay Cup*
On soil at edge of woodland tracks; common.
50, 51. SH 87, 96, 97; SJ 05, 13, 16, 17, 24–26, 35, 36. CC, CE, CF, CN, GW, LH, ML, MO, NF, PP. (20) 1880. (48, 49, 52)

Peziza cerea Sow. *Cellar Cup*
On very rotten elm logs; uncommon.
50, 51. SH 97; SJ 26. ML, MO. (2) 1978. (49, 52)

Peziza echinospora P. Karst. *Charcoal Cup*
On burnt ground; frequent.
50. Recorded only from Coed Coch (SH 87) in 1880. (48, 49, 52)

Peziza howsei Boud.
Peziza emileia Cooke
On limestone soil; uncommon.
51. Recorded only from the Alyn Valley woodlands (SJ 16) in 1983.

Peziza michelii (Boud.) Dennis
Peziza plebeia (Le Gal) Nannf.
On woodland soil; uncommon.
50. Recorded only from Tan-y-Gopa (SH 97) in 1999. (49, 52)

Peziza micropus Pers.
On rotten logs; common.
50, 51. SH 87, 97; SJ 06, 08, 26, 34, 36. ER, GW, ML, WW. (7) 1979.

Peziza petersii Berk. & M.A. Curt.
On site of bonfire; frequent.
51. Recorded only from Ddol Uchaf Nature Reserve (SJ 17) in 1978. (49, 52)

Peziza proteana (Boud.) Seaver *Bonfire Cauliflower*
On bonfire site; uncommon.
50, 51. SJ 17, 34. DU, ER. (2) 1977.

Peziza repanda Pers. *Palamino Cup*
On woody debris on forest floor; common.
50, 51. SH 97; SJ 07, 16, 17, 24, 26, 34, 36. CE, ER, GW, LH, LO, MA, ML, PP, WW. (11) 1976. (48, 49, 52)

Peziza rifai Moravec & Spooner
On calcareous soil; rare.
51. Recorded only from Ddol Uchaf Nature Reserve (SJ 17) in 1985. This was one of the first British records, and the first for Wales.

Species Accounts

Peziza saniosa Schrad.
On woodland soil; frequent.
50. Recorded only from Hafod Wood, near Betws-y-Coed (SH 85) in 1924.

Peziza succosa Berk. *Yellowing Cup*
On soil; frequent.
50, 51. SH 87, 97; SJ 16, 17, 26, 34. AV, CC, ER, LO, MA. (7) 1880. (49, 52)

Peziza vacinii (Velen.) Svrček
On calcareous soil; rare.
51. Recorded only from Ddol Uchaf Nature Reserve (SJ 17) in 1985. The first record for Wales.

Peziza varia (Hedw.) Fr. *Layered Cup*
On rich soil in woodland; common.
50, 51. SJ 16, 26, 34. AV, ER, MA. (3) 1977. (48)

Peziza vesiculosa Bull. *Blistered Cup*
On buried wood and rich soil; common.
50, 51. SJ 16, 17, 26, 34. DU, ER, LO, MO. (5) 1975. (48, 49, 52)

Peziza violacea Pers.
Peziza praetervisa Bres.
On sites of bonfire; frequent.
50, 51. SJ 17, 26, 35. AV, DU, MA. (4) 1978. (48, 49, 52)

Family Pyronemataceae

Aleuria aurantia (Pers.) Fuckel *Orange-peel Fungus*
On compacted soil at edge of tracks; common.
50, 51. SH 85, 97; SJ 07, 15-18, 24-27, 34-36. CE, ER, GW, HC, LH, LO, MF, NF, WW. (22) 1972. (48, 49, 52)

Aleuria luteonitens (Berk. & Broome) Gillet
On woodland soil; uncommon.
50. Recorded only from Fairy Glen (SH 85) in 1924. (49)

Anthracobia macrocystis (Cooke) Boud.
On bonfire sites; frequent.
50, 51. SH 97; SJ 07, 15. (3) 1981. (48, 52)

Anthracobia maurilabra (Cooke) Boud.
On bonfire site; uncommon.
51. Recorded only from Ddol Uchaf Nature Reserve (SJ 17) in 1985. (49, 52)

Anthracobia melaloma (Alb. & Schwein.) Boud.
On bonfire site; uncommon.
51. Recorded only from Ddol Uchaf Nature Reserve (SJ 17) in 1985. (48, 49, 52)

Cheilymenia fimicola (de Not. & Bagl.) Dennis
On cow dung; common.
50, 51. SJ 16, 24, 26, 27, 34. AV, ER, MO. (7) 1910. (48, 49, 52)

Cheilymenia granulata (Bull.) J. Moravec
Coprobia granulata (Bull.) Boud.
On dung; common.
50, 51. SH 85; SJ 16, 25, 26, 34, 43. CE, ER, NF. (6) 1910. (48, 49, 52)

Cheilymenia theleboloides (Alb. & Schwein.) Boud.
On cow and rabbit dung; frequent.
51. SJ 25, 26. HM, MO. (2) 1978. (48)

Cheilymenia vitellina (Pers.) Dennis
On very rotten wood and soil; common.
50, 51. SH 87; SJ 06, 16, 26, 34. AV, CC, ER, LO. (5) 1880. (49, 52)

Geopora sumneriana (Cooke) de la Torre *Cedar Cup*
Sepultaria sumneriana (Cooke) Massee
On a pile of cedar sawdust; uncommon. This is usually found sunk into the grass underneath living cedars in lawns.
51. Recorded only from Bodelwyddan (SH 97) in 1984.

Humaria hemisphaerica (J.H. Wigg.) Fuckel *Glazed Cup*
On soil; frequent.
50, 51. SH 97; SJ 17, 34. ER. (3) 1981. (48, 49, 52)

Lamprospora crec'hqueraultii (P. Crouan & H. Crouan) Boud.
Ramsbottomia crec'hqueraultii (P. Crouan & H. Crouan) Benkert & T. Schumach.
On soil in plant pot; uncommon.
51. Recorded only from Mold (SJ 26) in 1976.

Lamprospora crouani (Cooke) Seaver
In moss in dunes; uncommon.
51. Recorded only from Point of Ayr (SJ 18) in 1979. (49, 52)

Lamprospora wrightii (Berk. & M.A. Curt.) Seaver
Octospora wrightii (Berk. & M.A. Curt.) Moravec
On *Amblystegium* mosses; uncommon.
51. Recorded only from Bodelwyddan (SH 97) in 1864, the type locality.

Melastiza chateri (W.G. Sm.) Boud. *Orange Cup*
On damp soil; uncommon.
50, 51. SJ 15–17, 24, 35. AV, DU. (6) 1986. (49, 52)

Miladina lecithina (Cooke) Svrček
In foam in fast flowing rivers; common.
50, 51. SJ 16, 26. AV, LO, MO. (4) In Rivers Alyn and Terrig. 1981, as the *Actinosporella* anamorph.

Neottiella rutilans (Fr.) Dennis
In moss on sandy soil; uncommon.
50, 51. SJ 17, 27. (3) 1984. (48, 49)

Octospora humosa (Fr.) Dennis
On soil amongst *Polytrichum* mosses; uncommon.
50, 51. SH 85; SJ 26. LO. (2) 1977. (49)

Orbicula parietina (Schrader) Hughes
On rotten branches; frequent.
50, 51. SJ 07, 34, 36. ER, GW. (3) 1976.

Otidea alutacea (Pers.) Massee *Tan Ear*
In woodland litter; uncommon.
50. SH 77, 85, 97. ML. (3) 1979. (49, 52)

Otidea bufonia (Pers.) Boud. *Toad's Ear*
In woodland litter; frequent.
50, 51. SJ 05, 07, 24, 26. CF, PP, WW. (4) 1983. (49, 52)

Otidea cochleata (L.) Fuckel *Brown Ear*
In woodland litter on limestone; frequent.
50, 51. SH 85; SJ 16, 23, 26, 34. AV, WW. (5) 1902. (49, 52)

Otidea grandis (Pers.) Rehm
In woodland litter on limestone; uncommon.
51. SJ 16. AV. (2) 1977. (49)

Otidea leporina (Batsch) Fuckel *'Rabbit's Ear'*
In woodland litter; frequent.
50. Recorded only from Coed Coch (SH 87) in 1880. (52)

Otidea onotica (Pers.) Fuckel *Hare's Ear*
In woodland litter; frequent.
50. SH 85; SJ 05, 34. CF. (5) 1910. (49, 52)

Pulvinula cinnabarina (Fuckel) Boud.
On marl soil; uncommon.
51. Recorded only from Ddol Uchaf Nature Reserve (SJ 17) in 1985.

Pyronema omphalodes (Bull.) Fuckel
On bonfire sites; frequent.
51. SJ 26. LO, MO. (2) 1977. (49)

Scutellinia barlae (Boud.) Maire
On mossy bank at edge of woodland path; uncommon.
50. Recorded only from Big Covert, Maeshafn (SJ 26) in 1992. (52)

Scutellinia olivascens (Cooke) Kuntze
Scutellinia ampullacea in part
On damp soil; uncommon.
50. Recorded only from Erddig Park (SJ 34) in 2018.

Scutellinia scutellata (L.) Lambotte *Common Eyelash*
On rotten wood in damp woodland soil; common.
50, 51. SH 85, 87, 97; SJ 05, 07, 08, 13, 15–18, 24–27, 34–36. AV, BO, CC, CE, CN, DU, ER, GW, LO, MA, ML, NF, PP, WW. (39) 1880. (48, 49, 52)

Scutellinia trechispora (Berk. & Br.) Lambotte
On damp, rotten wood; frequent.
50, 51. SH 97; SJ 08, 15–17, 26, 34. AV, BO, DU, ER, LO. (7) 1910. (48, 52)

Scutellinia umbrorum (Fr.) Lambotte
On soil among rushes and sedges; uncommon.
50. Recorded only from Alyn Waters Country Park (SJ 35) in 1998. (49, 52)

Stephensia bombycina (Vitt.) Tul. & C. Tul.
A truffle, in soil under larch; uncommon.
51. Recorded only from Loggerheads Country Park (SJ 26) in 1978.

Tarzetta catinus (Holmsk.) Korf & Rogers
On soil under beech; frequent.
50, 51. SJ 16, 17, 26. LO, MF. (3) 1983. (52)

Tarzetta cupularis (L.) Lambotte *Toothed Cup*
On soil in woodland; uncommon.
50, 51. SH 77; SJ 17, 34. CN, DU, ER. (5) 1985. (49, 52)

Trichophaea woolhopeia (Cooke & W. Phill.) Boud.
On rotten wood; frequent.
50. Recorded only from Erddig (SJ 34) in 1985. (48, 49, 52)

Family Rhizinaceae

Rhizina undulata Fr. *Pine Firefungus*
On the ground in conifer plantation; uncommon. Another species typical of Scottish pine forest which is appearing further south in recent years.
51. Recorded only from Hawarden Woods (SJ 36) in 1972. (48, 49)

Family Sarcoscyphaceae

Sarcoscypha austriaca (Beck) Boud. var. **austriaca** *Scarlet Elfcup*
On rotten wood on the ground in winter. This and the next species can only be separated microscopically. This species is probably more common than the next but the older records cannot be assigned in the absence of specimens.
50, 51. SH 97; SJ 05, 08, 15–17, 24–27, 34, 35. AV, CE, CF, CN, DU, LO, MA, ML, MO, NF, PP, WW. (41) 1979. (49, 52) This species has become much more common during the last fifty years and was especially abundant in the Loggerheads Country Park in early 2006 and Ddol Uchaf in 2017.

Species Accounts

Sarcoscypha austriaca var. **lutea** Ruini & Ruedl
On rotten wood on the ground, in winter; rare.
50. SH 97; SJ 05. (2) Recorded only from Bont Newydd, Marli since 2003 and Lady Bagot's Drive, Rhewl in 2006. This yellow variety was described in 1999 and these are the first British records; it was found in 2008 in Somerset.

Sarcoscypha coccinea (Scop.) Lambotte *Scarlet Elfcup*
On rotten wood on the ground in winter. Most of the records are likely to refer to the previous species; only specimens from Pantymwyn (SJ 16) in 1960 and Marli (SH 97) in 1979, deposited at Kew, are definitely this species..
50, 51. SH 97; SJ 08, 16, 17, 25, 35. AV, HM, ML, NF. (12) 1813. (52)

Family Tuberaceae

Tuber aestivum Vitt. *Summer Truffle*
A truffle, in soil under beech trees; rare except in the south of England.
51. Recorded only from Bodelwyddan (SH 97) in 1983.

Tuber borchii Vitt.
A truffle, in soil under beech trees; uncommon.
51. Recorded only from Loggerheads Country Park (SJ 26) in 1990.

Tuber excavatum Vitt.
A truffle, in soil under beech trees; uncommon.
50, 51. SJ 26. LO, MA. (2) 1992.

Tuber maculatum Vitt.
A truffle, in soil under beech; rare.
50. Recorded only from Big Covert, Maeshafn (SJ 26) in 1992.

Tuber nitidum Vitt.
A truffle, in soil under trees; rare.
51. Recorded from Dyserth (SJ 08) by Higgins (1962) but the record is extremely doubtful.

Tuber rufum Pico *Red Truffle*
A truffle, in soil under birch; uncommon.
51. Recorded only from Prestatyn (SJ 08) in 1930.

Class SORDARIOMYCETES
Sub-class Hypocreomycetidae
Order HYPOCREALES
Family Bionectriaceae

Bionectria ralfsii (Berk. & Broome) Schroers & Samuels
Nectria ralfsii Berk. & Br.
On fallen branches; uncommon.
50. SH 87; SH 34. CC. (2) 1880. (48, 49)

Nectriopsis candicans (Plowr.) Maire
Nectria candicans (Plowr.) Samuels
On myxomycetes, esp. *Diderma globosum*; uncommon.
51. SJ 17, 36. GW, LH. (2) 1972.

Nectriopsis rexiana (Sacc.) Rossman, I. Crogghard & Crous
On myxomycetes, esp. *Metatrichia floriformis*; common.
51. SJ 17, 26. DU, LO, WW. (3) 1975. Only as the anamorph, *Verticillium rexianum* (Sacc.) Sacc.

Nectriopsis violacea (Schm.) Maire
Nectria violacea (Schm.) Fr.
On the myxomycete *Fuligo septica*; uncommon.
50, 51. SJ 23, 25, 26, 34. ER, LO, NF, NM. (5) 1977.

Roumegueriella rufula (Berk. & Br.) Malloch & Cain
On cow dung; uncommon.
51. Recorded only from Gwysaney Park, Mold (SJ 26) in 1978.

Family Clavicipitaceae

Berkelella stilbigera (Berk. & Broome) Sacc.
Byssostilbe stilbigera (Berk. & Br.) Petch.
Parasitic on myxomycetes, especially species of *Trichia*; common.
50, 51. SH 85; SJ 05, 26, 34, 36. CF, ER, GW, LO. (6) 1910. (48, 49, 52) Always as the anamorph *Polycephalomyces tomentosus* (Schrad.) Seifert.

Unassigned anamorph of **Berkellela**:

Polycephalomyces ramosus (Peck) Mains
On *Hirsutella* on dead fly; rare.
51. Recorded only from a cave in Tremeirchion (SJ 07) in 1954.

Claviceps purpurea (Fr.) Tul. *Ergot*
The 'ergot' of grasses and cereals; very common. **Poisonous.**
50, 51. SJ 05, 07, 13, 15, 16, 18, 23, 24, 26, 27, 34–36. AV, ER, GW, LO, MO, PA. (20) 1972. (48, 49, 52)

Epichloë baconii White
Causing 'choke' on stems of *Agrostis capillaris*; uncommon.
51. Recorded only from Point of Ayr (SJ 18) in 1991.

Epichloë clarkii White
Causing 'choke' on stems of *Holcus lanatus* and *H. mollis*; common.
51. Recorded only from Bagillt (SJ 27) in 1978. (48)

Epichloë festucae Leuchtm., Schardl & Siegel
Causing 'choke' on stems of species of *Festuca*; uncommon.
51. Recorded only from Meliden (SJ 07) in 1989.

Species Accounts

Epichloë sylvaticum Leuchtm. & Schardl
Causing 'choke' on stems of *Brachypodium sylvaticum*; rare.
50, 51. SJ 15, 26, 34. ER, MO. (3) 1974.

Epichloë typhina (Pers.) Tul. & C. Tul. *Choke*
The cause of 'choke' on stems of a variety of grasses; common.
50, 51. SJ 07, 16, 26, 36. AV, MO. (5) 1974. (48, 49, 52)

Unassigned anamorph of **Metacordyceps**:

Metarhizum marquandii (Massee) Kapler, S.A. Rohner & Hunser
Paecilomyces marquandii (Massee) S.J. Hughes
On gills of *Camarophyllus virgineus*, causing lilac patches; rare.
50, 51. SH 97; SJ 23. BO. (2) 2001.

Family Cordycipitaceae

Cordyceps bassiana Z.Z. Li, C.R. Li, B. Hueng & M.Z. Fan
Beauveria bassiana (Bals.-Criv.) Vuill.
Parasitic on insects, especially beetles; common.
50, 51. SJ 26, 34. ER, LO. (2) 1977. (48)

Cordyceps bifusispora O.E. Erikss.
Cordyceps tuberculata (Lebert) Maire
On buried moth pupae; rare.
51. Recorded only from Ysceifiog (SJ 17) in 1978. (52) **RDL/VU**

Cordyceps farinosa (Höhnel) Kepler, B. Shresh. & Spatafora
Paecilomyces farinosus (Höhnel) Mason & S.J. Hughes
On buried moth caterpillars and pupae; frequent.
50, 51. SH 85; SJ 08, 17, 24, 25, 34, 36. CE, ER, GW, NF. (11) 1924. (48, 49)

Cordyceps militaris (L.) Link *Scarlet Caterpillarclub*
On buried moth larvae; common.
50, 51. SH 85, 97; SJ 15, 16, 23, 24, 26, 35, 36. GW, LO, ML, MO. (11) 1924. (48, 49, 52)

Torrubiella arachnophila (Johnston) Mains
On linyphiid spiders; uncommon.
51. Recorded only from Nant-y-Ffrith (SJ 25) in 1976, as the *Hymenostilbe* anamorph. (49, 52)

Family Glomerellaceae

Glomerella cingulata (Stonem.) Spaulding & v. Shrenk
Colletotrichum gloeosporioides (Pers.) Penzig & Sacc.; *Gloeosporium rhododendri* Briosi & Cavara
Leaf spot on rhododendrons; common.
51. SJ 26, 36. GW. (2) 1972. (49)

Unassigned anamorphs of **Glomerella**:

Colletotrichum liliacearum (Wesyend.) Duke
Black spots on dead stems of *Hyacinthoides non-scriptus*; common.
50, 51. SJ 16, 26, 34, 35. AV, ER, LO. (4) 1977.

Colletotrichum trichellum (Fr.) Duke
Leaf spot on ivy; common.
51. SJ 26, 27, 36. GW, MO. (3) 1972. (49)

Family Hypocreaceae

Hypocrea alutacea (Pers.) Ces. & de Not. *Dogend*
Podostroma alutaceum (Pers.) Atk.
On soil under conifers; rare.
50, 51. SJ 15–17. BW. (4) 1994.

Hypocrea argillacea W. Phill. & Plowr.
On dead wood; uncommon.
51. SJ 07, 08. (2) 1981. (52)

Hypocrea aureoviridis Plowr. & Cooke
Chromocrea aureoviridis (Plowr. & Cooke) Petch
On fallen branches; uncommon.
51. Recorded only from Loggerheads Country Park (SJ 26) in 2002. (52)

Hypocrea citrina (Pers.) Fr.
On rotten coniferous wood; uncommon.
51. SJ 07, 08. (2) 1981. (49)

Hypocrea gelatinosa (Tode) Fr.
Creopus gelatinosus (Tode) Link
On rotten wood; frequent.
51. SJ 26. (2) 2002. (49, 52)

Hypocrea pulvinata Fuckel *Ochre Cushion*
On decaying brackets of *Piptoporus betulinus*; frequent.
50, 51. SJ 07, 08, 15, 17. (4) 1978. (48, 49, 52)

Hypocrea rufa (Pers.) Fr.
On fallen branches; common.
50, 51. SH 77, 87, 97; SJ 07, 08, 16–18, 25, 26, 34–36. BO, CC, DU, ER, GW, LO, MF, NF, PA, WW. (30) 1880. Usually the *Trichoderma* anamorph.

Hypomyces aurantius (Pers.) Tul.
On dead brackets of various polypores; common.
50, 51. SH 97; SJ 16, 17, 25, 26, 34. BO, ER, MA, MF, ML, MO, NF (12) 1864. (48, 49, 52)

Species Accounts

Hypomyces chrysospermus Tul. *Bolete Mould*
Apiocrea chrysosperma (Tul.) Sydow
On decaying boletes, *Paxillus* and earthballs; common.
50, 51. SH 77, 87; SJ 05, 07, 08, 15–17, 25, 26, 34–36. CC, CF, ER, GW, LH, LO, MF, NF, NM, WW. (23) 1880. Usually as the *Sepedonium* anamorph. (48, 49, 52)

Hypomyces luteovirens (Fr.:Fr.) Tul. & C. Tul.
Peckiella viridis (Alb. & Schwein.) Sacc.
On decaying fruit bodies of *Lactarius* species; uncommon.
51. Recorded only from Loggerheads Country Park (SJ 26) in 1976.

Hypomyces ochraceus (Pers.) Tul. & C. Tul.
On decaying fruit bodies of *Russula*; frequent.
50, 51. SJ 26, 34. ER, LO. (2) 1976. Only as the *Cladobotryum* anamorph.

Hypomyces rosellus (Alb. & Schwein.) Tul. & C. Tul.
On decaying fruit bodies of *Trametes versicolor*; frequent.
50. SH 87; SJ 34. CC, ER. (2) 1880. (48, 49, 52)

Sphaerostilbella aureonitens (Tul. & C. Tul.) Seifert et al.
Nectriopsis aureonitens (Tul. & C. Tul.) Maire
On decaying fruit bodies of *Stereum hirsutum*; uncommon.
50. SH 87. CC. (2) 1880. (48, 49)

Family Nectriaceae

Dialonectria episphaeria (Tode) Cooke
Nectria episphaeria (Tode) Fr.
On old pyrenomycetes on fallen branches; frequent.
50, 51. SJ 26, 34. AV, ER, LO. (3) 1910. (48, 49, 52)

Unassigned anamorph of **Gibberella**:

Fusarium tricinctum (Corda) Sacc. f. poae (Peck) Snyder & Hankin
On flowers of *Dactylis glomerata*; frequent.
51. Recorded only from Mold (SJ 26) in 1978.

Nectria cinnabarina (Tode) Fr. *Coral Spot*
On sticks and branches, often still attached; very common.
50, 51. SH 77, 85, 87, 97; SJ 05–08, 15–18, 23–27, 34–36, 43. AV, BO, BW, CC, CE, CN, DU, ER, GW, HC, LH, LO, MA, MF, ML, MO, NF, NM, PA, PP, WW. (102) 1880. Usually accompanied by the *Tubercularia* anamorph. (49, 49, 52)

Nectria peziza (Tode) Fr.
On rotten wood; frequent.
50, 51. SJ 15, 18, 34, 36. ER, GW. (4) 1976. (48, 49, 52)

Nectria pseudopeziza (Desm.) Rossman
Calonectria ochraceopallida (Berk. & Br.) Sacc.
On fallen wood; frequent.
51. Recorded only from Pontblyddyn (SJ 26) in 1973.

Nectria sinopica (Fr.) Fr.
On old ivy stems; uncommon.
50. Recorded from Coed Coch (SH 87) in 1880.

Unassigned anamorph of **Nectria**:

Flagellospora curvula Ingold
In foam in fast flowing rivers; common.
50, 51. SJ 16, 24–26, 34, 35. AV, LO, MO, NF. (9) In Rivers Alyn, Cegidog, Clywedog, Dee, Ffrith and Terrig. 1981. (49)

Neonectria coccinea (Pers.) Rossman & Samuels
Nectria coccinea (Pers.) Fr.
On beech trunks and branches; frequent.
50, 51. SJ 16, 24–26, 34. CN, ER, LO, MO, NF, PP. (8) 1910. (48, 49, 52)

Neonectria ditissima (Tul. & C. Tul.) Samuels & Rossman
Nectria ditissima Tul. & C. Tul.
On fallen beech branch; uncommon.
50. Recorded only from Erddig (SJ 34) in 1910.

Neonectria lugdunensis (Sacc. & Therry) L. Lomard & Crous
Nectria lugdunensis J. Webster
In foam in fast flowing rivers; common.
50, 51. SJ 16, 17, 24, 26, 34. AV, DU, LO, MO. (6) 1981. In Rivers Alyn, Clywedog, Dee and Wheeler. (48, 49) Always as the anamorph – *Heliscus lugdunensis* Sacc. & Therry.

Unassigned anamorph of **Neonectria**:

Fusidium griseum Link
Leaf spot on *Betula*; common.
50. SH 85; SJ 16, 34. ER, NA. (3) 1910. (48, 49, 52)

Unassigned anamorph of **Pseudonectria**:

Volutella ciliata (Alb. & Schwein.) Fr.
On the bark of living trees in moist chamber culture; common.
50, 51. SH 76, 97; SJ 24, 26, 53. BO, LO. (5) 2005. (48, 52)

Thelonectria mammoidea (Phill. & Plowr.) C. Salgado & Sanchez
Nectria mammoidea Phill. & Plowr.
On sticks and dead stems; uncommon.
50. Recorded only from Coed Coch (SH 87) in 1880. (48, 49, 52)

Thelonectria veuillotiana (Rostr. & Sacc.) P. Chaveni & C. Salgado
Nectria veuillotiana Rostr. & Sacc.
On fallen twigs; frequent.
51. Recorded only from Loggerheads Country Park (SJ 26) in 1983. (48)

Species Accounts

Thyronectria aquifolii (Fr.) Jaklitsch & Voglmayr
Nectria aquifolii (Fr.) Berk.
On holly bark; frequent.
51. Recorded only from Frith Mountain, Moel Famau (SJ 16) in 1978.

Unassigned anamorph of Nectriaceae:

Aphanocladium album (Preuss) Gams
On the myxomycete *Comatricha nigra*; frequent.
51. Recorded only from Mold (SJ 26) in 1977. (48, 49)

Family Ophiocordycipitaceae

Ophiocordyceps entomorrhiza (Dicks.) G.H. Sung, J.F. Sung, Hywel-Jones & Spatafora
Cordyceps entomorrhiza (Dicks.) Link
Parasitic on beetles; rare, considered extinct by some workers.
51. Recorded only from Loggerheads Country Park (SJ 26) in 1985, as the *Hirsutella* anamorph. (49)

Ophiocordyceps forquignonii (Quél.) G.H. Sung, J.F. Sung, Hywel-Jones & Spatafora
Cordyceps forquignonii Quél.
Parasitic on flies; uncommon.
50, 51. SH 85; SJ 17. (2) 1924. (48, 52)

Ophiocordyceps gracilis (Grev.) G.H. Sung, J.F. Sung, Hywel-Jones & Spatafora
Cordyceps gracilis (Grev.) Durieu & Mont.
Parasitic on buried moth larvae; uncommon.
50. SJ 15. (2) 1994. (48, 49, 52)

Unassigned anamorph of **Ophiocordyceps**:

Hirsutella guignardii (Maheis) Sampen
On dead fly in cave; rare.
51. Recorded only from Tremeirchion (SJ 07) in 1954.

Tolypocladium capitatum (Holmsk.) C.H. Quandt, Kepler & Spatafora
Elaphocordyceps capitata (Holmsk.) G.H. Sung, J.F. Sung & Spatafora
Cordyceps capitata (Holmsk.) Link
Parasitic upon the false truffle *Elaphomyces granulatus*; uncommon.
50. Recorded only from Hafod Wood (SH 85) in 1924. (48, 49, 52)

Tolypocladium microsporum (Jaap) Bisset
Sesquicillium microsporum (Jaap) Veenbaas Rijks & W. Gams
Parasitic on the myxomycete *Stemonitis fusca*; common.
51. Recorded only from Wepre Country Park (SJ 26) in 1993. (49)

Tolypocladium ophioglossoides (J.F. Gmel.) C.H. Quandt, Kepler & Spatafora
Snaketongue Truffleclub
Elaphocordyceps ophioglossoides (Pers.) G.H. Sung, J.F. Sung & Spatafora;
Cordyceps ophioglossoides (Pers.) Link
On the false truffle *Elaphomyces muricatus*; uncommon.
50. SH 76, 85. (3) 1924. (48, 49)
51. Recorded only from Bagillt (SJ 27) in 1978. (48)

Unassigned anamorphs of Hypocreales:

Calcarisporium arbuscula Preuss
Parasitic on *Geoglossum fallax*; uncommon.
51. Recorded only from Halkyn Common (SJ 17) in 2008. (52)

Myrothecium carmichaelii Grev.
On dead stems of *Epilobium hirsutum*; frequent.
51. Recorded only from Mold (SJ 26) in 1978.

Stilbella erythrocephala (Ditm.) Lindau
On rabbit dung; common.
51. SJ 08, 17, 26. BP, LH, MO. (3) 1976. (48)

Order MICROASCALES
Family Halosphaeriaceae

Unassigned anamorph of **Corollospora**:

Clavatospora longibrachiata (Ingold) S. Nilsson ex Marvanová & S. Nilsson
In foam in fast flowing rivers; frequent.
51. SJ 16, 25, 26. AV, MO, NF. (3) In Rivers Alyn and Ffrith. 1981. (48, 49)

Family Microascaceae

Unassigned anamorph of **Microascus**:

Wardomyces pulvinatus (Marchal) Dickinson
Isolated from saltmarsh soil; rare.
51. Recorded only from Flint (SJ 27) in 1962.

Unassigned anamorph of Microascaceae:

Cephalotrichum stemonitis (Pers.) Nees
Doratomyces stemonitis (Pers.) Morton & Smith
On the bark of living elder; widespread and common on herbaceous stems.
50. Recorded only from Bryn Euryn (SJ 88) in 2008.

Unassigned anamorph of Microascales:

Sphaeronaemella fimicola Marchal
On rabbit dung; common.
51. Recorded only from Mold (SJ 26) in 1976. (49)

Species Accounts

Sub-class Sordariomycetidae
Order BOLINIALES
Family Boliniaceae

Camaropsella lutea (Alb. & Schwein.) Lar N. Vassiljeva
Camarops lutea (Alb. & Schwein.) Nannf.
On dead alder stems; uncommon.
50, 51. SJ 17, 26, 34. DU, ER, LO. (3) 1972. (48, 49)

Order CHAETOSPHAERIALES
Family Chaetosphaeriaceae

Chaetosphaeria innumera Tul. & C. Tul.
On fallen beech branch; frequent.
50. Recorded only from Coed Coch (SH 87) in 1880.

Chaetosphaeria myriocarpa (Fr.) Booth
On fallen sycamore branches; frequent.
51. AV, LO. 26. (2) 1983. (48, 49)

Chaetosphaeria pulviscula (Currey) Booth
On fallen sycamore branches; uncommon.
51. Recorded only from Loggerheads Country Park (SJ 26) in 1983.

Zignoella paecilostoma (Berk. & Br.) Sacc.
On fallen gorse branches; rare.
50. Recorded only from Coed Coch (SH 87) in 1880.

Order CONIOCHAETALES
Family Coniochaetaceae

Coniochaeta pulveracea (Ehrenb.) Munk
On fallen sycamore branches; frequent.
51. Recorded only from Loggerheads Country Park (SJ 26) in 1983.

Coniochaeta scatigera (Berk. & Br.) Cain
On rabbit dung; common.
51. Recorded only from Mold (SJ 26) in 1978.

Order DIAPORTHALES
Family Diaporthaceae

Diaporthe arctii (Lasch) Nitschke
On dead herbaceous stems; common.
50. SH 87; SJ 16. CC, MA. (2) 1880. (48)

Diaporthe epilobii Fuckel
On dead stems of *Chamaerion angustifolium*; common.
51. Recorded only from Mold (SJ 26) in 1989.

Diaporthe eres Nitschke
On herbaceous stems and woody twigs; frequent.
50, 51. SH 87; SJ 26, 34. CC, LO. (3) 1880. (48, 49, 52)

Diaporthe nucleata (Currey) Cooke
On fallen gorse branches; frequent.
51. Recorded only from Hope Mountain (SJ 25) in 1978.

Diaporthe pardalota (Mont.) Nitschke ex Fuckel
On dead herbaceous stems; frequent.
51. Recorded only from Mold (SJ 26) in 1976.

Diaporthe pustulata (Desm.) Sacc.
On fallen sycamore branches; frequent.
51. Recorded only from Loggerheads Country Park (SJ 26) in 1983.

Diaporthe strumella (Fr.) Fuckel
On dead *Ribes rubrum* stems; uncommon.
50. Recorded only from Erddig (SJ 34) in 1985.

Diaporthe varians (Currey) Sacc.
On fallen sycamore branches; common.
51. Recorded only from Loggerheads Country Park (SJ 26) in 1978.

Unassigned anamorph of **Diaporthe**:

Phomopsis tamicola (Desm.) Trav.
On dead leaves of *Tamus communis*; rare.
51. Recorded only from Mold (SJ 26) in 1978.

Mazzantia angelicae (Berk.) Lar N. Vassiljeva
Diaporthopsis angelicae (Berk.) Wehmeyer
On dead *Heracleum sphondylium* stems; frequent.
51. Recorded only from Mold (SJ 26) in 1978.

Family Gnomoniaceae

Apiognomonia errabunda (Desm.) Höhn.
On fallen beech twigs; frequent.
50. Recorded only from Big Covert, Maeshafn (SJ 26) in 1992.

Cryptodiaporthe hystrix (Tode) Petrak
On fallen sycamore branches; uncommon.
51. Recorded only from Loggerheads Country Park (SJ 26) in 1983.

Cryptodiaporthe salicina (Pers.) Wehmeyer
On dead and dying sallow twigs; uncommon.
51. Recorded only from Point of Ayr (SJ 18) in 1985, as the *Cytospora* anamorph.

Species Accounts

Gnomonia gnomon (Tode) J. Schröt.
On fallen hazel leaves; frequent.
51. Recorded only from Hawarden (SJ 36) in 2014.

Gnomonia rostellata (Fr.) Wehm.
On bramble twigs; uncommon.
50. Recorded only from Fairy Glen (SH 85) in 1924. (48, 49)

Gnomiella tubaeformis (Tode) Sacc.
Leaf spot on alder; frequent.
50. Recorded only from Alyn Waters Country Park (SJ 35) in 2016.

Gnomiopsis comari (P. Karst.) Spooner
Gnomonia comari P. Karst.
On dead rosaceous stems; uncommon.
51. SJ 26, 27. MO. (2) 1950.

Ophiovalsa corylina (Tul. & C. Tul.) Petrak
Winterella corylina (Tul. & C. Tul.) O. Kuntze
On fallen hazel branches; uncommon.
50. Recorded only from Hafod Wood (SH 85) in 1924.

Family Melanconidaceae

Melanconis flavovirens (Otth) Wehmeyer
On fallen hazel branches; uncommon.
50. Recorded only from Coed Coch (SH 87) in 1880.

Melanconis stilbostoma (Fr.) Tul.
On fallen birch branches; uncommon.
50. Recorded only from Hafod Wood (SH 85) in 1924. (48, 49)

Prosthecium innesii (Currey) Wehmeyer
On fallen sycamore branches; uncommon.
51. Recorded only from Loggerheads Country Park (SJ 26) in 1983.

Family Sydowiellaceae

Sillia ferruginea (Pers.) P. Karst.
On hazel twigs; frequent.
50. SH 87; SJ 15. CC. (2) 1880. (48)

Family Valsaceae

Mamiania fimbriata (Pers.) Ces. & de Not.
Leaf spot on hornbeam; rare.
50. SJ 16, 34. LO. (2) 2012.

Mamianiella coryli (Batsch) Höhnel
Leaf spot on hazel; uncommon.
50. Recorded only from Loggerheads Country Park (SJ 16) in 2016.

Valsa ambiens (Pers.) Fr.
On fallen sycamore branches; frequent.
50, 51. SH 87; SJ 26, 34. CC, ER, LO. (3) 1880. (48) Includes records of the *Cytospora* anamorph.

Valsa ceratosperma (Tody) Maire
On fallen branches; frequent.
50. Recorded only from Fairy Glen (SH 85) in 1924. (48, 49)

Order OPHIOSTOMATALES
Family Ceratocystidaceae

Unassigned anamorphs of **Ceratocystis**:

Chalara pteridina Syd.
On bracken litter; common.
51. Recorded only from Hope Mountain (SJ 25) in 1978. (48, 49)

Thielaviopsis basicola (Berk. & Br.) Ferraris
Causing a stem and root rot of China asters; uncommon.
51. Recorded from Flintshire, without locality, by Moore (1959).

Family Ophiostomataceae

Ophiostoma novo-ulmi Brasier
The causal agent of 'Dutch elm disease' which devastated the British elm population in the 1970s; now much less common.
50, 51. SH 97; SJ 07, 08, 15–17, 25–27, 34–36. DU, ER, MO, WW. (15) 1974.

Order PHYLLACHORALES
Family Phyllachoraceae

Phyllachora dactylidis Delacr.
On living leaves and stems of *Dactylis glomerata*; common.
50, 51. SJ 16, 26. LO, MA, MO. (4) 1976. (49)

Phyllachora graminis (Pers.) Fuckel
On living grass leaves; common.
50, 51. SH 85; SJ 16, 18, 26, 45. LO, PA. (5) 1924. (48, 49, 52)

Phyllachora heraclei (Fr.) Fuckel
On living leaves and stems of *Heracleum sphondylium*; frequent.
51. Recorded only from Mold (SJ 26) in 1978, as the *Phleospora* anamorph.

Phyllachora junci (Alb. & Schwein.) Fuckel
On stems of living *Juncus* species; common.
51. SJ 18, 25, 36. GW, HM, NF, PA. (4) 1910. (48, 49)

Species Accounts

Order SORDARIALES
Family Bertiaceae

Bertia moriformis (Tode) de Not. *Wood Mulberry*
On fallen branches; frequent.
50, 51. SJ 08, 16, 25, 26. LO, NF. (4) 1978. (48, 49, 52)

Family Chaetomiaceae

Chaetomium elatum Schm. & Kunze
On decaying vegetable matter; common.
50, 51. SJ 16, 18, 26. LO, MO, PA. (3) 1976. (48)

Chaetomium globosum Kunze
On decaying vegetable matter in compost heap; frequent.
51. Recorded only from Mold (SJ 26) in 1976.

Family Lasiosphaeriaceae

Echinosphaeria strigosa (Alb. & Schwein.) Declercq
Lasiosphaeria strigosa (Alb. & Schwein.) Sacc.
On fallen sycamore branches; rare.
51. Recorded only from Loggerheads Country Park (SJ 26) in 1983.

Lasiosphaeria hirsuta (Fr.) Ces. & de Not.
On fallen branches; frequent.
51. SH 97; SJ 26. AV, BO. (2) 2001. (49, 52)

Lasiosphaeria ovina (Pers.) Ces. & de Not.
On fallen wood; common.
50, 51. SH 87; SJ 25, 26, 34, 36. CC, ER, GW, LO, NF. (6) 1880. (49, 52)

Ruzenia spermoides (Hoffm.) O. Hilber
Lasiosphaeria spermoides (Hoffm.) Ces. & de Not.
On fallen wood; common.
50, 51. SH 85; SJ 15–17, 26, 36. DU, LO. (5) 1975. (48, 49, 52)

Schizothecium conicum (Fuckel) Lundqvist
Podospora curvula (Niessl.) Winter
On rabbit dung; common.
51. Recorded only from Mold (SJ 26) in 1976. (52)

Family Nitschkiaceae

Nitschkia cupularis (Pers.) P. Karst.
On fallen branches; frequent.
50. Recorded only from Loggerheads Country Park (SJ 16) in 1995.

Nitschkia grevillei (Rehm) Nannf.
On fallen sycamore branches; uncommon.
51. Recorded only from Loggerheads Country Park (SJ 26) in 1983.

Family Sordariaceae

Neurospora sitophila Sher & B. Dodge
On rotting bread; frequent.
51. Recorded only from Mold (SJ 26) in 1975, as the anamorph, *Chrysonilia sitophila* (Mont.) v. Arx.

Sordaria fimicola (Desm.) Ces. & de Not.
On dung; common.
51. SJ 08, 25, 26. HM, MO. (3) 1964. (52)

Sub-class Xylariomycetidae
Order XYLARIALES
Family Amphisphaeriaceae

Monographella nivalis (Schaffnit) E. Müll. & v. Arx
On grass, especially after winter snow cover; frequent.
51. Recorded only from Loggerheads Country Park (SJ 26) in 1981.

Family Diatrypaceae

Anthostoma turgidum (Pers.) Nitschke
Lopadostoma turgidum (Pers.) Traverso
On fallen beech branches; frequent.
51. SJ 17, 36. DU, GW. (2) 1972. (49, 52)

Cryptosphaeria eunomia (Fr.) Fuckel
On fallen ash twigs; frequent.
51. Recorded only from Loggerheads Country Park (SJ 26) in 1977. (48, 49, 52)

Diatrype bullata (Hoffm.) Fr. *Willow Barkspot*
On fallen sallow twigs and branches; frequent.
50, 51. SJ 17, 18, 35, 43. DU, PA. (4) 1985. (48, 52)

Diatrype disciformis (Hoffm.) Fr. *Beech Barkcrust*
On fallen beech branches; common.
50, 51. SH 77, 85, 87; SJ 05, 15–17, 23–27, 34, 36. AV, CC, CE, ER, GW, LH, LO, MA, WW. (29) 1880. (49, 52)

Diatrype stigma (Hoffm.) Fr. *Common Tarcrust*
On fallen branches; common.
50, 51. SH 87, 97; SJ 06, 08, 15–17, 24–26, 34, 36. AV, BO, CC, DU, ER, GW, LH, LO, MA, NF. (19) 1880. (48, 49, 52)

Diatrypella favacea (Fr.) de Not.
On fallen branches; common.
50, 51. SH 85; SJ 05, 08, 15, 26. LO. (5) 1924. (48, 49, 52)

Diatrypella quercina (Pers.) Cooke
On fallen oak branches; frequent.
50, 51. SH 85, 87; SJ 15, 25–27, 34, 35. CC, ER, LO. (13) 1880. (48, 49, 52)

Species Accounts

Eutypa acharii Tul. & C. Tul.
On fallen sycamore wood; common.
50, 51. SH 87; SJ 15, 26, 27. CC, CE, LO. (5) 1880. (49, 52)

Eutypa lata (Pers.) Tul. & C. Tul.
On fallen wood; common.
50. SH 85, 87; SJ 34. CC, ER. (4) 1880. (48, 49)

Eutypa spinosa (Pers.) Tul. & C. Tul.
On fallen branches; frequent.
50, 51. SJ 26, 34, 36. ER, GW, LO. (3) 1972. (48, 49, 52)

Eutypa subtecta (Fr.) Fuckel
On fallen sycamore branches; frequent.
51. Recorded only from Loggerheads Country Park (SJ 26) in 1983.

Eutypella quaternata (Pers.) Rappaz
Quaternaria quaternata (Pers.) Schröter
On fallen beech branches; common.
50, 51. SJ 17, 25, 26, 34. DU, ER, LO, MA, NF, WW. (7) 1910. (48, 49)

Eutypella stellulata (Fr.) Sacc.
On fallen elm branches; uncommon.
50. Recorded only from Coed Coch (SH 87) in 1880.

Family Hyponectriaceae

Hyponectria buxi (DC.) Sacc.
On dead box leaves; rare.
50. Recorded only from Erddig (SJ 34) in 1910. (48)

Hyponectria cookeana (Auersw.) Barr
On oak leaves in litter; uncommon.
51. SJ 17. DU. (2) 1978.

Family Xylariaceae

Annulohypoxylon multiforme (Fr.) Ju, Rogers & Hsieh *Birch Woodwart*
Hypoxylon multiforme (Fr.) Fr.
On dead wood of alder and birch; common.
50, 51. SJ 06, 15-17, 25-27, 34-36. AV, CE, CN, ER, GW, LH, LO, MA, MF, NF, NM, WW. (35) 1910. (48, 49, 52)

Anthostomella lugubris (Desm.) Sacc.
On dead stems of marram grass; frequent.
51. Recorded only from Point of Ayr (SJ 18) in 1978. (52)

Anthostomella phaeosticta (Berk.) Sacc.
On dead marram stems; frequent.
51. Recorded only from Point of Ayr (SJ 18) in 1985. (48, 52)

Biscogniauxia nummularium (Bull.) O. Kuntze *Beech Tarcrust*
Hypoxylon nummularium Bull.
On fallen branches; uncommon.
50, 51. SJ 05, 15–17, 23, 25–27, 35, 36. AV, CE, CN, LH, LO, NF. (22) 1984. (48, 49)

Daldinia concentrica (Bolt.) Ces. & de Not. *Cramps Ball; King Alfred's Cakes*
On living and fallen ash trunks and branches, common; rare on beech.
50, 51. SH 97; SJ 06–08, 15–18, 24–27, 34–36. AV, BO, BW, CE, CN, DU, ER, GW, HC, LO, ML, NF, PP, WW. (73) 1909. (48, 49, 52)

Daldinia fissa Lloyd
Daldinia vernicosa (Schwein.) Ces. & de Not.
On burnt gorse stems; uncommon.
50, 51. SH 97; SJ 15–18, 25, 35. HC, HM, PA. (7) 1978. (49, 52)

Entoleuca mammata (Wahlenb.) Rogers & Ju
Hypoxylon mammatum (Wahlenb.) Miller
On living sallows, causing a serious disease; uncommon.
51. Recorded only from Ddol Uchaf Nature Reserve (SJ 17) in 1985. This is the first record from Wales and the first UK report as a pathogen.

Euepixylon udum (Pers.) Laessoe & Spooner
Hypoxylon udum (Pers.) Fr.
On rotten oak wood; rare.
50. Recorded only from Henllan (SJ 06) in 1979. (48)

Hypoxylon fragiforme (Scop.) Kickx *Beech Woodwart*
On fallen trunks and branches of beech; common.
50, 51. SH 77, 87, 97; SJ 05, 06, 15–17, 23–27, 34–36, 54. AV, CC, ER, GW, LH, LO, MA, ML, NF, NM. (37) 1880. (48, 49, 52)

Hypoxylon fuscum (Pers.) Fr. *Hazel Woodwart*
On fallen stems and branches of hazel; common.
50, 51. SH 97; SJ 06, 08, 15–17, 24–27, 34–36. AV, BO, BW, CE, CN, DU, ER, GW, LO, MA, NF, WW. (33) 1910. (48, 49, 52)

Hypoxylon howeanum Peck
On dead hazel stems; uncommon.
50, 51. SJ 07, 26, 34, 36. ER, GW. (4) 1978. (49, 52)

Hypoxylon petriniae Stadler & Fournier
On dead ash wood; uncommon.
51. Recorded only from Rhydymwyn Nature Reserve (SJ 26) in 2011.

Hypoxylon rubiginosum (Pers.) Fr. *Rusty Woodwart*
On dead ash and oak wood; common.
50, 51. SH 87; SJ 08, 15–26, 34, 36. AV, CC, CN, DU, ER, GW, LO. (10) 1880. (48, 49, 52) Some of the records on ash may refer to *H. petriniae* Stadler & Fournier but no material has been preserved. The records from the Alyn Valley, Coed y Felin and Hawarden are all true *rubiginosum*.

Species Accounts

Kretzschmaria deusta (Hoffm.) P.M.D. Martin *Brittle Cinder*
Ustulina deusta (Hoffm.) Lind
On beech stumps and logs; common.
50, 51. SH 77; SJ 06, 07, 15, 16, 18, 24, 26, 27, 34–36. ER, GW, LO, MF, MO. (22) 1910. (48, 49, 52)

Nemania confluens (Tode) Laessoe & Spooner
Hypoxylon confluens (Tode) Westend.
On rotten oak wood; frequent.
50. SH 87; SJ 15. CC. (2) 1880.

Nemania serpens (Pers.) Gray
Hypoxylon serpens (Pers.) Fr.
On fallen beech wood; frequent.
50, 51. SH 87; SJ 15–17, 26. AV, CC, LH, LO, MO. (8) 1880. (48, 49, 52)

Rosellinia aquila (Fr.) de Not.
On rotten wood; frequent.
50, 51. SH 87; SJ 17, 34, 36. CC, ER, GW. (4) 1880. (48, 49, 52)

Rosellinia britannica L.E. Petrini, Petrini & S.M. Francis
R. mammiformis auct.
On dead branches, especially of ivy; uncommon.
50, 51. SJ 16, 17, 26. AV, MA. (4) 1978. (52)

Xylaria carpophila (Pers.) Fr. *Beechmast Candlesnuff*
On fallen beech mast in litter; common.
50, 51. SH 85; SJ 13, 15, 16, 25–27, 34–36. AV, CE, CN, ER, GW, LO, MA, MO. (17) 1910. (48, 49, 52)

Xylaria hypoxylon (L.) Grev. *Candlesnuff Fungus*
On dead wood of all kinds; very common. The infected wood is stained black.
50, 51. SH 77, 84, 85, 87, 95, 97; SJ 02, 05–08, 15–18, 23–27, 34–36, 43, 54. AV, BO, BP, BW, CC, CE, CF, CN, DU, ER, GW, HC, LH, LO, MA, MF, ML, MO, NF, NM, PA, PP, WW. (122) 1880. (48, 49, 52)

Xylaria longipes Nitschke *Dead Moll's Fingers*
On fallen sycamore wood; common.
50, 51. SJ 05–08, 15–17, 24–27, 34–36. AV, CE, CF, DU, ER, GW, HC, LO, MA, NF, PP, WW. (32) 1972. (48, 49, 52)

Xylaria polymorpha (Pers.) Grev. *Dead Man's Fingers*
On fallen trunks and stumps; common.
50, 51. SH 77, 97; SJ 05–08, 15–18, 24–27, 34–36. AV, BO, BW, CE, CF, CN, DU, ER, GW, HC, LH, LO, MA, ML, MO, NF, PP, WW. (55) 1910. (48, 49, 52)

Unassigned anamorph of Xylariaceae:

Nodulisporium cecidiogenes J. Koch
Galling *Coniophora puteana*; rare.
50. Recorded only from Cilygroeslwyd Wood (SJ 15) in 1994. This is the first Welsh record, only published as new to Britain in 2000.

(Family uncertain)

Phomatospora dinemasporium Webster
On grass litter; frequent.
51. Recorded only from Mold (SJ 26) in 1978, as the *Dinemasporium* anamorph. (48, 49, 52)

(Order uncertain)
Family Apiosporaceae

Apiospora montagnei Sacc.
On dead marram stems; frequent.
51. Recorded only from Point of Ayr (SJ 18) in 1985.

Family Magnoporthaceae

Gaeumannomyces graminis (Sacc.) v. Arx & Olivier
On living *Dactylis glomerata*; common.
51. Recorded only from Mold (SJ 26) in 1978. (52)

(Sub-class uncertain)
Order MELIOLALES
Family Meliolaceae

Appendiculella calostroma (Desm.) Höhnel
On stems of raspberry; frequent.
50, 51. SH 85; SJ 17. LH. (3) 1977. (49, 52)

Order TRICHOSPHAERIALES
Family Chaetosphaerellaceae

Chaetosphaerella phaeostroma (Durieu & Mont.) E. Muller & C. Booth
On fallen branches; common.
50, 51. SH 87, 97; SJ 06, 07, 15–17, 24–27, 34, 36. AV, CW, ER, GW, LO, MA, MO, NF, NM, WW. (25) 1880. (49, 52)

Crassochaete fusispora (Sivan.) Réblové
Chaetosphaella fusispora Sivan.
On fallen sycamore branch; uncommon.
51. Recorded only from Loggerheads Country Park (SJ 26) in 1973.

Helminthosporium clavariarum (Tul.) Fuckel
Parasitic on *Clavulina cinerea*; frequent.
51. SJ 07, 26. LO. (2) 1975.

(Order uncertain)
Family Abrothallaceae

Abrothallus parmeliarum (Sommerf.) Arnold
Parasitic on *Parmelia sulcata*; frequent.
51. Recorded only from Ffrith Mountain, Moel Famau (SJ 16) in 1978. (48)

Species Accounts

Family Myxotrichaceae

Myxotrichum chartarum (Nees) Kuntze
On rabbit dung; frequent.
51. Recorded only from Mold (SJ 26) in 1977.

Unassigned anamorphs of Pezizomycotina:

Bactridium flavum Kuntze
On very rotten wood; common.
50, 51. SH 97; SJ 26, 34. AV, BO, ER. (3) 1985. (48, 49, 52)

Bispora antennata (Pers.) Mason
On cut ends of tree stumps; frequent.
51. Recorded only from Nant-y-Ffrith (SJ 25) in 1910. (49)

Botryosporium pulchrum Corda
On decaying parts of herbaceous plants; common.
50, 51. SJ 16, 17, 26, 34. CN, DU, ER, MO. (4) 1910.

Camposporium pellucidum (Grove) S.J. Hughes
In foam in fast flowing rivers; uncommon.
51. Recorded only from the River Alyn in Mold (SJ 26) since 1981.

Cryptostroma corticale Gregory & Waller
The cause of 'sooty bark disease' on sycamore, spread by grey squirrels, and very damaging; becoming common.
50, 51. SJ 16, 25–27, 34–36. ER, MO. (9) 1977.

Dendrospora erecta Ingold
In foam in fast flowing rivers; uncommon.
51. Recorded only from the River Alyn in Mold (SJ 26) since 1981. (48, 49)

Flabellospora acuminata Descals
In foam in fast flowing rivers; frequent.
50, 51. SJ 16, 25, 26, 35. AV, LO, MO, NF. (5) In Rivers Alyn and Ffrith. 1981. (49)

Haplariopsis fagicola Oud.
On fallen beech cupules; frequent.
51. Recorded only from Loggerheads Country Park (SJ 26) in 1985.

Heliscella stellata (Ingold & Cox) Marvanová & S. Nilsson
Clavatospora stellata (Ingold & Cox) S. Nilsson
In foam in fast flowing rivers; frequent.
51. SJ 16, 26, 35. AV, MO. (3) In River Alyn. 1985. (48, 49)

Hymenopsis typhae (Peck) Sutton
On leaves and stems of *Typha latifolia*; rare.
51. Recorded only from Ddol Uchaf Nature Reserve (SJ 17) in 1985. (48)

Lateriramulosa uni-inflata Matsushima
In foam in fast flowing rivers; uncommon.
51. Recorded only from the River Alyn in Mold (SJ 26) since 1986. (48)

Septocyta ruborum (Lib.) Petrak
Leaf spot on *Rubus* species; common.
51. SJ 26. MO. (2) 1976.

Spondylocladiopsis cupulicola M.B. Ellis
On fallen beech cupules; frequent.
51. Recorded only from Loggerheads Country Park (SJ 26) in 1985.

Tumularia aquatica (Ingold) Descals & Marvanová
Dactylella aquatica (Ingold) Ranzoni
In foam in fast flowing rivers; uncommon.
50, 51. SJ 16, 26, 35. AV, LO, MO. (4) 1985. (49)

Volucrispora aurantiaca Haskins
Tricellula aurantiaca (Haskins) v. Arx
In foam in fast flowing rivers; uncommon.
51. Recorded only from the River Alyn in Mold (SJ 26) in 1986.

Volucrispora graminea Ingold, McDougal & Dann
Lambdasporium gramineum (Ingold, McDougall & Dann) Descals & Marvanová
In foam in fast flowing rivers; uncommon.
50, 51. SJ 16, 26, 35. LO, MO. (4) In River Alyn. 1981.

Sub-phylum SACCHAROMYCOTINA
Class SACCHAROMYCETES
Order SACCHAROMYCETALES
Family Dipodascaceae

Dipodascus geotrichum (E.E. Butler & L.J. Petersen) Arx
Isolated from garden soil; common.
51. Recorded only from Mold (SJ 26) in 1978, as the *Geotrichum* anamorph.

Unassigned anamorph of Saccharomycetales:

Candida albicans (Robin) Berkh.
The cause of 'thrush' in Man; common.
Recorded for generations but not recorded for mycological purposes. From personal experience it has occurred in Mold!

Sub-phylum TAPHRINOMYCOTINA
Class TAPHRINOMYCETES
Order TAPHRINALES
Family Protomycetaceae

Burenia inundata (Dangeard) M. Reddy & Kramer
Forming blisters on leaves of *Apium nodiflorum*; uncommon.
51. Recorded only from Mold (SJ 26) in 1977. (49, 52)

Species Accounts

Protomyces macrosporus Unger
In warts on leaves of *Aegopodium podagraria* and *Anthriscus sylvestris*; common.
50, 51. SJ 12, 16, 26, 35, 36, 43. AV, LO, MO. (8) 1969. (48, 49, 52)

Taphridium umbelliferarum (Rostr.) Lagerh. & Juel
In coarse leaf spots on *Heracleum sphondylium*; common.
50, 51. SJ 16, 26, 34. CN, ER, MO. (3) 1977. (52)

Family Taphrinaceae

Taphrina alni (Berk. & Br.) Gjaerum *Alder Tongue*
Taphrina amentacea (Sadeb.) Rostr.
Producing red 'tongues' from female alder catkins; once rare but now spreading rapidly. The causes for its spread since the early 1990s may be the planting of imported alders, mostly *Alnus incana* and *A. glutinosa*, or climate change. The species is common in central and southern Europe.
50, 51. SJ 16, 25, 26, 34, 36. AV, ER, LO, MO, WW. (9) 2000. (49, 52)

Taphrina betulae (Fuckel) Johanson
Causing yellow-brown leaf spots on birch leaves; uncommon.
50, 51. SJ 16, 26. LO, MO. (2) 1991.

Taphrina betulina Rostr. *Birch Besom*
On deformed leaves of witches' broom galls on birch branches; common. This is the more likely cause of the brooms: mycoplasmas and mites have been implicated but all the galls examined in this survey have carried the fungus.
50, 51. SH 95–97; SJ 07, 08, 16–18, 24–27, 34–36, 43. AV, BO, DU, ER, GW, LH, LO, MA, MO, NF, NM, PP. (25) 1972. (48, 49)

Taphrina caerulescens (Desm.) Tul. & C. Tul.
Forming raised spots on oak leaves; uncommon.
51. SJ 25, 26. MO, NM. (2) 1993. (48, 49)

Taphrina deformans (Berk.) Tul. & C. Tul.
The cause of 'peach-leaf curl' on almonds, peaches and apricots; common.
50, 51. SJ 24, 26, 35, 36. MO. (4) 1975. (49, 52)

Taphrina filicina Rostr. ex Johanson
Causing frond-tip curl and browning in *Dryopteris* species; uncommon.
50. Recorded only from Erddig (SJ 34) in 2002.

Taphrina padi (Jacz.) Mix
Forming 'pocket plums' on bird cherry; uncommon except in Scotland.
51. SJ 26, 36. MO. (2) 1997. (52)

Taphrina populina Fr.
Forming conspicuous blister galls on leaves of poplars; common.
51. SJ 18, 26. MO, PA, WW. (3) 1976. (49)

Taphrina pruni Tul. & C. Tul. *Pocket Plum*
Forming 'pocket plums' and leaf crinkles on blackthorn; becoming more common.
50, 51. SH 87; SJ 26, 27. MO. (3) 1993. (48, 49, 52)

Taphrina sadebackii Johanson
Forming small, yellow blister galls on alder leaves; frequent.
50, 51. SJ 08, 16, 25, 26, 36. AV, LO, MO. (8) 1991.

Taphrina tormentillae Rostr.
T. potentillae (Farlow) Johanson
Forming yellow swellings on stems and leaves of *Potentilla erecta*; frequent, especially in the west of the British Isles.
50, 51. SJ 15, 25. NM. (2) 1995. (48)

Taphrina tosquinetii (Westend.) Magnus
Forming large, pale blister galls on leaves of alder; common.
50, 51. SJ 14, 16, 25, 26, 34. AV, LO, MO. (8) 1974. (48, 52)

Taphrina ulmi (Fuckel) Johanson
Forming leaf spots, and possibly witches' brooms, on elms; rare.
51. Recorded only from a hedge in Bagillt (SJ 27) in 1978.

Taphrina wiesneri (Rathay) Mix
Forming witches' brooms and leaf curls on wild cherry, *Prunus avium*; common.
50, 51. SJ 06, 16, 26. AV, WW. (4) 1978. (48, 49)

Species Accounts

Phylum BASIDIOMYCOTA
Sub-phylum AGARICOMYCOTINA
Class AGARICOMYCETES
Order AGARICALES
Family Agaricaceae

Agaricus arvensis Schaeff. *Horse Mushroom*
incl. *Agaricus leucotrichus* (F.H. Møll.) F.H. Møll.
In grass in manured fields, dune meadows and on roadside verges; common.
50, 51. SH 87, 95, 97; SJ 06, 08, 15-18, 23, 25, 26, 34, 36. BO, BP, CC, ER, GW, LO, MO, PA. (22) 1880. (48, 49, 52)

Agaricus augustus Fr. *The Prince*
In grass on roadsides, parks and woodland margins; uncommon.
50, 51. SH 77; SJ 05, 07, 15, 26. LO. (6) 1962. (49, 52)

Agaricus bernardii Quél.
In base-rich turf; usually coastal, rare inland.
50, 51. SH 97; SJ 27. ML. (2) 1980. (52)

Agaricus bisporus (J.E. Lange) Imbach *Cultivated Mushroom*
Agaricus brunnescens auct.
In garden soil, especially among potatoes; frequent.
50, 51. SJ 16, 25, 26. MO. (3) 1976.

Agaricus bitorquis (Quél.) Sacc.
On bare soil at edge of paths and roads; uncommon.
50, 51. SH 97; SJ 15, 16, 23, 25, 26, 35. LO, NM. (9) 1978.

Agaricus campestris L. *Field Mushroom*
In unploughed grassland where inorganic fertiliser has not been used for several years; common in good years.
50, 51. SH 87, 96, 97; SJ 05-08, 15, 16, 18, 23-27, 34-36, 43. AV, BO, BP, CC, CE, ER, GW, LO, MA, MF, ML, MO, PA, WW. (56) 1880. (48, 49, 52)

Agaricus capellianus Hlaváček
Agaricus pseudovillaticus Rauschart; *A. vaporarius* (Pers.) Capelli
On grassy banks and under hedges; uncommon.
50, 51. SJ 07, 16, 17. LO. (3) 1994.

Agaricus comtulus Fr.
In grass at edge of woodland; uncommon.
50, 51. SH 87; SJ 26, 34, 36. CC, LO. (4) 1880. (48)

Agaricus crocodilinus *Macro Mushroom*
Agaricus urinascens (F.H. Møll. & Schaeff.) Sing.; *A. macrosporus* (F.H. Møll. & Schaeff.) Pilát
In grassland at edge of woodland; uncommon.
50, 51. SH 97; SJ 08, 15, 16, 25, 26. BP, LO, ML. (8) 1983. (48, 49, 52)

Agaricus devoniensis P.D. Orton
In sand dune turf; frequent.
51. SJ 08, 18. PA. 1979. (48, 52)

Agaricus dulcidulus S. Schulz. *Rosy Wood Mushroom*
Agaricus semotus auct.
In grassland on limestone; uncommon.
50, 51. SH 97; SJ 06, 08, 15, 34. BO, BP, ER, ML. (7) 1974.

Agaricus fuscofibrillosus (F.H. Møll.) Pilát
In woodland soil; uncommon.
50. SH 76, 97; SJ 15, 34. ER, ML. (4) 1979. (49)

Agaricus impudicus (Rea) Pilát
Agaricus variegans F. Møll.
In broad-leaved woodland litter; uncommon.
50, 51. SH 97; SJ 26, 34. ER, LO, ML. (3) 1979. (49, 52)

Agaricus langei (F.H. Møll.) F.H. Møll. *Scaly Wood Mushroom*
In mixed woodland soil; frequent.
50, 51. SH 97; SJ 07, 08, 25–27, 34. AV, BO, BP, CE, ER, LO, ML. (12) 1973. (48, 49, 52)

Agaricus litoralis (Wak. & Pears.) Pilát
Agaricus spissicaulis F.H. Møll.
In sandy grassland; uncommon.
51. SH 97. BO. 1998. (52)

Agaricus macrocarpus (F.H. Møll.) F.H. Møll.
In woodland soil; uncommon.
50. Recorded only from Glan-yr-Afon, Marli (SH 97) in 1979.

Agaricus moelleri Wasser *Inky Mushroom*
Agaricus praeclaresquamosus A.E. Freeman; *A. placomyces* auct.
In soil in base-rich woodland; frequent.
50, 51. SH 97; SJ 05, 06, 15, 17, 26, 35. MO. (7) 1978. (48, 49, 52)

Agaricus osecanus Pilát
Agaricus nivescens (F.H. Møll.) F.H. Møll.
In meadows grazed by horses and cattle; frequent.
50, 51. SH 97; SJ 07, 08, 16, 23. ML. (5) 1979.

Agaricus phaeolepidotus (F.H. Møll.) F.H. Møll.
In garden soil under trees; rare.
51. Recorded only from Broughton (SJ 36) in 2002.

Agaricus porphyrizon P.D. Orton *Lilac Mushroom*
In soil in broad-leaved woodland; uncommon.
50. Recorded only from Glan-yr-Afon, Marli (SH 97) in 1979. (52)

Species Accounts

Agaricus porphyrocephalus F.H. Møll.
In grassland on basic soil; uncommon.
50, 51. SH 97; SJ 17. DU, ML. (2) 1986. (49)

Agaricus silvaticus Schaeff. *Blushing Wood Mushroom*
incl. *Agaricus haemorrhoidarius* S. Schulz.
In woodland litter, usually of broad-leaves; common.
50, 51. SH 87, 97; SJ 05-08, 15-17, 24-27, 34-36. AV, BO, BP, CE, CN, ER, GW, HC, LO, ML, MO, NF, PP, WW. (29) 1910. (48, 49, 52)

Agaricus silvicola (Vitt.) Peck *Wood Mushroom*
incl. *Agaricus abruptibulbis* sensu F.H. Møll.
In conifer litter; common.
50, 51. SH 97; SJ 08, 15-17, 23, 24, 26, 34, 36. BO, BP, ER, GW, LH, LO, MA, ML, PP, WW. (17) 1910. (49, 52)

Agaricus subperonatus (J.E. Lange) Singer
On bare soil at edge of fields; uncommon.
50. Recorded only from Pant-yr-Ochain, Gresford (SJ 35) in 2005. (52)

Agaricus xanthodermus Genev. *Yellow Stainer*
In grassland and woodland edges; frequent. **Poisonous.**
50, 51. SH 77; SJ 06-08, 18, 25-27, 35, 36. AV, GW, MA, MO, PA, WW. (17) 1962. (49, 52)

Chamaemyces fracidus (Fr.) Donk *Dewdrop Dapperling*
In rich woodland soil; uncommon.
50, 51. SH 97; SJ 08, 15, 17, 25. (5) 1993. (49)

Chlorophyllum brunneum (Farl. & Burt.) Vellinga
Macrolepiota rhacodes var. *bohemica* (Wichansky) Bellù & Lanzoni
On rich soil in gardens; rare.
50, 51. SJ 15, 26. (2) 1986.

Chlorophyllum olivieri (Barla) Wasser
In woodland soil; uncommon.
51. Recorded only from Hawarden (SJ 36) in 2019.

Chlorophyllum rhacodes (Vitt.) Vellinga *Shaggy Parasol*
Macrolepiota rhacodes (Vitt.) Sing.; *Lepiota rhacodes* (Vitt.) Quél.
In rich soil in roadsides and woods; common.
50, 51. SH 77, 87, 97; SJ 05, 06, 08, 15-17, 23, 24, 26, 34-36. BO, BP, CC, CE, CF, CN, ER, GW, MA, MF, ML, MO, PP, WW. (48) 1880. (48, 49, 52)

Coprinus comatus (O.F. Müll.) Pers. *Shaggy Inkcap*
In enriched soil of roadside verges, playing fields, parks and field margins; common.
50, 51. SH 77, 87, 97; SJ 05, 06, 08, 15-18, 23-27, 34-36. AV, BO, CC, CE, DU, ER, GW, HC, LH, LO, ML, MO, NF, PP, WW. (67) 1880. (49, 52)

Cystolepiota bucknallii (Berk. & Br.) Sing. & Clém. *Lilac Dapperling*
Lepiota bucknallii (Berk. & Br.) Sacc.
In damp woodland; frequent.
50, 51. SH 97; SJ 15, 17. BO, ML. (5) 1974. (49, 52) Easily recognised by its smell of coal gas.

Cystolepiota hetieri (Boud.) Sing.
In basic woodland soils; uncommon.
50, 51. SH 97; SJ 16, 17, 35, 36. GW, LO, ML. (5) 1979. (49, 52)

Cystolepiota moelleri Knudsen
Lepiota rosea Rea
On clay banks in hedgerows; rare.
50, 51. SH 97; SJ 17, 26. ML, WW. (4) 1978. (52)

Cystolepiota seminuda (Lasch) M. Bon *Bearded Dapperling*
Cystolepiota sistrata auct.; *Lepiota sistrata* auct.
On base-rich woodland soils; frequent.
50, 51. SH 85, 87, 97; SJ 05-07, 15-17, 24, 26, 27, 34-36. AV, CC, CN, DU, ER, GW, LO, MA. (27) 1880. (49, 52)

Echinoderma asperum (Pers.) M. Bon *Freckled Dapperling*
Lepiota aspera (Pers.) Quél.; *Cystolepiota aspera* (Pers.) M. Bon;
Lepiota acutesquamosa (Weinm.) P. Kumm.; *L. friesii* (Lasch) Quél.
In rich woodland soil, often in nettle beds; common.
50, 51. SH 77, 97; SJ 05, 07, 08, 15-17, 26, 34-36. AV, BO, BP, CN, DU, ER, GW, LO, MA, ML, MO, WW. (19) 1863. (49, 52)

Echinoderma pseudoasperulum (Knudsen) M. Bon
Lepiota pseudoasperula (Knudsen) Knudsen; *Lepiota eriophora* sensu D.A. Reid
In soil in limestone woods; uncommon.
50, 51. SH 97; SJ 16. CN, ML. (2) 1979.

Lepiota boudieri Bres. *Girdled Dapperling*
Lepiota fulvella Rea
On clay banks in hedgerows; uncommon.
50, 51. SH 97; SJ 15, 16, 26, 35. AV, CN, ML, WW. (6) 1979. (49, 52)

Lepiota brunneoincarnata Chod. & Mart. *Deadly Dapperling*
In beech litter; rare. **Poisonous.**
50, 51. SH 97; SJ 26. LO, ML. (2) 1973. **PF**

Lepiota castanea Quél. *Chestnut Dapperling*
incl. *Lepiota ignicolor* Bres.
In woodland litter; frequent.
50, 51. SH 97; SJ 05, 07, 24, 26, 34, 35. ER, LO, ML, PP. (10) 1979. (49, 52)

Lepiota clypeolaria (Bull.) P. Kumm. *Shield Dapperling*
In grassy edges of woodland paths; uncommon.
51. SJ 16, 25. AV, NF. (2) 1978. (49, 52)

Species Accounts

Lepiota cortinarius J.E. Lange
On soil in deciduous woodland; uncommon.
50. Recorded only from Pen Parc Llwyd, Henllan (SJ 06) in 2004.

Lepiota cristata (Bolt.) P. Kumm. *Stinking Dapperling*
In vegetation at the edge of paths, often among nettles; common. **Poisonous.**
50, 51. SH 77, 85, 87, 97; SJ 05–08, 15–18, 23–27, 34–36. AV, BO, BP, CC, CE, DU, ER, GW, HC, LH, LO, MA, ML, MO, NF, NM, PA, PP, WW. (54) 1880. (48, 49, 52)

Lepiota erminea (Fr.) P. Kumm.
Lepiota alba (Bres.) Sacc.
In base-rich grassland, esp. in sand dunes; frequent.
50, 51. SH 87; SJ 08, 17, 18, 26. CC, PA, WW. (5) 1880. (48, 49, 52)

Lepiota felina (Pers.) P. Karst.
In woodland litter; frequent.
50, 51. SJ 08, 16, 23, 25, 26, 34. CE, ER, LO, NF. (7) 1982. (49)

Lepiota fourquignonii Quél.
In woodland litter, especially on limestone; rare.
51. SJ 16, 17. BW. (2) 1990.

Lepiota fuscovinacea F.H. Møll. & J.E. Lange
In soil in limestone woods; uncommon.
50. Recorded only from Glan-yr-Afon, Marli (SH 97) in 1979. (49, 52)

Lepiota grangei (Eyre) Kühner *Green Dapperling*
In rich woodland soil; rare.
50. Recorded only from Big Wood, Erddig (SJ 34) in 1999. (52) **PF**

Lepiota ignivolvata Bousset & Joss.
In litter in limestone woods; uncommon.
50, 51. SH 97; SJ 07, 15, 16, 26. AV, MA, ML. (5) 1990. (49, 52) **PF**

Lepiota lilacea Bres.
In woodland litter; rare.
51. Recorded only from Greenfield Valley Country Park (SJ 17) in 2002.

Lepiota locquinii Bon
In woodland soil; rare.
50. Recorded only from Glan-yr-Afon, Marli (SH 97) in 1979.

Lepiota magnispora Murrill
Lepiota ventriospora D. Reid
In limestone woodland litter; uncommon.
50, 51. SH 97; SJ 15, 16, 26. AV, MA, ML. (6) 1979. (49, 52)

Lepiota oreadiformis Velen.
In grassland; uncommon.
51. Recorded only from Point of Ayr (SJ 18) since 1995. (48, 52)

Lepiota pseudolilacea Huijsman
Lepiota pseudohelveola Hora
In limestone grassland; uncommon.
50, 51. SH 97; SJ 08. ML. (2) 1979.

Lepiota subalba P.D. Orton
In litter in base-rich woods; uncommon.
50, 51. SH 97; SJ 26, 34. ER, LO, ML. (3) 1978. (52)

Lepiota subgracilis Kühner
Under beech in limestone woods; uncommon.
51. Recorded only from Y Graig Nature Reserve, Tremeirchion (SJ 07) in 2003.

Lepiota subincarnata J.E. Lange *Fatal Dapperling*
In mixed woodland litter; uncommon. **Poisonous.**
50, 51. SH 97; SJ 26, 34. ER, ML, WW. (3) 1979.

Lepiota xanthophylla P.D. Orton
In limestone woods; rare.
50. Recorded only from Glan-yr-Afon, Marli (SJ 97) in 1979. (52) **PF**

Leucoagaricus badhamii (Berk. & Br.) Sing. *Blushing Dapperling*
In limestone woods; uncommon.
50, 51. SH 97; SJ 15, 26. ML. (3) 1979. (52)

Leucoagaricus georginae (W.G. Smith) Candusso
On acid woodland soil; rare.
51. Recorded only from Wepre Woods (SJ 26) in 2006; apparently new to Wales.

Leucoagaricus leucothites (Vittad.) Wasser *White Dapperling*
In grass at edge of woodland; uncommon.
50, 51. SJ 05, 06, 26, 27. MA. (4) 1994. (48, 49, 52)

Leucoagaricus melanotrichus (Malençon & Bertault) Trimbach
In woodland litter; rare.
50. Recorded only from Glan-yr-Afon, Marli (SH 97) in 1981.

Leucoagaricus meleagris (Sow.) Sing.
In heated greenhouses an introduced alien, uncommon.
50. Recorded only from Coed Coch (SH 87) in 1880.

Leucoagaricus nympharum (Kalchbr.) M. Bon
Macrolepiota puellaris (Fr.) M.M. Moser
Under shrubs at woodland edge; uncommon.
50, 51. SH 87, 97; SJ 07. CC, ML. (4) 1880. (52)

Leucoagaricus pilatianus (Demoulin) Bon & Boiffard
In litter in limestone woods; rare.
50. SJ 16, 26. LO, MA. (2) 2000. First records from Wales. **RDL/NT**

Species Accounts

Leucoagaricus roseilividus (Murrill) E. Ludw.
Leucoagaricus marriagei (D.A. Reid) Bon
In limestone woods; rare.
50. Recorded only from Glan-yr-Afon, Marli (SH 97) in 1980. **RDL/NT**

Leucoagaricus serenus (Fr.) Bon & Boiffard
Lepiota serena (Fr.) Quél.
In short vegetation at edge of woodland paths; frequent.
50, 51. SJ 06, 07, 15–18, 35. AV. (7) 1978.

Leucoagaricus sericifer (Locq.) Vellinga
Lepiota sericata Kühn. & Romagn.
On bare soil in basic woods; frequent.
50, 51. SJ 05, 26, 34, 35. CF, ER, LO. (4) 1982.

Leucoagaricus tener (P.D. Orton) Bon
In woodland soil; rare.
50. Recorded only from Glan-yr-Afon, Marli (SH 97) in 1981.

Leucoprinus birnbaumii (Corda) Sing. *Plantpot Dapperling*
In pots of house-plants; frequent. This bright yellow, tropical species is becoming widespread in hothouses, and with house-plants, throughout the temperate regions.
51. SJ 26, 36. MO. (2) 1989.

Leucocoprinus brebissonii (Godey) Locq. *Skullcap Dapperling*
In rich soil at edge of woodland paths; frequent.
50, 52. SJ 05, 34, 36. ER, GW. (3) 1985. (49, 52)

Leucocoprinus cepistipes (Sow.) Pat.
In heated greenhouses; an increasing alien occasionally found outside in gardens.
50. Recorded only from Coed Coch (SH 87) in 1880. (49)

Leucocoprinus straminellus (Bagl.) Narducci & Caroti
In pots of house plants; an uncommon alien.
50. Recorded only from Llangollen (SJ 24) in 2003.

Macrolepiota excoriata (Schaeff.) Wasser
Lepiota excoriata (Schaeff.) P. Kumm.
In grassland on sand and clay; frequent.
50, 51. SJ 07, 08, 15, 18, 25, 26, 34, 36. GW, LO, PA, WW. (11) 1976. (48, 52)

Macrolepiota konradii (P.D. Orton) M.M. Moser
In open woodland; uncommon.
50, 51. SH 97; SJ 26, 35. WW. (3) 1989. (48, 49, 52)

Macrolepiota mastoidea (Fr.) Sing. *Slender Parasol*
Macrolepiota gracilenta (Krombh.) M.M. Moser
In open woodland; frequent.
50, 51. SJ 06, 26, 34, 35. ER, MA, WW. (5) 1996. (48, 49, 52)

Macrolepiota procera (Scop.) Sing. *Parasol*
Lepiota procera (Scop.) Gray
In tall vegetation at roadsides, woodland edges and dunes; common.
50, 51. SH 77, 96, 97; SJ 05–08, 15–18, 23–27, 34–36. AV, BO, BP, CE, ER, GW, LH, LO, MF, ML, MO, PA, PP. (43) 1951. (48, 49, 52)

Melanophyllum eyrei (Mass.) Sing. *Greenspored Dapperling*
In litter in limestone woods; uncommon.
50, 51. SJ 15–17, 26. LO, MA. (6) 1978. (52) **PF**

Melanophyllum haematospermum (Bull.) Kriesel *Redspored Dapperling*
In rich woodland soil; uncommon.
50. SH 97; SJ 17, 26, 34. ER, MA, ML. (4) 1980. (49, 52) **PF**

Family Amanitaceae

Amanita ceciliae (Berk. & Br.) Bas *Snakeskin Grisette*
Amanita inaurata Secr.; *A. strangulata* auct.
Mycorrhizal with oak; uncommon.
50, 51. SH 97; SJ 07. BO, ML. (3) 1974. (49, 52)

Amanita citrina (Schaeff.) Pers. var. **citrina** *False Deathcap*
Mycorrhizal with oak; common.
50, 51. SH 76, 85, 87, 97; SJ 05, 07, 08, 16–18, 23, 26, 34–36. AV, BO, CC, ER, GW, LH, ML, WW. (22) 1880. (48, 49, 52)

Amanita citrina var. **alba** (Gilb.) Gilb.
Mycorrhizal with oak; frequent.
50, 51. SJ 05, 06, 16–18, 26, 34, 36. AV, CE, ER, GW, WW. (12) 1981. (48, 49, 52)

Amanita crocea (Quél.) Sing. *Orange Grisette*
Mycorrhizal with birch; frequent.
50, 51. SH 96; SJ 05, 17. CF, LH. (11) 1990. (48, 49, 52)

Amanita excelsa (Fr.) Bertilloni var. **excelsa** *Grey Spotted Amanita*
Mycorrhizal with oak; frequent.
50, 51. SH 85, 87, 95, 97; SJ 05, 16, 17, 26, 34, 36. CC, CE, CF, ER, GW, LH, LO, ML, WW. (16) 1880. (48, 49, 52)

Amanita excelsa var. **spissa** (Fr.) Neville & Poumerat *'Tall Amanita'*
Amanita spissa (Fr.) Opiz
Mycorrhizal with oak; uncommon.
50, 51. SH 77, 96; SJ 05, 13, 26. AV, CF. (5) 1990.

Amanita franchetii (Boud.) Fayod
Amanita aspera auct.
Mycorrhizal with beech on limestone; rare.
50. Recorded only from Glan-yr-Afon, Marli (SH 97) in 1981.

Species Accounts

Amanita fulva (Schaeff.) Pers. *Tawny Grisette*
Mycorrhizal with birch; common.
50, 51. SH 85, 95–97; SJ 05, 17, 18, 24–26, 34, 36, 43. BO, CE, CF, GW, LH, LO, ML, NF, NM. (23) 1962. (48, 49, 52)

Amanita junquillea Quél. *Jewelled Amanita*
Amanita gemmata (Fr.) Bertilloni
Mycorrhizal with birch; uncommon.
50, 51. SJ 17, 26. (3) 2000. The specimen from Rhosesmor (SJ 26) was found in May 2006, exceptionally early even for this species.

Amanita muscaria (L.) Lam. *Fly Agaric*
Mycorrhizal with birch, occasionally with conifers such as pines and silver fir; common. **Poisonous.**
50, 51. SH 76, 77, 85, 87, 95–97; SJ 05–07, 15–17, 23–27, 34–36, 43. AV, BO, CC, CE, CF, ER, GW, LH, LO, MF, ML, NF, PP, WW. (59) 1880. (48, 49)

Amanita pantherina (DC.) Krombh. *Panthercap*
Mycorrhizal with beech; uncommon. **Poisonous.**
50, 51. SH 77, 85; SJ 05, 16, 17, 26. CE, CF, LH, LO, MA. (8) 1962. (48, 49, 52)

Amanita phalloides (Fr.) Link *Deathcap*
Mycorrhizal with beech and oak; frequent. **Poisonous.**
50, 51. SH 87, 97; SJ 05–07, 15–18, 26, 34, 36. AV, BO, CC, CF, ER, GW, LH, MA, MF, ML. (22) 1880. (48, 49, 52)

Amanita porphyria (Alb. & Schwein.) Mlady *Grey Veiled Amanita*
Mycorrhizal with conifers; uncommon.
50, 51. SH 76, 85; SJ 05, 17, 26. AV. (5) 1924. (48, 49)

Amanita rubescens Pers. var. **rubescens** *Blusher*
Mycorrhizal with oak and beech; common.
50, 51. SH 76, 77, 85, 87, 95–97; SJ 05, 07, 13, 15–18, 24–27, 34, 36, 43. AV, BO, CC, CE, CF, CN, ER, GW, HC, LH, LO, MA, MF, ML, NF, NM, PP, WW. (55) 1880. (48, 49, 52)

Amanita rubescens var. **annulosulphurea** Gillet
Mycorrhizal with oak and beech; uncommon.
50, 51. SH 85, 95; SJ 05, 07, 17, 26, 34. CE, ER, LH. (9) 1924. (48, 49)

Amanita strobiliformis (Paul) Bertilloni *Warted Amanita*
Mycorrhizal with beech and willows; rare.
51. SH 97; SJ 17. BO, DU. 1980.

Amanita submembranacea (Bon) Gröger
Mycorrhizal with birch; uncommon.
51. Recorded only from Nant-y-Ffrith (SJ 25) in 2003.

Amanita vaginata (Bull.) Lam. *Grisette*
Mycorrhizal with birch; common.
50, 51. SH 76, 87, 95, 97; SJ 05, 07, 13, 16, 17, 24–26, 34, 36. BO, CE, CF, ER, GW, LH, MF, ML, NF, NM. (22) 1902. (48, 49, 52)

Amanita virosa (Fr.) Bertilloni *Destroying Angel*
Mycorrhizal with oaks; uncommon.
50, 51. SJ 18, 24. PP. (2) 1991.

Limacella delicata (Fr.) Earle var. **glioderma** (Fr.) Gminder
In leaf litter in limestone woodland; rare.
50. Recorded only from Glan-yr-Afon, Marli (SH 97) in 1979.

Limacella guttata (Pers.) Konrad & Maubl.
In leaf litter under broad-leaves; uncommon.
50, 51. SH 87, 97; SJ 26. BO, CC, MA. (3) 1880.

Family Biannulariaceae

Ripartites tricholoma (Alb. & Schwein.) P. Karst. *Bearded Seamine*
In woodland litter; uncommon.
50, 51. SH 97; SJ 16, 23, 26. AV, LO, MF, ML. (6) 1978. (48, 49)

Family Bolbitiaceae

Bolbitius reticulatus (Pers.) Ricken *Netted Fieldcap*
Bolbitius aleuriatus (Fr.) Sing.; *Pluteolus aleuriatus* (Fr.) P. Karst.
On sawdust heap; uncommon.
50, 51. SH 87; SJ 26. AV, CC, WW. (3) 1880. (49, 52)

Bolbitius titubans (Bull.) Fr. *Yellow Fieldcap*
Bolbitius vitellinus (Pers.) Fr.
In enriched grassland and on dung; common.
50, 51. SH 77, 87, 97; SJ 05–08, 13, 15–18, 23–26, 34–36, 43. AV, BO, BP, CC, CE, CN, DU, ER, GW, HC, LH, LO, MA, ML, NF, NM, PA, WW. (52) 1880. (48, 49, 52)

Conocybe ambigua Watling
In lawns; uncommon.
51. Recorded only from a garden in Gronant (SJ 08) in 1990.

Conocybe antipus (Lasch) Fayod
In mosses in woodland; uncommon.
51. Recorded only from Bishop Wood, Prestatyn (SJ 08) in 1990.

Conocybe apala (Fr.) Arnolds *Milky Conecap*
Conocybe lactea (J. Lange) Métrod
In lawns, roadside grass verges and grassland generally; common.
50, 51. SJ 06, 08, 15–18, 23, 25, 26, 34–36. AV, BP, ER, LH, LO, MA, MF, MO, PA. (22) 1973. (48, 49, 52)

Conocybe brunneola (Kühn.) Kühn. & Watling
In lawns; uncommon.
50, 51. SH 97; SJ 06, 08, 15, 26, 35. BO, CE, MA. (7) 1989. (48)

Conocybe dunensis T.J. Wallace *'Dune Conecap'*
In fixed dune turf; frequent.
51. SJ 08, 18. PA. (2) 1978. (48, 52)

Conocybe juniana (Velen.) Hauskn. & Svrček
Conocybe magnicapitata P.D. Orton
In short turf and on roadside verges; frequent.
50, 51. SJ 02, 17, 26. MA. (3) 1978. (48, 49)

Conocybe macrocephala Kühn. & Watling
In woodland soil; uncommon.
51. Recorded only from Gwysaney, Mold (SJ 26) in 1978. (52)

Conocybe pilosella (Pers.) Kühn.
In woodland soil; uncommon.
50, 51. SH 87, 97; SJ 26, 35. BO, CC, WW. (5) 1880. (52)

Conocybe pubescens (Gillet) Kühn.
Conocybe pseudopilosella sensu auct.
In lawns and short grassland; frequent.
50, 51. SJ 07, 16, 18, 24–26. BW, PA, WW. (6) 1985. (49, 52)

Conocybe pulchella (Velen.) Hauskn. & Svrček
In short grass in open woodland; uncommon.
50. Recorded only from Alyn Waters Country Park (SJ 35) in 2016. (52)

Conocybe pygmaeoaffinis (Fr.) Kühner
In lawns and short grassland; uncommon.
51. Recorded only from Pengwern College (SJ 07) in 2007.

Conocybe rickeniana P.D. Orton
In grass at edge of woodland paths; frequent.
50, 51. SJ 24, 26. AV, MA, PP. (4) 1985.

Conocybe rickenii (Jul. Schaeff.) Kühn.
On old cow dung; frequent.
51. Recorded only from Wepre Woods Country Park (SJ 26) in 1983. (52)

Conocybe siennophylla (Berk. & Br.) Sing.
On manured soil at edge of woodland; uncommon.
50. Recorded only from Foxhall, Denbighshire (SJ 06) in 1996.

Conocybe subovalis Kühn. & Watling
In woodland soil; common.
50, 51. SH 95; SJ 06–08, 15–18, 24–26, 34–36, 43. AV, DU, ER, GW, LH, LO, MF, MO, NF, PA, WW. (27) 1974. (48, 49, 52)

Conocybe tenera (Schaeff.) Fayod
In short turf; common.
50, 51. SH 85, 87, 97; SJ 05, 08, 15, 16, 18, 23, 25, 26, 34, 36. AV, CC, ER, GW, MA, ML, PA. (16) 1880. (48, 49, 52)

Conocybe velata (Velen.) Watling
Conocybe appendiculata Watling
In woodland soil; uncommon.
51. Recorded only from Rhydymwyn Nature Reserve (SJ 26) in 2009.

Conocybe velutipes (Velen.) Hauskn. & Svrček
Conocybe kuhneriana Sing.; *C. ochracea* sensu auct.
In short turf on limestone; common.
51. SJ 07, 08, 26. LO. (3) 1982. (49, 52)

Pholiotina aporos (Kits van Wav.) Clémençon
Conocybe aporos Kits van Wav.
In rich calcareous woodland soil; frequent.
50, 51. SJ 15, 16, 26. CN. (4) 1994. (48, 49, 52)

Pholiotina arrhenii (Fr.) Singer *Ringed Conecap*
Conocybe arrhenii (Fr.) Kits van Wav.
In grassland; frequent.
50, 51. SH 97; SJ 06, 07, 15, 25, 26. AV, BO, NF. (11) 1994. (48, 49, 52)

Pholiotina blattaria (Fr.) Fayod
Conocybe blattaria (Fr.) Kühn.
In woodland soil; uncommon.
50, 51. SH 88, 97; SJ 26. AV, BO. (3) 1996.

Pholiotina exannulata (Kühner & Watling) M. Moser
Conocybe exannulata Kühner & Watling
In woodland soil; rare.
51. Recorded only from Rhydymwyn Nature Reserve (SJ 26) in 2007.

Pholiotina filaris (Fr.) Singer *Fool's Conecap*
Conocybe filaris (Fr.) Kühner
In grassland, especially on limestone; frequent.
50, 51. SH 95, 97; SJ 08, 15–18, 25, 26, 35, 36. AV, CE, CN, GW, ML, NF, PA. (18) 1979. (49, 52)

Pholiotina tenerioides (J.E. Lange) Singer
Conocybe percincta P.D. Orton
In woodland soil; frequent.
50, 51. SJ 07, 08, 24, 26, 36. AV, GW, PP. (5) 1981.

Pholiotina vestita (Fr.) Singer
Conocybe vestita (Fr.) Kühn.
In rich woodland soil; rare.
50. Recorded only from Loggerheads Country Park (SJ 16) in 2000.

Family Clavariaceae

Camarophyllopsis phaeophylla (Romagn.) Arnolds
In unimproved grassland; rare.
50. Recorded only from Minera (SJ 25) in 2004. This species is not in the *Checklist of Basidiomycetes* and may be new to Britain.

Species Accounts

Camarophyllopsis schulzeri (Bres.) Herink
Hygrophorus schulzeri Bres.
In unimproved, mossy grassland; uncommon.
51. Recorded only from Point of Ayr (SJ 18) in 1978. (48)

Clavaria acuta Sow. *Pointed Club*
Clavaria asterospora Pat.; *C. falcata* Pers.
In damp woodland soil; frequent.
50, 51. SH 85, 97; SJ 04, 08, 15–17, 23, 25, 26, 34. AV, BO, ER, HC, MA, NF. (13) 1976. (48, 49, 52)

Clavaria argillacea Pers. *Moor Club*
On peat in moorland, also in gardens; frequent.
50, 51. SJ 16, 17, 24–26. MF, MO. (6) 1910. (48, 49, 52)

Clavaria flavipes Pers.
Clavaria straminea Cotton
In unimproved grassland; uncommon.
50. SJ 04, 23. (2) 2004. (48, 49, 52)

Clavaria fragilis Holmsk. *White Spindles*
Clavaria vermicularis Sow.
In unimproved grassland; frequent.
50, 51. SH 85, 95, 97; SJ 05, 06, 08, 16, 17, 23–27, 34–36. AV, BO, BP, ER, GW, HC, LO, MA, ML. (24) 1902. (48, 49, 52)

Clavaria fumosa Pers. *Smoky Spindles*
In unimproved grassland; uncommon.
50, 51. SH 97; SJ 07, 17, 25, 26, 34, 35. BO, ER, ML, MO. (9) 1978. (48, 49)

Clavaria incarnata Weinm.
In unimproved grassland; rare.
50, 51. SH 87; SJ 23. BO. (2) 2001. (49, 52)

Clavaria tenuipes Berk. & Broome
Clavaria kriegelsteineri Kajan & Grauw.
In mossy, sandy soil; rare.
50. Recorded only from Marford Quarry Nature Reserve (SJ 35) in 2000.

Clavaria zollingeri Lév. *Violet Coral*
In ancient lawns; rare. This is an indicator of an ancient piece of grassland.
51. SJ 25, 26. (2) 1973. (48, 49, 52)

Clavulinopsis corniculata (Fr.) Corner *Meadow Coral*
In grassland; common.
50, 51. SH 97; SJ 04, 06–08, 15–17, 23–26, 34, 36. BO, CE, ER, HC, LO, MA, ML, PP. (25) 1910. (48, 49, 52)

Clavulinopsis fusiformis (Sow.) Corner *Golden Spindles*
In unimproved grassland; frequent.
50, 51. SH 85, 97; SJ 07, 15, 16, 23–26, 35, 36. AV, BO, GW, ML, NF, NM, PP. (16) 1910. (48, 49, 52)

Clavulinopsis helvola (Pers.) Corner *Yellow Club*
Clavaria inaequalis auct.
In grassland; common.
50, 51. SH 77, 85, 87, 95, 97; SJ 05, 07, 16–18, 23–26, 34–36. AV, BO, CC, CE, CF, ER, GW, LO, ML, NF, NM, PP, WW. (37) 1880. (48, 49, 52)

Clavulinopsis laeticolor (Berk. & M.A. Curt.) R.H. Petersen *Handsome Club*
Clavulinopsis pulchra (Peck) Corner
In unimproved grassland; uncommon.
50, 51. SH 77, 85, 97; SJ 07, 08, 17, 23, 25, 26, 34. BO, ER, HC, MA. (13) 1980. (48, 49, 52)

Clavulinopsis luteoalba (Rea) Corner *Apricot Club*
In limestone grassland; frequent.
50, 51. SH 97; SJ 17, 23, 24, 26, 35. HC, LO, ML, PP. (6) 1979. (48, 49, 52)

Clavulinopsis umbrinella (Sacc.) Corner *Beige Coral*
Clavaria cinereoides (G.F. Atk.) Corner
In base-rich, unimproved grassland; uncommon.
50. SH 85, 97; SJ 15, 23. ML. (4) 1924. (49, 52)

Hodophilus foetens (W. Phill.) Berkebak & Adamčik
Camarophyllopsis foetens (W. Phill.) Arnolds; *Hygrophorus foetens* W. Phill.
On soil under trees; rare.
50, 51. SH 87; SJ 17. CC, HC. (2) 1880. (49)

Hodophilus hymenocephalus (A.H. Sm. & Hesler) Berkebak & Adamčik
Dark Fanvault
Camarophyllopsis hymenocephala (A.H. Sm. & Hesler) Arnolds
On soil under trees; rare.
51. Recorded only from Pengwern College (SJ 07) in 2007.

Hodophilus micaceus (Berk. & Broome) Berkebak & Adamčik
Camarophyllopsis micacea (Berk. & Br.) Arnolds; *Hygrophorus micaceus* Berk. & Br.
On calcareous soil in woodland; rare.
50. Recorded only from Coed Coch (SH 87) in 1880. (52)

Ramariopsis biformis (G.F. Atk.) R.H. Petersen
In woodland soil; rare.
50. Recorded only from Glan-yr-Afon, Marli (SH 97) in 1981.

Ramariopsis kunzei (Fr.) Corner *Ivory Coral*
In woodland soil and grassland; frequent.
50, 51. SJ 08, 23, 26, 34. LO, MO. (6) 1910. (48, 49, 52)

Ramariopsis subtilis (Pers.) R.H. Petersen
Clavulinopsis subtilis (Pers.) Corner
In limestone grassland; rare.
51. SJ 08, 26. LO. (2) 2004. (48)

Species Accounts

Ramariopsis tenuiramosa Corner
In base-rich soil in a garden, near to a yew hedge; uncommon.
50. Recorded only from Rhyd-y-Creuau Field Centre (SH 85) in 1988.

Family Clitocybaceae

Clitocybe agrestis Harm.
In grassy woodland; rare.
51. Recorded only from Coed-y-Garth (SJ 18) in 1991.

Clitocybe brumalis (Fr.) Gillet
In woodland litter; rare.
50. Recorded only from Coed Coch (SH 87) in 1880. (52)

Clitocybe diatreta (Fr.) P. Kumm.
In conifer litter; rare.
50. Recorded only from Coed Coch (SH 87) in 1880.

Clitocybe ditopus Fr. *Mealy Frosted Funnel*
In woodland litter; frequent.
50, 51. SJ 05, 07, 08, 15–18, 24, 25, 34, 36. AV, CE, CN, ER, GW, MF, WW. (17) 1910. (48, 49, 52)

Clitocybe ericetorum Quél.
In heathy soil; rare.
50. Recorded only from Coed Coch (SH 87) in 1880.

Clitocybe fragrans (With.) P. Kumm. *Fragrant Funnel*
Clitocybe suavolens (Schum.) P. Kumm.; *C. decepta* H.E. Bigelow
In woodland litter; common.
50, 51. SH 77, 87, 95, 97; SJ 04–08, 15–18, 24–27, 34–36. AV, BO, BP, CC, CE, CF, CN, DU, ER, GW, HC, LH, LO, MA, MF, ML, NF, NM, PP, WW. (68) 1880. (48, 49, 52)

Clitocybe fuscosquamula J.E. Lange
Clitocybe parilis sensu Rea
In woodland litter; very rare and misunderstood.
50. Reported only from Coed Coch (SH 87) in 1880.

Clitocybe metachroa (Fr.) P. Kumm.
Clitocybe dicolor (Pers.) Murrill; *C. decembris* Sing.
In woodland litter; frequent.
50, 51. SH 95; SJ 05–07, 15–18, 24–26, 35. AV, CE, MA, MF, NF, PP, WW. (18) 1910. (48, 49, 52)

Clitocybe nebularis (Batsch) P. Kumm. *Clouded Funnel*
Lepista nebularis (Batsch) Harmaja
In woodland litter; common.
50, 51. SH 77, 85, 87, 96, 97; SJ 05–08, 15–18, 23–27, 34–36. AV, BO, BP, CC, CE, CN, ER, GW, HC, LH, LO, MA, MF, ML, MO, NF, PP, WW. (62) 1880. (48, 49, 52)

Clitocybe odora (Bull.) P. Kumm. *Aniseed Funnel*
In woodland litter; frequent.
50, 51. SH 87, 97; SJ 05, 07, 15–17, 25–27, 34, 36, 43. AV, CC, CE, ER, GW, LH, LO, MA, MF, NF, WW. (23) 1880. (48, 49, 52)

Clitocybe phyllophila (Pers.) P. Kumm. *Frosty Funnel*
incl. *Clitocybe cerussata* (Fr.) P. Kumm.
In woodland litter; common.
50, 51. SH 87, 97; SJ 05–07, 15–17, 24–26, 34–36. AV, CC, CE, DU, ER, LH, LO, MA, MO, NF, NM, PP, WW. (36) 1880. (48, 49, 52)

Clitocybe rivulosa (Pers.) P. Kumm. *Fool's Funnel*
C. dealbata auct.
In lawns, short turf and grassy places in woods; common. **Poisonous.**
50, 51. SH 87, 97; SJ 05, 08, 15, 16, 18, 23–27, 34–36. AV, BO, BP, CC, CE, ER, GW, LO, MA, MO, PA, WW. (39) 1880. (48, 49, 52)

Clitocybe vibecina (Fr.) Quél. *Mealy Funnel*
Clitocybe langei Hora
In woodland litter; common.
50, 51. SJ 05–07, 15–17, 23–26, 35, 36. AV, CE, CF, GW, LH, LO, MF, PP, WW. (20) 1976. (48, 49, 52)

Lepista irina (Fr.) H.E. Bigelow *Flowery Blewit*
In woodland leaf litter; uncommon.
50, 51. SH 97; SJ 06, 08, 15–17, 24, 26. BP, HC, LO, ML, PP. (9) 1977.

Lepista nuda (Fr.) Cooke *Wood Blewit*
In woodland litter; common.
50, 51. SH 96, 97; SJ 05–08, 15–18, 23–27, 34, 36, 54. AV, BO, BP, BW, CE, CF, DU, ER, GW, HC, LH, LO, MA, MF, ML, MO, NF, NM, PA, PP, WW. (67) 1910. (48, 49, 52)

Lepista panaeolus (Fr.) P. Karst.
Lepista luscina auct.
In woodland litter; uncommon.
51. SJ 17, 25. LH. (3) 1977. (48, 49, 52)

Lepista saeva (Fr.) P.D. Orton *Field Blewit*
Lepista personata (Fr.) Cooke
In base-rich grassland; frequent.
50, 51. SH 97; SJ 07, 08, 14–16, 18, 24–27, 34, 36. AV, BO, BP, ER, LP, PA. (19) 1962. (48, 49, 52)

Lepista sordida (Fr.) Sing.
In woodland litter; uncommon.
50, 51. SH 77, 87, 97; SJ 07, 15–17, 24–27, 34, 35. AV, CC, ER, HC, LH, LO, MA. (22) 1880. (48, 49, 52)

Species Accounts

Paralepista flaccida (Sow.) *Tawny Funnel*
Lepista flaccida (Sow.) Pat.; *Clitocybe flaccida* (Sow.) P. Kumm.;
Lepista inversa (Scop.) Pat.
In woodland litter; common.
50, 51. SH 77, 85, 87, 95–97; SJ 05–08, 15–18, 23–26, 34–36. AV, BP, BW, CC, CE, CN, ER, GW, LH, LO, MA, MF, ML, NF, NM, PA, PP, WW. (59) 1880. (48, 49, 52)

Singerocybe phaeophthalma (Pers.) Harmaja *Chicken Run Funnel*
Clitocybe phaeophthalma (Pers.) Kuyp.; *C. hydrogramma* auct.
In woodland litter; uncommon.
50, 51. SH 87, 97; SJ 07, 26. AV, CC, LO, ML, WW. (6) 1880. (49, 52)

Family Cortinariaceae

Cortinarius acutus (Pers.) Fr.
Mycorrhizal with conifer; uncommon.
50. SH 85, 95; SJ 34. ER. (3) 1910. (48, 49, 52)

Cortinarius alboviolaceus (Pers.) Fr. *Pearly Webcap*
Mycorrhizal with beech; frequent.
50, 51. SH 85; SJ 16, 26, 27, 34, 35. AV, ER, LO, MA, MO. (8) 1910. (48, 49)

Cortinarius alnetorum (Velen.) M.M. Moser
Mycorrhizal with alder; uncommon.
50. Recorded only from Alyn Waters Country Park (SJ 35) in 2016. (48, 49, 52)

Cortinarius amoenolens P.D. Orton
Cortinarius anserinus auct.
Mycorrhizal with beech; uncommon.
50, 51. SJ 08, 16, 24. AV, PP. (3) 1902.

Cortinarius anomalus (Fr.) Fr. *Variable Webcap*
Cortinarius lepidopus Cooke; *C. azureovelatus* P.D. Orton
Mycorrhizal with beech; frequent.
50, 51. SH 85, 87, 97; SJ 05, 08, 16, 17, 24, 26, 34–36. AV, BO, CC, ER, GW, LO, PP. (19) 1880. (48, 49, 52)

Cortinarius anthracinus Fr.
Mycorrhizal with beech and oak; uncommon.
50, 51. SH 87, 97; SJ 06. BO, CC. (3) 1880. (49, 52)

Cortinarius armillatus (Fr.) Fr. *Red Banded Webcap*
Mycorrhizal with birch; frequent.
50, 51. SH 76, 87; SJ 17, 25, 26. CC, CE, LH, NF. (5) 1880. (49)

Cortinarius bibulus Quél.
Cortinarius pulchellus J.E. Lange
Mycorrhizal with alder; uncommon.
50. Recorded only from Erddig (SJ 34) in 1985. (48, 49, 52)

Cortinarius bivelus (Fr.) Fr.
Mycorrhizal wih birch; uncommon.
50. Recorded only from Coed Coch (SH 87) in 1880.

Cortinarius bolaris (Pers.) Fr. *Dappled Webcap*
Mycorrhizal with birch; frequent.
50, 51. SH 85; SJ 05, 16, 17, 26, 34. AV, CF, ER, LH, LO, MO. (8) 1924. (48, 49, 52)

Cortinarius brunneus (Pers.) Fr.
Mycorrhizal with conifers; rare.
50. Recorded only from Fairy Glen (SH 85) in 1924. (48, 49)

Cortinarius bulbosus (Sow.) Fr.
Mycorrhizal with broad-leaved trees; rare and poorly known.
50. Recorded only from Erddig (SJ 34) in 1910.

Cortinarius caerulescens (Schaeff.) Fr. *Mealy Bigfoot Webcap*
Mycorrhizal with beech; rare.
50. Recorded only from Erddig (SJ 34) in 1910.

Cortinarius cagei Melot
Mycorrhizal with a variety of tree species; rare.
50. Recorded only from Coed Coch (SH 87) in 1880.

Cortinarius callisteus (Fr.) Fr.
Mycorrhizal with conifers; rare.
50. Recorded only from Glan-yr-Afon, Marli (SH 97) in 1979. This is the only Welsh record.

Cortinarius calochrous (Pers.) Gray
Mycorrhizal with beech; uncommon.
50, 51. SH 97; SJ 07. (3) 1994. (49, 52)

Cortinarius caninus (Fr.) Fr.
Mycorrhizal with broad-leaved trees; frequent.
50, 51. SH 85, 87; SJ 05, 07, 16, 25, 26, 35. AV, MF, NF, NM, WW. (11) 1924. (48, 49, 52)

Cortinarius cinnabarinus Fr.
Dermocybe cinnabarina (Fr.) Wünsche
Mycorrhizal with beech; rare.
50, 51. SH 85; SJ 05, 36. CF, GW. (4) 1902. (48, 49)

Cortinarius cinnamomeus (L.) Gray *Cinnamon Webcap*
Dermocybe cinnamomea (L.) Wünsche
Mycorrhizal with conifers and birch; common.
50, 51. SH 85, 87; SJ 07, 16, 17, 25, 26, 34–36. AV, CC, DU, ER, GW, LH, LO, NM. (14) 1880. (48, 49, 52)

Cortinarius croceocaeruleus (Pers.) Fr.
Mycorrhizal with beech; rare.
50. SH 97; SJ 34. ER. (3) 1979. (52)

Species Accounts

Cortinarius croceus (Schaeff.) Gray
Cortinarius cinnamomeobadius R. Henry
Mycorrhizal with conifers and, occasionally, with *Helianthemum*; uncommon.
50. SH 95; SJ 15, 16, 25. MF. (4) 1981. (48, 49, 52)

Cortinarius decipiens (Pers.) Fr. *Sepia Webcap*
Mycorrhizal with a variety of deciduous trees; common.
50, 51. SH 85, 97; SJ 05, 08, 15–17, 24–26, 34–36. AV, BO, LO, MA, NF, PP. (19) 1924. (48, 49, 52)

Cortinarius delibutus Fr. *Yellow Webcap*
Mycorrhizal with birch; frequent.
50, 51. SH 85; SJ 17, 25, 26. CE, NF. (4) 1924. (48, 49, 52)

Cortinarius diasemospermus Lamoure
Mycorrhizal with birch; probably widespread.
51. Recorded only from Rhydymwyn Nature Reserve (SJ 26) in 2007. (48, 49, 52) A recent segregate from *C. flexipes*.

Cortinarius elatior Fr. *Wrinkled Webcap*
Cortinarius lividoochraceus (Berk.) Berk.
Mycorrhizal with beech; frequent.
50, 51. SH 85; SJ 06, 16, 17, 34. ER, LO. (6) 1924. (48, 49, 52)

Cortinarius elegantissimus Rob. Henry
Cortinarius aureoturbinatus J.E. Lange
Mycorrhizal with beech over limestone; rare.
50. SJ 15, 34. CG. (2) 1990.

Cortinarius erythrinus (Fr.) Fr.
Mycorrhizal with conifers; rare.
50. SH 85, 87; SJ 34. CC, ER. (5) 1880. (48, 49) Not seen in North Wales since 1950 until its rediscovery in Big Wood at Erddig in 1996.

Cortinarius evernius (Fr.) Fr.
Mycorrhizal with conifers and birch; rare.
50, 51. SJ 16, 26. CE, MF. (2) 1981. (52) These Welsh records must be treated as doubtful in the absence of herbarium specimens.

Cortinarius flexipes (Pers.) Fr. var. **flexipes**
Mycorrhizal with birch; common.
50, 51. SJ 05, 16, 17, 25, 26, 34–36. AV, CE, CF, ER, LH, LO, NF, WW. (12) 1976. (48, 49, 52)

Cortinarius flexipes var. **flabellus** (Fr.) Lindst. & Melot *Pelargonium Webcap*
Cortinarius paleaceus auct.
Mycorrhizal with birch; common.
50, 51. SH 76, 85, 87, 95; SJ 06, 08, 15–17, 26, 34, 35. AV, CC, ER, LH, MA, WW. (15) 1880. (48, 49, 52)

Cortinarius fulvescens Fr.
Cortinarius fasciatus auct.
Mycorrhizal with pines; rare.
50. SH 87; SJ 05. CC, CF. (2) 1880. Not seen between 1880 and 2002 in North Wales.

Cortinarius fulvoochraceus Rob. Henry var. **cyanophyllus** Rob. Henry
Mycorrhizal with oak; very rare.
50. Recorded only from Colwyn Bay (SH 87) in 1988. This is the only British record.

Cortinarius glandicolor (Fr.) Fr.
Cortinarius brunneus var. *glandicolor* (Fr.) Lindstr. & Melot
Mycorrhizal with birch; common.
50, 51. SJ 05, 16, 17, 26, 27, 34. AV, CE, CF, ER, HC, LH, WW. (7) 1977. (48)

Cortinarius helvelloides (Fr.) Fr.
Mycorrhizal with alder; uncommon.
50, 51. SJ 08, 35. (2) 1998. (48, 49, 52)

Cortinarius hemitrichus (Pers.) Fr. *Frosty Webcap*
Mycorrhizal with birch; frequent.
50, 51. SH 76, 85; SJ 08, 15, 16, 25, 26, 34, 35. AV, ER, NF, WW. (13) 1924. (48, 49, 52)

Cortinarius hinnuleus (Sow.) Fr. *Earthy Webcap*
Mycorrhizal with beech; frequent.
50. SH 85, 87; SJ 16, 34. CC, ER, MA, MF. (5) 1880. (49)

Cortinarius illibatus Fr.
Mycorrhizal with birch; very rare.
50. Recorded only from Coed Coch (SH 87) in 1880.

Cortinarius illuminus Fr.
Mycorrhizal with deciduous trees; very rare.
50. Reported only from Coed Coch (SH 87) in 1880.

Cortinarius incisus (Pers.) Fr.
Mycorrhizal with deciduous trees; very rare.
50. Reported only from Coed Hafod near Betws-y-Coed (SH 85) in 1924. (48)

Cortinarius infractus (Pers.) Fr. *Bitter Webcap*
Mycorrhizal with beech; uncommon.
50, 51. SJ 16, 26. AV, MA. (2) 1983. (52)

Cortinarius ionophyllus M.M. Moser
Mycorrhizal with pine; very rare.
51. Recorded only from Coed y Felin Nature Reserve (SJ 16) in 1993. This was originally recorded in the British Mycological Society foray report as being associated with beech. There are several pines in the wood and growing with them are several other species of northern fungi. This species is otherwise Recorded only from Scotland.

Species Accounts

Cortinarius largus Fr.
Cortinarius nemorensis (Fr.) J.E. Lange
Mycorrhizal with birch and beech, especially on limestone; frequent.
50, 51. SH 85, 96; SJ 06, 16, 25, 26, 34. AV, ER, LO, NM. (9) 1924. (48, 49, 52)

Cortinarius limonius (Fr.) Fr. *Sunset Webcap*
Mycorrhizal with conifers; uncommon.
51. Recorded only from Prestatyn (SJ 08) in 1951. (48)

Cortinarius malicorius Fr.
Cortinarius croceifolius Peck
Mycorrhizal with pines; rare.
50. Recorded only from Maes Mynan (SJ 17) in 1994.

Cortinarius mucifluoides (Rob. Henry) Rob. Henry *Purple Stocking Webcap*
Cortinarius pseudosalor J.E. Lange; *Cortinarius stillatitius* auct.
Mycorrhizal with beech; frequent.
50, 51. SH 87, 95; SJ 05, 24, 26, 34. CC, CE, CF, ER, PP. (8) 1880.

Cortinarius obtusus (Fr.) Fr.
Mycorrhizal with broad-leaved trees; uncommon.
50, 51. SH 85, 87; SJ 16, 17, 34, 36. AV, CC, ER, GW, MF. (8) 1880. (48, 49, 52)

Cortinarius ochroleucus (Schaeff.) Fr.
Mycorrhizal with beech; uncommon.
50. Recorded only from Coed Coch (SH 87) in 1880. (52)

Cortinarius orellanus Fr. *Fool's Webcap*
Mycorrhizal with oak; rare. **Very poisonous.**
50. Recorded only from Coed Coch (SH 87) in 1880. (49) Not seen in North Wales since 1950.

Cortinarius cf. **parvannulatus** Kühner
Mycorrhizal with alder and willow; rare.
51. Recorded only from Rhydymwyn Nature Reserve (SJ 26) in 2007.

Cortinarius pholideus (Fr.) Fr. *Scaly Webcap*
Mycorrhizal with birch; uncommon.
50, 51. SH 85; SJ 26. LO. (3) 1924. (49)

Cortinarius porphyropus (Alb. & Schwein.) Fr.
Mycorrhizal with broad-leaved trees; uncommon.
50. Recorded only from Coed Coch (SH 87) in 1880. (48)

Cortinarius purpurascens (Fr.) Fr. *Bruising Webcap*
Mycorrhizal with beech and oak; rare.
50, 51. SJ 07, 34. ER. (2) 1980. (48)

Cortinarius purpureus (Pers.) Fuckel
Mycorrhizal with conifers; rare.
50. Recorded only from Coed Coch (SH 87) in 1880. (49) Not seen in North Wales since 1950.

Cortinarius raphanoides (Pers.) Fr.
Cortinarius betuletorum M.M. Moser
Mycorrhizal with birch; uncommon.
50, 51. SJ 16, 26, 35. AV, MO, WW. (4) 1983. (48, 49)

Cortinarius rigens (Pers.) Fr.
Cortinarius duracinus Fr.
Mycorrhizal with conifers; uncommon.
50, 51. SH 95, 97; SJ 26. LO. (3) 1999. (52)

Cortinarius salor Fr.
Mycorrhizal with different trees; very rare.
50. Recorded only from Coed Coch (SH 87) in 1880.

Cortinarius sanguineus (Wulf.) Fr. *Bloodred Webcap*
Dermocybe sanguinea (Wulf.) Wünsche
Mycorrhizal with conifers; uncommon.
50, 51. SJ 05, 26. CE. (2) 1984. (48, 49, 52)

Cortinarius saniosus (Fr.) Fr.
Mycorrhizal with birch; uncommon.
50, 51. SJ 16, 17, 26, 35, 36. AV, GW, LH, LO. (6) 1977. (48)

Cortinarius saturninus (Fr.) Fr.
Cortinarius cohabitans P. Karst.
Mycorrhizal with deciduous trees; uncommon.
50, 51. SH 85; SJ 17, 25, 26, 34. AV, ER, MO, NF. (6) 1924. (48, 49, 52)

Cortinarius semisanguineus (Fr.) Gillet *Surprise Webcap*
Dermocybe semisanguinea (Fr.) M.M. Moser
Mycorrhizal with conifers; frequent.
50, 51. SJ 05, 16, 25, 26. CE, CF, MF, NM. (5) 1976. (48, 49, 52)

Cortinarius spilomeus (Fr.) Fr.
Mycorrhizal with oak; rare.
50. Recorded only from Hafod Wood (SH 85) in 1924.

Cortinarius subbalaustinus Rob. Henry
Mycorrhizal with birch; uncommon.
51. SJ 26, 36. CE. (2) 1981.

Cortinarius subferrugineus (Batsch) Fr.
Mycorrhizal with conifers; rare and not accepted as British in the absence of authentic specimens.
50. SH 85, 87. Reported only from Coed Coch in 1880 and Fairy Glen in 1924. (49) Listed because the Coed Coch fungi were recorded by the leading mycologists of the day; not reported from North Wales since the 1924 foray.

Species Accounts

Cortinarius tabularis (Fr.) Fr.
Cortinarius decoloratus (Fr.) Fr.
Mycorrhizal with birch; uncommon.
50. Recorded only from Coed Hafod, near Betws-y-Coed (SH 85) in 1924.

Cortinarius torvus (Fr.) Fr. *Stocking Webcap*
Mycorrhizal with beech; uncommon.
50, 51. SH 85, 87; SJ 05, 16, 24, 34. AV, CC, ER. (6) 1880. (48, 49, 52)

Cortinarius traganus (Fr.) Fr. *Gassy Webcap*
Mycorrhizal with birch; uncommon.
50, 51. SH 96; SJ 36. GW. (2) 1974.

Cortinarius triumphans Fr. *Birch Webcap*
Cortinarius crocolitus Quél.
Mycorrhizal with birch; uncommon.
50, 51. SH 76, 77, 96; SJ 17, 26. CE, LH. (5) 1977. (48, 49, 52)

Cortinarius trivialis J.E. Lange *Girdle Webcap*
Mycorrhizal with willow; uncommon.
50. Recorded only from Llyn Syberi (SH 76) in 1979. (48, 49)

Cortinarius umbrinolutens P.D. Orton
Cortinarius rigidus auct.
Mycorrhizal with broad-leaved trees; uncommon.
50, 51. SH 85; SJ 25, 26. MA, NF. (3) 1924. (49, 52)

Cortinarius variicolor (Pers.) Fr. *Contrary Webcap*
Cortinarius varius auct.
Mycorrhizal with conifers; uncommon.
50, 51. SJ 08, 34. ER. (2) 1910. (49) Not seen in North Wales since 1960.

Cortinarius violaceus (L.) Gray *Violet Webcap*
Mycorrhizal with birch; uncommon.
51. SJ 26. AV. (2) 2004. **RDL/NT.** This rare species seems to be increasing, especially in disturbed sites which are being colonised by birch.

Family Crepidotaceae

Crepidotus applanatus (Pers.) P. Kumm. *Flat Oysterling*
On woody debris in litter; common.
50, 51. SH 97; SH 05, 08, 16, 17, 24–26, 34, 36. AV, BO, CE, DU, ER, LO, MA, MF, ML, PP, WW. (23) 1985. (48, 49, 52)

Crepidotus autochthonus J.E. Lange
On soil at edge of woodland paths; rare.
51. Recorded only from Celyn Woods (SJ 26) in 1986. (49)

Crepidotus carpaticus Pilát
Crepidotus wakefieldiae Pilát
On fallen branches; uncommon.
50. Recorded only from the Aberduna Nature Reserve, Maeshafn (SJ 26) in 2001. (48, 49)

Crepidotus cesati (Rab.) Sacc.
On twigs; common.
50, 51. SJ 16, 17, 24–26, 34–36. AV, CN, DU, ER, GW, MA, NF. (9) 1978. (48, 49, 52)

Crepidotus epibryus (Fr.) Quél.
Crepidotus herbarum (Peck) Sacc.; *Pleurotellus graminicola* Fayod
On grass litter and mosses on woodland floor; uncommon.
50, 51. SH 87; SJ 07, 16, 18, 35. AV, CC, CN, PA. (6) 1880. (48, 49, 52)

Crepidotus lundellii Pilát
On small rotten sycamore branches; uncommon.
50, 51. SJ 26, 35. AV. (2) (49, 52)

Crepidotus luteolus (Lamb.) Sacc. *Yellowing Oysterling*
On herbaceous stems in litter; frequent.
50, 51. SH 85; SJ 07, 16, 26, 34. AV, CE, ER, LO. (6) 1984. (48, 49, 52)

Crepidotus mollis (Schaeff.) Staude *Peeling Oysterling*
On fallen hardwood trunks and branches; common.
50, 51. SH 85, 87, 97; SJ 05–07, 16, 17, 23, 26, 27, 34, 35. BO, CC, CE, CN, ER, LO, MA, MF, WW. (25) 1880. (48, 49, 52)

Crepidotus variabilis (Pers.) P. Kumm. *Variable Oysterling*
On twigs in litter and in bramble clumps; common.
50, 51. SH 97; SJ 05, 07, 15–18, 23–27, 34–36. AV, BO, CE, CF, CN, DU, ER, GW, LH, LO, MA, MF, ML, MO, NF, NM, PP, WW. (56) 1910. (48, 49, 52)

Simocybe cetunculus (Fr.) P. Karst.
On rotten trunk; uncommon.
50. Recorded only from the grounds of Chirk Castle (SJ 23) in 1993. (49, 52)

Simocybe haustellaris (Fr.) Watling
On rotten twigs and sticks; uncommon.
50, 51. SJ 24, 26, 34. AV. (3) 1910. (48, 49, 52)

Family Cyphellaceae

Baeospora myosura (Fr.) Sing. *Conifercone Cap*
On half-buried cones of a variety of conifers; common.
50, 51. SH 77; SJ 26, 34, 36. ER, LO, WW. (6) 1910. (49, 52)

Species Accounts

Chondrostereum purpureum (Pers.) Pouzar *Silverleaf Fungus*
Stereum purpureum Pers.
Parasitic on a wide range of deciduous trees, then on dead wood, also the causal agent of *Silverleaf* disease of fruit trees; common.
50, 51. SH 77, 87; SJ 06–08, 15–17, 24, 26, 27, 34–36. AV, BW, CC, CE, CN, DU, ER, GW, LO, MA, MO, PP, WW. (27) 1880. (48, 49, 52)

Granulobasidium vellereum (Ellis & Cragin) Jülich
Hypochnicium vellereum (Ellis & Cragin) Parmasto
On rotten bark; uncommon.
51. Recorded only from Point of Ayr (SJ 18) in 1985.

Family Cyphellopsidaceae

Flagelloscypha minutissima (Burt) Donk
On dead herbaceous stems; frequent.
51. SJ 26, 36. AV, GW. (2) 2007. (48, 52)

Merismodes anomala (Pers.) Sing.
Cyphellopsis anomala (Pers.) Donk
On fallen branches; uncommon.
50, 51. SH 85; SJ 25, 26, 34. CE, ER, NF. (5) 1910. (48, 49, 52)

Merismodes fasciculata (Schwein.) Donk
On fallen branches; rare.
51. Recorded only from Nant-y-Ffrith (SJ 25) in 1910. (49)

Family Cystodermataceae

Cystoderma amianthinum (Scop.) Fayod *Earthy Powdercap*
In woodland litter; common.
50, 51. SH 76, 85, 95, 97; SJ 04, 05, 07, 15–17, 23–26, 34–36. AV, CF, ER, GW, HC, LH, MA, MF, NF, NM, WW. (42) 1910. (48, 49, 52)

Cystoderma carcharias (Pers.) Fayod *Pearly Powdercap*
In woodland litter; uncommon.
50. SH 87, 95; SJ 24, 34. CC, ER. (4) 1880. (49)

Cystoderma granulosum (Batsch) Fayod
In acidic woodland litter; uncommon.
50. Recorded only from Coed Coch (SH 87) in 1880. (49)

Phaeolepiota aurea (Matt.) Maire ex Konrad & Maubl. *Golden Bootleg*
In grassy places in open woods and grass verges; uncommon.
51. SJ 17, 25. DU, NM. (2) 1985.

Family Entolomataceae

Clitocella popinalis (Fr.) Kluting, T.J. Baroni & Bergemann
Rhodocybe popinalis (Fr.) Sing.
In dune grassland and in grassy woodland; frequent.
50, 51. SH 87; SJ 06, 18. CC, PA. (3) 1880. (48, 49, 52)

Clitopilus hobsonii (Berk.) P.D. Orton
On rotten wood; uncommon.
50. Recorded only from Erddig (SJ 34) since 1985. (48, 49, 52)

Clitopilus prunulus (Scop.) P. Kumm. *The Miller*
In grassy places in woods; common.
50, 51. SH 77, 87, 95–97; SJ 05–08, 15–17, 24, 26, 34–36. AV, BO, CC, CF, ER, GW, LO, MF, ML, PP, WW. (33) 1880. (48, 49, 52)

Entoloma aethiops (Scop.) G. Stev.
Leptonia aethiops (Scop.) Gillet
In acid forest soil; rare.
50, 51. SH 85, 87; SJ 26. CC, LO. (3) 1880. (49)

Entoloma ameides (Berk. & Br.) Sacc.
Nolanea ameides (Berk. & Br.) P.D. Orton
In meadows; rare.
50, 51. SH 97; SJ 04, 25. BO. (3) 1863, at Bodelwyddan from where it was originally described. (49, 52)

Entoloma anatinum (Lasch) Donk
Leptonia anatina (Lasch) P. Kumm.
In limestone grassland; uncommon.
51. SJ 17, 26. LO. (2) 1978. (48, 49)

Entoloma aprile (Britz.) Sacc.
Under hawthorn hedges, in spring; uncommon.
51. Recorded only from Ddol Uchaf Nature Reserve (SJ 17) in 1978.

Entoloma araneosum (Quél.) M.M. Moser
On basic soil under *Mercurialis*; rare.
50. Recorded only from Glan-yr-Afon, Marli (SH 97) in 1979.

Entoloma asprellum (Fr.) Fayod
In grassland; uncommon.
50, 51. SH 97; SJ 23, 25, 35. BO. (5) 1997. (49, 52)

Entoloma atrocaeruleum Noordel.
In unimproved grassland; rare.
50. Recorded only from Bryn Alyn (SJ 25) in 2003.

Species Accounts

Entoloma bloxamii (Berk. & Br.) Sacc. *Big Blue Pinkgill*
Entoloma madidum auct.
In unimproved grassland; rare.
50, 51. SH 87; SJ 17. CC, HC. (2) 1880. (49, 52)

Entoloma caesiocinctum (Kühner) Noordel.
In unimproved grassland; rare.
51. SJ 17, 26. AV. (2) 2003. (49, 52)

Entoloma catalaunicum (Singer) Noordel.
In unimproved grassland; rare.
50, 51. SJ 17, 25, 27. HC. (3) 2003. These are the first records from Wales.

Entoloma cetratum (Fr.) Noordel. *Honey Pinkgill*
Nolanea cetrata (Fr.) P. Kumm.
In grassland, both acid and basic; common.
50, 51. SJ 16, 17, 24–26, 34. AV, CE, ER, MF, NF, PP. (9) 1976. (48, 49, 52)

Entoloma chalybaeum (Pers.) Noordel. var. **chalybaeum** *Indigo Pinkgill*
Leptonia chalybaea (Pers.) P. Kumm.
In unimproved grassland; frequent.
50, 51. SH 85, 87; SJ 08, 23, 26, 34. BP, CC, MA. (6) 1880. (48)

Entoloma chalybaeum var. **lazulinum** (Fr.) Noordel.
Leptonia lazulina (Fr.) Quél.
In unimproved grassland; frequent.
50, 51. SJ 08, 15, 16, 25, 26, 34, 35. AV, LO. (7) 1983. (48, 49, 52)

Entoloma clandestinum (Fr.) Noordel.
In unimproved grassland; rare.
50, 51. SJ 17, 25. HC. 2003. These are the first records from Wales.

Entoloma clypeatum (L.) P. Kumm. *Shield Pinkgill*
Under rosaceous shrubs; common.
50, 51. SH 97; SJ 07, 15–17, 26. BO, ML, MO, WW. (11) 1973. (48, 49, 52)

Entoloma conferendum (Britz.) Noordel. *Star Pinkgill*
Nolanea staurospora Bres.
In grassland of all kinds; common.
50, 51. SH 85, 87, 97; SJ 04–06, 15–17, 23, 25, 26, 34, 35. BO, CC, CE, DU, ER, HC, ML, NM. (22) 1880. (48, 49, 52)

Entoloma corvinum (Kühn.) Noordel.
Leptonia corvina (Kühn.) P.D. Orton
In limestone grassland; frequent.
50, 51. SH 97; SJ 06, 17, 25, 26. AV, HC, LO, MA. (10) 1978. (49, 52)

Entoloma depluens (Batsch) Hesler
On decayed wood; rare.
50. Recorded only from Coed Coch (SH 87) in 1880.

Entoloma dysthales (Peck) Sacc.
On base-rich soil under *Mercurialis*; frequent.
50. Recorded only from Nant Gwyn (SJ 06) in 1997.

Entoloma elodes (Fr.) P. Kumm.
Trichopilus elodes (Fr.) P.D. Orton
In *Sphagnum*; uncommon.
50, 51. SH 87; SJ 25. CC, HM. (2) 1880.

Entoloma euchroum (Pers.) Donk
On hazel stumps; uncommon.
50. Recorded only from Glan-yr-Afon, Marli (SH 97) in 1980.

Entoloma exile (Fr.) Hesler
In unimproved grassland; rare.
50. Recorded only from Minera (SJ 25) in 2003. (48, 49, 52)

Entoloma formosum (Fr.) Noordel.
In unimproved grassland; rare.
51. Recorded only from Halkyn Common (SJ 17) in 2003. This is the first record from Wales.

Entoloma glaucobasis Noordel.
In unimproved grassland; rare.
51. Recorded only from Halkyn Common (SJ 17) in 2003. This appears to be the first record from the United Kingdom of species described from northern Italy.

Entoloma griseocyaneum (Fr.) P. Kumm. *Felted Pinkgill*
Leptonia griseocyanea (Fr.) P.D. Orton
In unimproved grassland; rare.
50. SH 87; SJ 25, 34. CC. (3) 1880. (49, 52)

Entoloma hebes (Romagn.) Trimbach *Pimple Pinkgill*
Nolanea tenuipes P.D. Orton; *N. hebes* (Romagn.) P.D. Orton
In grassland; uncommon.
50, 51. SH 87, 97; SJ 26. AV, BO, CC. (3) 1880. (49, 52)

Entoloma hirtipes (Schum.) M.M. Moser
Nolanea hirtipes (Schum.) P. Kumm.
In coniferous litter; frequent.
50, 51. SH 95, 97; SJ 07, 08, 15–17, 24–27, 35. BP, CE, CN, DU, LH, LO, ML, WW. (21) 1981. (52)

Entoloma incanum (Fr.) Hesler *Mousepee Pinkgill*
Leptonia incana (Fr.) Gillet
In basic grassland; frequent.
50, 51. SH 97; SJ 08, 15, 24–26, 35. BO, MA. (10) 1983. (49, 52)

Entoloma infula (Fr.) Noordel.
Nolanea infula (Fr.) Gillet
In unimproved grassland; uncommon.
50, 51. SH 87; SJ 17, 26. AV, CC, HC. (5) 1880. (49)

Species Accounts

Entoloma jubatum (Fr.) P. Karst. *Sepia Pinkgill*
Trichopilus jubatus (Fr.) P.D. Orton
In heathland; frequent.
50, 51. SH 87; SJ 16. CC, MF. (2) 1880. (48, 49, 52)

Entoloma langei Noordel. & T. Borgen
In unimproved grassland; rare.
51. Recorded only from Halkyn Common (SJ 17 and 27) in 2003. These are the first records from Wales.

Entoloma longisetum (Peck) Noordel. var. **sarcitulum** (P.D. Orton) Noordel.
Leptonia sarcitula P.D. Orton
In short grass at roadside; uncommon.
50. SH 95, 97. (2) 1997. (48, 49)

Entoloma lucidum (P.D. Orton) M.M. Moser
Nolanea lucida P.D. Orton
In grassy clearings in woodland; uncommon.
51. Recorded only from Celyn Woods (SJ 26) in 1986.

Entoloma minutum (P. Karst.) Noordel.
In unimproved grassland; rare.
51. Recorded only from Halkyn Common (SJ 17 and 27) in 2003. These are the first records from Wales.

Entoloma mougeotii (Fr.) Hesl.
In limestone grassland; uncommon.
51. Recorded only from Craig Fawr (SJ 08) in 1983.

Entoloma nigellum (Quél.) Noordel.
Eccilia nigella Quél.; *Claudopus nigella* (Quél.) P.D. Orton
In dune grassland; frequent.
51. Recorded only from Point of Ayr (SJ 18) in 1978.

Entoloma nigroviolaceum (P.D. Orton) Hesler
In unimproved grassland; rare.
50, 51. SJ 17, 25. HC. (2) 2003. These are the first records from Wales.

Entoloma niphoides Noordel.
On soil under hawthorn in late spring; rare.
50. Recorded only from Glan-yr-Afon, Marli (SH 97) in 1979.

Entoloma nitidum Quél.
In coniferous litter; frequent.
51. Recorded only from Loggerheads Country Park (SJ 26) in 1978. (52)

Entoloma ochromicaceum Noordel. & Liiv
In unimproved grassland; rare.
50. Recorded only from Bryn Alyn (SJ 25) in 2003. This is the first record from Wales.

Entoloma papillatum (Bres.) Dennis *Papillate Pinkgill*
Nolanea papillata Bres.
In unimproved grassland; frequent.
50, 51. SH 85; SJ 06, 08, 15–17, 23, 25. BP, HC, MA. (11) 1924. (48, 49, 52)

Entoloma pleopodium (DC.) Noordel. *Aromatic Pinkgill*
Nolanea icterina (Fr.) P. Kumm.
In conifer litter; frequent.
50, 51. SH 97; SJ 07, 16, 26, 35. BO, CE, MA, MF, WW. (7) 1976. (52)

Entoloma poliopus (Romagn.) Noordel.
In unimproved grassland; rare.
50, 51. SJ 17, 25. HC. (2) 2003.

Entoloma politum (Pers.) Donk
Eccilia polita (Pers.) P. Kumm.
In damp woodland; uncommon.
50. SH 76; SJ 15. (2) 1979. (48, 49, 52)

Entoloma porphyrophaeum (Fr.) P. Karst. *Lilac Pinkgill*
Trichopilus porphyrophaeus (Fr.) P.D. Orton
In pastures; common.
50, 51. SH 97; SJ 05, 08, 15–17, 23, 25, 26, 34, 36. AV, BO, BP, ER, HM, ML, MO. (18) 1973. (48, 49, 52)

Entoloma pratulense Noordel.
In unimproved grassland; rare.
50, 51. SH 85; SJ 17, 27. HC. (3) 2003. These are the first records from Wales.

Entoloma prunuloides (Fr.) Quél. *Mealy Pinkgill*
In limestone grassland; common.
50, 51. SH 15, 17, 25, 26. HC, LO. (5) 1978. (48, 49, 52)

Entoloma pulvereum Rea
On litter in oak woodlands; rare.
50. Recorded only from Erbistock (SJ 34) in 1910, where it was determined by Rea.

Entoloma rhodopolium (Fr.) P. Kumm. f. **rhodopolium** *Wood Pinkgill*
In litter in deciduous woodland; common.
50, 51. SH 76; SJ 05, 06, 08, 15–17, 26, 34, 35. AV, BP, CF, ER, LO, MA, WW. (13) 1983. (48, 49, 52)

Entoloma rhodopolium f. **nidorosum** (Fr.) Noordel. *'Nitrous Pinkgill'*
Entoloma nidorosum (Fr.) Quél.
In woodland litter; common.
50, 51. SH 85, 87, 97; SJ 05, 08, 15, 16, 18, 26, 34. BO, CC, ER, MF, MA, MO, WW. (15) 1880. (48, 49, 52)

Entoloma scabiosum (Fr.) Quél.
In litter in deciduous woodland; rare.
50. Recorded only from Cefn Rofft (SJ 04) in 2003. (52)

Species Accounts

Entoloma saepium (Noules & Dassier) Richon & Roze
Under cherry and hawthorn in scrub, in spring; uncommon.
50, 51. SJ 16, 17. AV, DU, LO. (3) 1978. (52)

Entoloma sericatum (Britz.) Sacc.
In boggy woodland; frequent.
50, 51. SH 97; SJ 16, 17, 23, 26. AV, BO, WW. (5) 1924. (48)

Entoloma sericellum (Fr.) P. Kumm. *Cream Pinkgill*
Leptonia sericella (Fr.) Barbier; *Alboleptonia sericella* (Fr.) P.D. Orton
In grassland; common.
50, 51. SH 85, 95, 97; SJ 04, 07, 13, 16, 17, 23, 25, 26, 34, 35. BO, CE, HC, MA, MF, MO, WW. (22) 1974. (48, 49, 52)

Entoloma sericeum (Bull.) Quél. *Silky Pinkgill*
Nolanea sericea (Bull.) P.D. Orton
In grassland; common, by far the commonest species of the genus.
50, 51. SH 85, 97; SJ 04, 05, 07, 08, 13, 15–18, 23, 25, 26, 34–36. AV, BO, BP, CE, CF, ER, GW, HC, LO, MA, ML, MO, PA. (35) 1910. (48, 49, 52)

Entoloma serrulatum (Fr.) Hesler *Blue Edge Pinkgill*
Leptonia serrulata (Fr.) P. Kumm.
In grassy woodland; common.
50, 51. SH 97; SJ 07, 08, 15, 17, 23, 25, 26. AV, BO, MA. (13) 1978. (48, 49, 52)

Entoloma sinuatum (Pers.) P. Kumm. *Livid Pinkgill*
incl. *Entoloma lividum* (Bull.) Quél.
In deciduous woods; common. **Poisonous.**
50, 51. SJ 06, 08, 16, 25, 26, 34–36. CE, ER, GW, LO, MA, NF, WW. (12) 1978. (49, 52)

Entoloma sodale Noordel.
Leptonia lampropus auct.
In grassland; frequent.
50, 51. SJ 08, 16, 17, 26, 34, 35. AV, ER, LO, MF, MO. (7) 1975. (48, 49, 52)

Entoloma solstitiale (Fr.) Noordel.
In unimproved grassland; rare.
50. Recorded only from Minera (SJ 25) in 2003.

Entoloma tibiicystidiatum Arnolds & Noordel.
In unimproved grassland; rare.
51. Recorded only from Halkyn Common (SJ 27) in 2003. This is the first record from Wales.

Entoloma tjallingiorum Noordel.
On decaying wood in mixed woodland; rare.
50, 51. SH 87; SJ 26. CC, MO. 1880.

Entoloma turbidum (Fr.) Quél.
In sandy soil with pine and birch; rare.
50. Recorded only from Coed Coch (SH 87) in 1880. (48, 49)

Entoloma turci (Bres.) M.M. Moser
In unimproved grassland; rare.
50, 51. SJ 15, 17, 25. HC. (4) 2003. (52)

Entoloma undatum (Fr.) Gillet
Entoloma sericeonitidum (P.D. Orton) Arnolds; *Eccilia sericeonitida* P.D. Orton; *Claudopus sericeonitidus* (P.D. Orton) P.D. Orton
In mossy woods; rare.
50, 51. SH 76; SJ 18, 25, 27. HC, PA. (4) 1979. (48, 52)

Entoloma vernum S. Lundell
Nolanea cucullata (J. Favre) P.D. Orton; *N. verna* (S. Lundell) P.D. Orton
In grassy places in conifer woodland; uncommon.
50, 51. SH 85; SJ 17, 34. ER, LH. (3) 1910. (48, 49)

Entoloma versatile (Gillet) M.M. Moser
In soil under broad-leaved trees; rare.
51. Recorded only from the Rhydymwyn Nature Reserve (SJ 26) in 2007. (52)

Entoloma wynnei (Berk. & Br.) Sacc.
On soil in coniferous woodland; rare.
50. Recorded only from Coed Coch (SH 87) in 1872, from where it was originally described; it has never been refound.

Entoloma xanthochroum (P.D. Orton) Noordel.
In unimproved grassland; rare.
51. Recorded only from Halkyn Common (SJ 17) in 2003. This is the first record from Wales.

Rhodocybe gemina (Fr.) Kuyp. & Noordel.
Rhodocybe truncata auct.
In limestone grassland; uncommon.
50, 51. SH 97; SJ 07. ML. (2) 1979.

Rhodocybe hirneola (Fr.) P.D. Orton
On soil under conifers; rare.
50. Recorded only from Coed Coch (SH 87) in 1879; this is the only Welsh record.

Family Fayodiaceae

Bonomyces sinopica (Fr.) Vizzini
Clitocybe sinopica (Fr.) P. Kumm.
On bonfire site in woodland; rare.
51. SJ 18, 28. WW. (2) 1992.

Gamundia striatula (Kühner) Raithelh.
On forest litter; rare.
50. Recorded only from Coed Coch (SH 87) in 1865 and 1879.

Species Accounts

Gerronema prescotii (Weinm.) Redhead
Cantharellus albidus auct.
On mossy woodland soil; rare.
50. Recorded only from Coed Coch (SH 87) in 1880.

Infundibulicybe costata (Kühner & Romagn.) Harmaja
Clitocybe costata Kuhn. & Romagn.; *C. incilis* auct.
In beech litter; uncommon.
50, 51. SH 87; SJ 07, 16. CC, MF. (4) 1880. (52)

Infundibulicybe geotropa (Bull.) Harmaja *Trooping Funnel*
Clitocybe geotropa (Bull.) Quél.
In woodland litter; common.
50, 51. SH 87, 97; SJ 05, 15, 24–27, 34, 35. AV, BO, BW, CC, CE, CF, ER, LO, MA, ML, MO, PP, WW. (27) 1880. (48, 49, 52)

Infundibulicybe gibba (Pers.) Harmaja *Common Funnel*
Clitocybe gibba (Pers.) P. Kumm.; *C. infundibuliformis* auct.
In woodland litter; common.
50, 51. SH 85, 87, 97; SJ 05, 07, 08, 15–17, 25, 26, 34–36. AV, CC, ER, GW, LO, MA, MO, WW. (25) 1880. (48, 49, 52)

Leucocortinarius bulbiger (Alb. & Schwein.) Sing. *White Webcap*
Under larch and pine in a mixed wood (with beech); rare.
51. Recorded only from Loggerheads Country Park (SJ 26) in 1993.

Myxomphalia maura (Fr.) Hora
On burnt ground and bonfire sites; frequent.
50, 51. SH 87; SJ 07, 16, 26, 34. CC, ER, LO, WW. (5) 1880. (52)

Omphalina pyxidata (Pers.) Quél.
On damp, sandy soil with lichens; frequent.
50, 51. SJ 25, 26, 35. LO, NF. (3) 1978. (48, 49, 52)

Pseudoomphalina graveolens (S. Petersen) Sing.
In soil in grassland; rare.
50. Recorded only from Glan-yr-Afon, Marli (SH 97) in 1981.

Pseudoomphalina pachyphylla (Fr.) Knudsen
On soil under conifers; rare.
50. Recorded only from Coed Coch (SH 87) in 1880.

Family Hydnangiaceae

Laccaria amethystina Cooke *Amethyst Deceiver*
Laccaria amethystea auct.
Mycorrhizal with beech and oak, characteristic of ancient woodland; woodland.
50, 51. SH 76, 77, 85, 95, 97; SJ 05–07, 15–18, 24–26, 34–36. AV, BO, BW, CE, CF, CN, ER, GW, LH, LO, MA, MF, ML, MO, NM, PP, WW. (47) 1910. (48, 49, 52)

Laccaria bicolor (R. Maire) P.D. Orton *Bicoloured Deceiver*
Mycorrhizal with broad-leaves; common.
50, 51. SH 95; SJ 13, 16, 17, 24, 26. AV, CE, LH, LO, WW. (10) 1977. (48, 49, 52)

Laccaria laccata (Scop.) Cooke *Deceiver*
Mycorrhizal with numerous tree species in all kinds of ecosystems; very common.
50, 51. SH 77, 85, 87, 95, 97; SJ 04–08, 13, 15–18, 23–27, 34–36, 43. AV, BO, BP, CC, CE, CF, CN, ER, GW, HC, LH, LO, MA, MF, ML, MO, NF, NM, PA, PP, WW. (87) 1880. (48, 49, 52)

Laccaria proxima (Boud.) Pat. *Scurfy Deceiver*
In wet grassland, often with *Sphagnum*; uncommon.
50, 51. SH 95, 97; SJ 05, 15–17, 25, 34, 35. BO, CF, ER, LH, MF, ML, NM. (11) 1974. (48, 49, 52)

Laccaria purpureobadia D.A. Reid
Under alder in wet woodland; rare.
50, 51. SJ 17, 26. CE, LH. (3) 1994.

Laccaria tortilis (Bolton) Cooke *Twisted Deceiver*
On soil under willow and alder; uncommon.
50, 51. SJ 05, 17. CF. (2) 1993. (48, 49, 52)

Family Hygrophoraceae

Ampullocybe clavipes (Pers.) Redhead, Lutzoni, Moncalvo & Vilgary *Club Foot*
Clitocybe clavipes (Pers.) P. Kumm.
In woodland litter; common.
50, 51. SH 87, 98; SJ 07, 16, 17, 23–26, 34, 36, 43. AV, CC, CE, ER, GW, LH, LO, MF, ML, MO, PP, WW. (21) 1880. (48, 49, 52)

Arrhenia acerosa (Fr.) Kühn. *Moss Oysterling*
Pleurotellus acerosus (Fr.) Konrad & Maubl.; *Omphalina acerosa* (Fr.) M. Lange
On mossy wood; uncommon.
50, 51. SH 85; SJ 08, 25. NM. (3) 1924. (48, 49, 52)

Arrhenia obscurata (D.A. Reid) Redhead, Lutzoni, Moncalvo & Vilgalys
Omphalina obscurata D.A. Reid
In short turf on sand; rare.
50. Recorded only from Marford Quarry Nature Reserve (SJ 35) in 1998.

Arrhenia rickenii (Hora) Watling
On mossy soil; rare.
51. Recorded only from the Rhydymwyn Nature Reserve (SJ 26) in 2007. New to North Wales.

Arrhenia rustica (Fr.) Redhead, Lutzoni, Moncalvo & Vilgalys
Omphalina rustica (Fr.) Quél.
Among mosses and lichens on sandy soil; rare.
50. SH 87; SJ 24. CC. (2) 1880.

Species Accounts

Arrhenia spathulata (Fr.) Redhead
Leptoglossum muscigenum (Bull.) P. Karst.
In mossy, damp grassland; uncommon.
50. Recorded only from Glan-yr-Afon, Marli (SH 97) in 1999. (48, 49, 52)

Arrhenia sphagnicola (Berk.) Redhead, Lutzoni, Moncalvo & Vilgary
Omphalina sphagnicola (Berk.) P. Karst.; *O. gerardiana* (Peck) Sing.
In *Sphagnum* in boggy areas of heathland; uncommon.
51. SJ 25. HM, NM. (2) 1978.

Cantharellula umbonata (J.F. Gmel.) Sing. *The Humpback*
In grassy places in woods; uncommon.
51. SJ 17 (2) 1991. (48, 49)

Chrysomphalina grossula (Pers.) Norvell, Redhead & Ammirati
Omphalina wynniae (Berk. & Br.) S. Ito
On decayed conifer wood; rare.
50. Recorded only from Coed Coch (SH 87) in 1880.

Cuphophyllus berkeleyi (P.D. Orton) Bon *Pale Waxcap*
Hygrocybe berkeleyi (P.D. Orton) P.D. Orton; *Hygrocybe pratensis* var. *pallida* (Cooke) Arnolds
In unimproved grassland; frequent.
50, 51. SH 97; SJ 08, 26. BO, BP, LO, MA. (4) 1978. (48, 49, 52)

Cuphophyllus colemannianus (A. Bloxam) Bon *Toasted Waxcap*
Hygrocybe colemanniana (A. Bloxam) P.D. Orton & Watling
In unimproved grassland; uncommon.
50, 51. SJ 06, 08, 15, 18, 25–27, 35. BP, MA, PA. (13) 1973. (48, 49, 52)

Cuphophyllus flavipes (Britzelm.) Bon
Hygrocybe flavipes (Britzelm.) Arnolds
In unimproved grassland; uncommon.
50, 51. SJ 17, 23. HC. (2) 2003. (48, 49, 52)

Cuphophyllus fornicatus (Fr.) Lodge, Padam. & Vizzini *Earthy Waxcap*
Hygrocybe fornicata (Fr.) Singer
In unimproved grassland; uncommon.
50, 51. SJ 04, 08, 15–17, 23, 25. HC, MA. (7) 2003. (48, 49, 52)

Cuphophyllus lacmus (Schumach.) Bon *Grey Waxcap*
Hygrocybe lacmus (Schumach.) P.D. Orton & Watling
In unimproved grassland; rare.
50. SH 85, 87. CC. (2) 1880. (48, 49)

Cuphophyllus pratensis (Pers.) Bon *Meadow Waxcap*
Hygrocybe pratensis (Pers.) Murrill
In unimproved grassland; common.
50, 51. SH 85, 87, 97; SJ 04, 05, 08, 15–17, 23–27, 34–36. AV, BO, BP, CC, CE, ER, GW, HC, HM, LO, MA, ML, PP. (45) 1880. (48, 49, 52)

Cuphophyllus russocoriaceus (Berk. & T.K. Mill.) Bon *Cedarwood Waxcap*
Hygrocybe russocoriacea (Berk. & T.K. Mill.) P.D. Orton & Watling
In unimproved grassland; frequent.
50, 51. SH 97; SJ 04, 08, 15, 17, 23, 25, 26. AV, HC, MA, ML. (10) 1980. (48, 49, 52)

Cuphophyllus virgineus (Wulf.) Kovalenko var. **virgineus** *Snowy Waxcap*
Hygrocybe virginea (Wulf.) P.D. Orton & Watling; incl. *Hygrocybe nivea* (Scop.) Murrill
In unimproved grassland; common.
50, 51. SH 85, 87, 95, 97; SJ 04–08, 15–18, 23–27, 34–36. AV, BO, BP, CC, CE, CF, ER, GW, HC, LO, MA, MF, ML, MO, PP, WW. (56) 1880. (48, 49, 52)

Cuphophyllus virgineus var. **fuscescens** (Bres.) Bon
Hygrocybe virginea var. *fuscescens* (Bres.) Arnolds
In unimproved grassland; frequent.
50, 51. SJ 16, 17, 24, 25, 35. HC, LO, MF, PP. (7) 1999. (49, 52)

Cuphophyllus virgineus var. **ochraceopallidus** (P.D. Orton) Bon
Hygrocybe virginea var. *ochraceopallida* (P.D. Orton) Boertm.
In unimproved grassland; frequent.
50, 51. SJ 15–17, 23, 25, 26. AV, HC, MA. (8) (48, 49, 52)

Dictyonema interruptum (Hook.) Parmasto
Forming a lichen association with terrestrial woodland mosses; rare.
50. Recorded only from Coed Coch (SH 87) in 1866.

Gliophorus laetus (Pers.) Herink *Heath Waxcap*
Hygrocybe laeta (Pers.) P. Kumm.
In grassland; frequent.
50, 51. SH 85; SJ 06, 15–17, 24–26, 34. HC, LO, MA, PP. (12) 1910. (48, 49, 52)

Gliophorus perplexus (A.H. Sm. & Hesler) Kovalenko
Hygrocybe psittacina var. *perplexa* (A.H. Sm. & Hesler) Boertm.;
H. perplexa (A.H. Sm. & Hesler) Arnolds; *H. sciophana* auct.
In limestone grassland; uncommon.
50, 51. SH 87; SJ 26. CC, LO. (2) 1880. (48, 49, 52)

Gliophorus psittacinus (Schaeff.) Herink *Parrot Waxcap*
Hygrocybe psittacina (Schaeff.) P. Kumm.
In unimproved grassland, especially on basic soils; common.
50, 51. SH 85, 87, 97; SJ 04, 06–08, 15–18, 23–27, 34–36. AV, BO, BP, CC, CE, ER, GW, HC, LO, MA, MF, ML, PA, PP. (52) 1880. (48, 49, 52)

Gliophorus unguinosus (Pers.) Kovalenko *Slimy Waxcap*
Hygrocybe irrigata (Pers.) Bon; *Hygrocybe unguinosa* (Fr.) P. Karst.
In unimproved grassland; uncommon.
50, 51. SH 85, 97; SJ 16, 17, 24–26, 34. BO, CE, ER, HC. (12) 1910. (48, 49, 52)

Glioxanthomyces vitellinus (Fr.) Lodge, Vizzini, Erole & Boertm.
Hygrocybe vitellina (Fr.) P. Karst.
In unimproved grassland; frequent.
50, 51. SH 97; SJ 07, 15, 16, 26, 34, 35. CE, ER, MA, MF, ML. (11) 1973. (48, 49, 52)

Species Accounts

Hygrocybe acutoconica (Clem.) Sing. *Persistent Waxcap*
Hygrocybe persistens (Britz.) Sing.; *H. langei* Kühn.
In limestone grassland; frequent.
50, 51. SH 97; SJ 08, 15, 17, 24–26. AV, BP, HC, LO, MA, ML, PP. (10) 1978. (48, 49, 52)

Hygrocybe aurantiosplendens R. Haller Aar. *Orange Waxcap*
In unimproved grassland; uncommon.
50, 51. SJ 04, 08, 17, 23, 25, 26. HC, MA. (7) 1980. (48, 49)

Hygrocybe cantharellus (Schwein.) Murrill *Goblet Waxcap*
Hygrocybe lepida Arnolds
On wet peat in plantations and moorland; frequent.
50, 51. SH 97; SJ 16, 17, 25, 34. ER, HC, MF, NM. (5) 1981. (48, 49)

Hygrocybe ceracea (Wulfen) P. Kumm. *Butter Waxcap*
In unimproved grassland; frequent.
50, 51. SH 85, 97; SJ 08, 16, 17, 23–26, 34–36. BO, ER, GW, HC, LO, ML, MO, PP. (19) 1902. (48, 49, 52)

Hygrocybe chlorophana (Fr.) Wünsche *Golden Waxcap*
Hygrocybe flavescens sensu auct.
In unimproved grassland; common.
50, 51. SH 85, 97; SJ 04–06, 13, 16, 17, 23–26, 34–36. AV, BO, CF, ER, GW, HC, HM, MA, MF, ML, MO, PP. (29) 1910. (48, 49, 52)

Hygrocybe citrinovirens (J.E. Lange) Jul. Schäff. *Citrine Waxcap*
In unimproved grassland; uncommon.
50. SH 97; SJ 24. MA, PP. (2) 1999. (48, 49, 52)

Hygrocybe coccinea (Schaeff.) P. Kumm. *Scarlet Waxcap*
In unimproved grassland; common.
50, 51. SH 85, 97; SJ 04, 05, 07, 08, 15–17, 23–26, 34, 36. AV, BO, BP, CF, ER, GW, HC, LO, MA, MF, ML, MO, NF, PP. (36) 1898. (48, 49, 52)

Hygrocybe coccineocrenata (P.D. Orton) Moser
Hygrocybe turunda auct.
In *Sphagnum* bogs; frequent.
51. SJ 17, 25. HM, LH. (2) 1977. (49)

Hygrocybe conica (Schaeff.) P. Kumm. *Blackening Waxcap*
incl. *Hygrocybe nigrescens* auct.; *H. olivaceonigra* (P.D. Orton) M.M. Moser
In grassland; this is the commonest species of the genus in most kinds of grassland.
50, 51. SH 85, 87, 95, 97; SJ 04–08, 13, 15–18, 23–27, 34–36. AV, BO, BP, CC, CE, CF, DU, ER, GW, HC, LO, MA, MF, ML, MO, NM, PA, PP. (72) 1880. (48, 49, 52) This is now known to be a complex of species and the synonyms may be reinstated as distinct in future.

Hygrocybe conicoides (P.D. Orton) P.D. Orton & Watling *Dune Waxcap*
In dune grassland; frequent.
51. Recorded only from Point of Ayr (SJ 18) since 1978. (48, 49, 52)

Hygrocybe glutinipes (J.E. Lange) R. Haller Aar. *Glutinous Waxcap*
Hygrocybe citrina sensu auct.
In unimproved grassland; uncommon.
50, 51. SH 85, 97; SJ 06, 16, 23, 26, 35. AV, CE, MA, ML. (10) 1986. (48, 49, 52)

Hygrocybe insipida (J.E. Lange) M.M. Moser *Spangle Waxcap*
Hygrocybe subminutula sensu auct.
In unimproved grassland; uncommon.
50, 51. SH 85, 97; SJ 04, 06, 08, 15, 17, 23, 25, 26. AV, BO, HC, LO, ML. (14) 1979. (48, 49, 52)

Hygrocybe intermedia (Pass.) Fayod *Fibrous Waxcap*
In unimproved grassland; rare.
50, 51. SH 97; SJ 34. BO, ER. (2) 1910. (48, 49, 52)

Hygrocybe marchii (Bres.) Singer
In unimproved grassland; uncommon.
50, 51. SJ 17, 25. HC. (2) 2003. (48, 49, 52)

Hygrocybe miniata (Fr.) P. Kumm. *Vermilion Waxcap*
Hygrocybe strangulata (P.D. Orton) Svrček
In acid grassland; frequent.
50, 51. SH 85, 87; SJ 08, 15–17, 23, 25, 26, 34. AV, CC, ER, HC, LO, MF, NM. (19) 1880. (48, 49, 52)

Hygrocybe mucronella (Fr.) P. Karst. *Bitter Waxcap*
Hygrocybe reai (Maire) J.E. Lange
In unimproved grassland; uncommon.
50, 51. SJ 15, 17, 23, 26, 34. ER, HC, MA. (5) 1910. (48, 49, 52)

Hygrocybe punicea (Fr.) P. Kumm. *Crimson Waxcap*
In unimproved grassland; frequent.
50, 51. SH 85, 87, 97; SJ 04, 08, 15, 16, 23–26, 34, 35. BO, CC, ER, HM, LO, MA, ML, MO. (22) 1880. (48, 49, 52)

Hygrocybe quieta (Kühn.) Sing. *Oily Waxcap*
Hygrocybe obrussea sensu auct.
In lawns and short grassland; frequent.
50, 51. SH 85, 97; SJ 04, 07, 08, 15, 17, 23, 25, 26, 34. BO, CE, HC, ML. (15) 1973. (48, 49, 52)

Hygrocybe reidii Kühn. *Honey Waxcap*
In unimproved grassland; common.
50, 51. SH 97; SJ 15–17, 23–26, 35. BO, HC, LO, MA, ML. (13) 1973. (48, 49, 52)

Hygrocybe splendidissima (P.D. Orton) M.M. Moser *Splendid Waxcap*
In unimproved grassland; uncommon.
50, 51. SH 97; SJ 24–26. HM, MA, ML. (4) 1985. (48, 49, 52)

Hygrocybe turunda (Fr.) P. Karst.
In acid grassland and heath; rare.
50. Recorded only from Coed Coch (SH 87) in 1880. (48, 49)

Species Accounts

Hygrophorus agathosmus (Fr.) Fr. *Almond Woodwax*
Mycorrhizal with conifers; rare.
50. Recorded only from Coed Coch (SH 87) in 1880.

Hygrophorus arbustivus Fr.
Mycorrhizal with beech; rare.
51. Recorded only from Loggerheads Country Park (SJ 26) in 1985. (49)

Hygrophorus chrysodon (Batsch) Fr. *Gold Flecked Woodwax*
Mycorrhizal with beech; uncommon.
50. Recorded only from Chirk Castle (SJ 23) in 1980.

Hygrophorus discoxanthus (Fr.) Rea *Yellowing Woodwax*
Hygrophorus chrysaspis Métrod; *H. cossus* sensu auct.
Mycorrhizal with beech; frequent.
50, 51. SJ 06, 15, 16, 26, 34, 35. AV, ER, LO, MA. (11) 1910. (49, 52)

Hygrophorus eburneus (Bull.) Fr. *Ivory Woodwax*
Mycorrhizal with beech; uncommon.
50, 51. SH 97; SJ 05, 06, 15, 16, 23, 26, 34. AV, ER, LO, MA. (12) 1978. (48, 49, 52)

Hygrophorus hypothejus (Fr.) Fr. *Herald of Winter*
Mycorrhizal with conifers; frequent.
50, 51. SH 87; SJ 07, 16, 24, 36. CC, GW, MF. (6) 1880. (49, 52)

Hygrophorus nemoreus (Pers.) Fr. *Oak Woodwax*
Mycorrhizal with oak and hazel; rare.
51. Recorded only from Bishop's Wood, Prestatyn (SJ 08) in 1990.

Hygrophorus persoonii Arnolds
Hygrophorus olivaceoalbus auct.
Mycorrhizal with oak; rare.
50. Recorded only from Glan-yr-Afon, Marli (SH 97) in 1980.

Hygrophorus unicolor Gröger
Mycorrhizal with beech; rare.
50. SJ 24, 26. MA. (2) 1910.

Lichenomphalia umbellifera (L.) Redhead, Lutzoni, Moncalvo & Vilgalys
Heath Navel
Omphalina ericetorum (Pers.) H.E. Bigelow; *Phytoconis ericetorum* (Pers.) Redhead & Kuyp.
On peaty soil in woodland, heath and moorland; common.
50, 51. SH 85, 87, 97; SJ 16–18, 25, 35, 43. CC, HM, MF, NM, PA. (10) 1880. (48, 49, 52) This is a lichenised basidiomycete, associated with a *Botrydina* stage.

Lichenomphalia velutina (Quél.) Redhead, Lutzoni, Moncalvo & Vilgalys
Omphalina velutina (Quél.) Quél.
On peaty soil in moorland; uncommon.
51. Recorded only from Nercwys Mountain (SJ 25) since 1978. (48, 49)

Neohygrocybe nitrata (Pers.) Herink *Nitrous Waxcap*
Hygrocybe nitrata (Pers.) Wünsche
In unimproved grassland; uncommon.
50, 51. SJ 07, 24. (2) 2000. (48, 49, 52)

Omphaliaster asterophorus (J.E. Lange) Lamoure
Hygroaster asterophorus (J.E. Lange) Singer; *Clitocybe asterophora* (J.E. Lange) Moser
In woodland litter; rare.
50, 51. SJ 15, 26. CE. (2) 1984.

Porpolomopsis calyptriformis (Berk.) Bresinsky *Pink Waxcap*
Hygrocybe calyptriformis (Berk.) Fayod
In unimproved grassland; uncommon.
50, 51. SH 87, 97; SJ 07, 15, 16, 24–26, 34, 43. BO, CC, HM, ML, MO, PP. (14) 1910. (48, 49, 52) This species is declining disastrously throughout Continental Europe, so the British populations are of international importance. A population has been successfully introduced to a lawn in a private garden near Mold.

Family Hymenogastraceae

Galerina badipes (Fr.) Kühn.
In conifer litter; uncommon.
50. Recorded only from Coed Coch (SH 87) in 1880. (49)

Galerina calyptrata P.D. Orton
On mosses on woodland floor; common.
51. SH 97; SJ 25. BO. This is probably under-recorded as it is regularly misidentified as *G. hypnorum*.

Galerina clavata (Velen.) Kühn.
In mossy turf; frequent.
50, 51. SJ 07, 15, 26. AV, MA. (4) 1992. (48, 49) Often common in lawns and certainly under-recorded.

Galerina hypnorum (Schrank) Kühn. *Moss Bell*
On mosses, especially *Mnium hornum*; common.
50, 51. SH 85, 87, 97; SJ 05, 07, 15–18, 23–26, 34, 36, 43. AV, CC, CF, CN, DU, ER, GW, HM, LH, LO, MF, ML, MO, NF, PP, WW. (29) 1880. (48, 49, 52) Probably over-recorded for similar species of mossy cushions.

Galerina laevis (Pers.) Sing.
In mossy turf; common.
50, 51. SJ 16, 35. MA, MF. (2) 1997.

Galerina marginata (Batsch) Kühn. *Funeral Bell*
Galerina unicolor (Vahl) Sing.
On stumps; common.
50, 51. SH 97; SJ 05, 16, 25, 26, 34. AV, ER, HM, LO, ML, MF, WW. (9) 1976. (49, 52)

Galerina mniophila (Lasch) Kühn.
On mosses, especially *Mnium hornum*; uncommon.
50, 51. SH 85; SJ 15, 16, 24–26, 34. AV, CE, ER, HM, LO, MA, NF, PP. (12) 1910. (48, 49, 52)

Galerina paludosa (Fr.) Kühn. *Bog Bell*
In *Sphagnum* bogs; common.
50, 51. SH 76; SJ 17, 25. HM, LH, NF, NM. (5) 1976. (48, 49, 52)

Galerina pseudomycenopsis Pilát
Among mosses on sandy soil; uncommon.
50, 51. SJ 26, 35. AV. (2) 1997.

Galerina pumila (Pers.) Sing. *Dwarf Bell*
Galerina mycenopsis (Fr.) Kühn.
In terrestrial mosses; frequent.
50, 51. SH 97; SJ 05, 16–18, 24, 25, 35. CF, HM, MF, ML, PA, WW. (12) 1978. (48, 49)

Galerina sphagnorum (Pers.) Kühn.
In *Sphagnum*; uncommon.
51. Recorded only from Hope Mountain (SJ 25) in 1978. (48, 49)

Galerina tibiicystis (G.F. Atk.) Kühn.
In *Sphagnum*; frequent.
51. SJ 16, 17, 25. AV, HM, LH. (3) 1977. (48, 49)

Galerina vittiformis (Fr.) Sing. *Hairy Leg Bell*
In woodland mosses; common.
50, 51. SH 95; SJ 05, 17, 24–26. AV, HM. (6) 1978. (48, 49, 52)

Gymnopilus fulgens (J. Favre & Maire) Singer
On sandy soil among lichens and mosses; rare.
50. Recorded only from Marford Quarry Nature Reserve (SJ 35) in 2000.

Gymnopilus hybridus (Sow.) Maire
On rotten wood on the ground, more frequently on hardwoods but also on conifer.
50, 51. SJ 15, 16, 26, 36. CE, GW, LO. (4) 1986. May be confused with *G. penetrans* but lacks its spotted gills.

Gymnopilus junonius (Fr.) P.D. Orton *Spectacular Rustgill*
Gymnopilus spectabilis auct.
On dead and dying trees and stumps; common.
50, 51. SH 76, 87, 96, 97; SJ 05, 16–18, 24, 26, 34–36. AV, BO, CC, CE, ER, GW, LO, MF, ML, MO, PP, WW. (20) 1880. (48, 49, 52)

Gymnopilus penetrans (Fr.) Murrill *Common Rustgill*
On small branches on the ground or just buried, especially of conifers; common.
50, 51. SH 76, 96, 97; SJ 05, 07, 13, 15–17, 23–27, 34, 36, 43. AV, CE, CF, CN, ER, GW, LH, LO, MA, MF, ML, MO, NF, NM, PP, WW. (32) 1972. (48, 49, 52)

Gymnopilus sapineus (Fr.) Maire *Scaly Rustgill*
On rotten wood of conifers; uncommon.
50, 51. SH 85, 97; SJ 26, 34. ER, ML, MO, WW. (5) 1924. Probably under-recorded because of confusion with *G. hybridus* and *G. penetrans*.

Hebeloma birrus (Fr.) Sacc.
incl. *Hebeloma anthracophilum* R. Maire; *H. pumilum* J.E. Lange
On bonfire sites and with birch; frequent.
50, 51. SJ 08, 17, 26, 34. DU, ER, LO, MO. (6) 1973. (49, 52)

Hebeloma crustuliniforme (Bull.) Quél. *Poisonpie*
Mycorrhizal with a range of broad-leaved trees and shrubs, in woods and gardens; common. **Poisonous.**
50, 51. SH 95, 97; SJ 05, 07, 08, 15–18, 24–27, 34–36. AV, BO, CE, ER, GW, HC, LH, LO, MF, MO, NF, PA, WW. (42) 1910. (48, 49, 52)

Hebeloma helodes J. Favre
In *Sphagnum* in mixed woodland; uncommon.
50, 51. SJ 17, 34. ER, LH. (2) 1974. (49)

Hebeloma hiemale Bres.
not *Hebeloma fragilipes* Romagn. *fide* Vesterholt
Mycorrhizal with broad-leaves in open woodland; frequent.
51. SJ 17, 26. AV. (3) 1986. (48)

Hebeloma laterinum (Batsch) Vesterh.
Hebeloma sinuosum auct. *H. edurum* Bon; *H. senescens* Sacc.
Mycorrhizal with deciduous trees, especially beech; frequent.
50, 51. SH 76, 97; SJ 05, 06, 08, 16, 17, 23, 24, 26, 34, 36. AV, CF, ER, GW, LO, MO, WW. (15) 1880. (49, 52)

Hebeloma mesophaeum (Pers.) Quél. *Veiled Poisonpie*
Hebeloma strophosum (Fr.) Sacc.
Mycorrhizal with birch and willow; common.
50, 51. SH 87, 95, 97; SJ 05–08, 15, 16, 18, 24–27, 34–36, 43. AV, BW, CC, CE, ER, LO, MA, MF, MO, NF, NM, PP, WW. (43) 1880. (48, 49, 52)

Hebeloma pusillum J.E. Lange
Mycorrhizal with willows; frequent.
50, 51. SH 95; SJ 07, 08, 17, 26, 35. MA. (8) 1997. (48, 49, 52)

Hebeloma radicosum (Bull.) Ricken *Rooting Poisonpie*
On soil, over the remains of small mammals; rare.
50. SH 76, 87. CC. (2) 1880. (49, 52)

Hebeloma sacchariolens Quél. *Sweet Poisonpie*
Hebeloma pallidoluctuosum Gröger & Zschiesch
In damp, broad-leaved woodland, usually on acid soils; common.
50, 51. SH 97; SJ 05, 07, 08, 15–18, 24, 26, 34, 35. AV, CE, DU, LO, PP, WW. (19) 1977. (48, 49, 52)

Species Accounts

Hebeloma sinapizans (Paulet) Gill *Bitter Poisonpie*
In deciduous woods; common.
50, 51. SH 97; SJ 05, 07, 08, 15–18, 25, 26, 34, 35. AV, ER, MA, ML, NM, WW. (17) 1979. (49, 52)

Hebeloma sordidum Maire
Hebeloma mesophaeum var. *crassipes* Vesterh.
Mycorrhizal with conifers; uncommon.
50. SH 85, 87; SJ 24. CC. (3) 1867. The last record from Britain was the 1924 report from Coed Hafod near Betws-y-Coed, during the British Mycological Society foray.

Hebeloma theobrominum Quadr.
Hebeloma truncatum auct.
Mycorrhizal with birch; uncommon.
50, 51. SJ 26, 35. CE. (2) 1986. (52)

Hebeloma velutipes Bruchet
Hebeloma leucosarx P.D. Orton; *H. longicaudum* auct.
Mycorrhizal with birch; frequent.
50, 51. SH 76, 87; SJ 05, 07, 08, 15–17, 25, 26, 34–36. AV, BO, BP, CC, ER, LH, LO, MF, MO, NF. (21) 1880. (48, 49, 52)

Hymenogaster hessei Soehner
In soil under beech, a 'false truffle'; rare.
51. Recorded only from Loggerheads Country Park (SJ 26) in 1993. (49)

Hymenogaster olivaceus Vittad.
In soil under beech; uncommon.
50, 51. SJ 26. LO, MA. (2) 1993.

Hymenogaster sulcatus R. Hesse
In soil under beech; rare.
50. Recorded only from Nantglyn (SJ 06) in 1880.

Hymenogaster tener Berk. & Broome
In soil under beech; uncommon.
51. Recorded only from Coed y Felin Nature Reserve (SJ 16) in 1993. (49)

Hymenogaster vulgaris Tul. & C. Tul.
In soil under beech; uncommon.
50, 51. SJ 26. LO, MA, WW. (3) 1990. (49)

Naucoria amaescens Quél.
In damp soil under willows; rare.
51. Recorded only from Rhydymwyn Nature Reserve (SJ 26) in 2007.

Naucoria bohemica Velen.
Mycorrhizal with alder; uncommon.
51. Recorded only from Cilcain in the Alyn Valley (SJ 16) in 1990. (48, 49, 52)

Naucoria escharioides (Fr.) P. Kumm. *Ochre Aldercap*
Mycorrhizal with alder; common.
50, 51. SH 85; SJ 08, 17, 25, 26, 24. CE, ER, LH, NF, WW. (13) 1910. (48, 49, 52)

Naucoria scolecina (Fr.) Quél.
Mycorrhizal with alder; frequent.
50, 51. SJ 25, 26, 34, 35. ER, WW. (4) 1981 (48, 49).

Naucoria striatula P.D. Orton *Striate Aldercap*
Mycorrhizal with alder; frequent.
51. Recorded only from Celyn Woods (SJ 26) in 1986. (48, 49)

Phaeocollybia lugubris (Fr.) R. Heim *Russet Rootshank*
In conifer litter; rare.
50. Reported only from Coed Coch (SH 87) in 1880; now confined to Scotland.

Family Inocybaceae

Inocybe adaequata (Britz.) Sacc.
Inocybe jurana (Pat.) Sacc.
Mycorrhizal with beech; frequent.
50, 51. SJ 15, 16, 26. AV, LO. (4) 1977. (52)

Inocybe aghardii (N. Lund) P.D. Orton
Mycorrhizal with willows in dune slacks; frequent.
51. Recorded only from Point of Ayr (SJ 18) since 1985. (48, 49)

Inocybe assimilata Britz.
Inocybe umbrina Bres.
Mycorrhizal with broad-leaves; frequent.
50, 51. SH 95, 97; SJ 08, 13, 16. AV, BP, ML. (5) 1979. (48, 49)

Inocybe asterospora Quél. *Star Fibrecap*
Mycorrhizal with oak and other trees; frequent.
50, 51. SH 85, 97; SJ 16, 24, 26, 34. AV, BO, CE, MF, ML. (9) 1910. (48, 49, 52)

Inocybe bongardii (Weinm.) Quél. *Fruity Fibrecap*
Mycorrhizal with beech; common.
50, 51. SH 87, 95; SJ 05, 08, 16, 17, 24, 26, 34. AV, CC, ER, LO. (10) 1880. (48, 52)

Inocybe calamistrata (Fr.) Gillet *Greenfoot Fibrecap*
Mycorrhizal with conifers; frequent.
50. SH 85, 97. ML. (3) 1979. (48, 49, 52)

Inocybe calospora Quél.
Mycorrhizal with birch; uncommon.
50, 51. SJ 16, 26. AV, LO. (2) 1992. (49, 52)

Inocybe cervicolor (Pers.) Quél.
Mycorrhizal with conifers; uncommon.
51. Recorded only from Cefn (SJ 07) in 1980.

Species Accounts

Inocybe cincinnata (Fr.) Quél. var. **cincinnata** *Collared Fibrecap*
Inocybe phaeocomis (Pers.) Kuyper
Mycorrhizal with beech and oak; common.
50, 51. SH 85, 87, 97; SJ 26. BO, CC, LO, MA. (6) 1880. (52)

Inocybe cincinnata var. **major** (S. Petersen) Kuyper
I. obscura auct.
Mycorrhizal with beech; uncommon.
50, 51. SJ 05, 07, 08, 24, 26. BP, WW. (5) 1990.

Inocybe cookei Bres. *Straw Fibrecap*
Mycorrhizal with conifers and broad-leaved trees; common.
50, 51. SH 87; SJ 07, 16, 26, 34. AV, CC, CE, ER, LO, MF. (8) 1880. (48, 49, 52)

Inocybe corydalina Quél. *Greenflash Fibrecap*
Mycorrhizal with hazel; uncommon.
50, 51. SH 97; SJ 26, 34. AV, CE, ER, ML. (4) 1979. (52)

Inocybe dulcamara P. Kumm.
Mycorrhizal with birch in damp woodland and willows in dunes, also on sandy heaths and old industrial sites; common.
50, 51. SJ 07, 16–18, 26, 35. AV, BW, PA. (7) 1977. (48, 49, 52)

Inocybe dunensis P.D. Orton
Mycorrhizal with willows in dunes; frequent.
51. SJ 08, 18. PA. (2) 1978. (48, 52)

Inocybe erubescens A. Blytt *Deadly Fibrecap*
Inocybe patouillardii Bres.
Mycorrhizal with beech on limestone; common. **Poisonous.**
50, 51. SJ 16, 23, 26. AV, LO. (3) 1973. (49)

Inocybe fibrosoides Kühn.
Mycorrhizal with deciduous trees; rare.
50. Recorded only from Erddig (SJ 34) in 1985. Material was checked at the Herbarium of the Edinburgh Botanic Garden.

Inocybe flocculosa Sacc. *Fleecy Fibrecap*
Inocybe gausapata Kühn.
Mycorrhizal with broad-leaves and conifers; uncommon.
50, 51. SH 87; SJ 07, 16, 24, 26, 35. AV, CC, LP, MA, MF, MO, PP. (12) 1880. (48, 49, 52)

Inocybe fraudans (Britz.) Sacc. *Pear Fibrecap*
Inocybe pyriodora auct.
Mycorrhizal with beech; common.
50, 51. SH 95, 97; SJ 06, 07, 16, 26, 34. AV, CC, CE, ER. (8) 1880. (48, 49, 52)

Inocybe fuscidula Velen.
Inocybe brunneo-atra (R. Heim) P.D. Orton
Mycorrhizal with beech; uncommon.
51. Recorded only from Hawarden Woods (SJ 36) in 1972. (48, 49, 52)

Inocybe geophylla (Fr.) P. Kumm. *White Fibrecap*
Mycorrhizal with broad-leaves; common.
50, 51. SH 76, 85, 87, 97; SJ 05, 06, 08, 15–18, 23–26, 34–36. AV, BO, CC, CE, ER, GW, LO, MA, ML, MO, NF, PP, WW. (38) 1880. (48, 49, 52)

Inocybe glabripes Rick.
Inocybe microspora J.E. Lange
Mycorrhizal with oak in damp woodland; uncommon.
50, 51. SJ 05, 17, 26, 34. ER, LH, WW. (4) 1977. (49, 52)

Inocybe godeyi Gillet
Mycorrhizal with beech; uncommon.
50. Recorded only from Lady Bagot's Drive, Rhewl (SJ 15) in 2001. (49, 52)

Inocybe grammata Quél.
Mycorrhizal with deciduous trees; rare.
50. Recorded only from Llyn Syberi (SH 76) in 1980. (49)

Inocybe griseolilacina J.E. Lange *Lilac Leg Fibrecap*
Mycorrhizal with broad-leaves; frequent.
50, 51. SH 97; SJ 07, 08, 16, 17, 26, 35. AV, BO, BW, LO, ML. (11) 1974. (49, 52)

Inocybe haemacta (Berk. & Cooke) Sacc.
Mycorrhizal with hazel; uncommon.
51. Recorded only from Ddol Uchaf Nature Reserve (SJ 17) in 1985. (49, 52)

Inocybe hirtella Bres.
Mycorrhizal with beech and hazel; frequent.
50, 51. SJ 08, 24, 26. BP, CE. (3) 1984. (52)

Inocybe hystrix (Fr.) P. Karst. *Scaly Fibrecap*
Mycorrhizal with beech; frequent.
50, 51. SH 85; SJ 05, 24, 26. CF, WW. (4) 1979. (48, 49)

Inocybe lacera (Fr.) P. Kumm. *Torn Fibrecap*
Mycorrhizal with conifers; common.
50, 51. SJ 05, 07, 16, 18, 24–26, 34, 35. AV, CE, CF, ER, LO, MF, NM, PA. (15) 1977. (48, 49, 52)

Inocybe lanuginosa (Bull.) P. Kumm. *Woolly Fibrecap*
Inocybe longicystis G.F. Atk.
Mycorrhizal with conifers; common.
50, 51. SH 97; SJ 16, 25, 26. AV, LO, ML, NF. (5) 1978. (48, 49, 52)

Inocybe lilacina (Peck) Kauffman *Lilac Fibrecap*
Inocybe geophylla var. lilacina (Peck) Gillet
Mycorrhizal with broad-leaves; common.
50, 51. SH 97; SJ 05–07, 15–17, 23–26. AV, BO, BW, CE, ER, LO, MA, ML, MO, NM. (28) 1973. (48, 49, 52)

Inocybe maculata Boud. *Frosty Fibrecap*
Mycorrhizal with beech and oak; common.
50, 51. SJ 07, 16, 17, 25, 26, 34, 36. AV, BO, ER, GW, LO, MF, MO, NF, WW. (11) 1975. (48, 49, 52)

Inocybe margaritispora (Cooke) Sacc.
Mycorrhizal with beech; rare.
50, 51. SH 97; SJ 26. AV, ML. (2) 1979. (49, 52)

Inocybe napipes J.E. Lange *Bulbous Fibrecap*
Mycorrhizal with birch and oak; common.
50, 51. SH 76; SJ 05, 16, 17, 26. AV, CE, WW. (6) 1976. (48, 49, 52)

Inocybe nitidiuscula (Britz.) Sacc.
Mycorrhizal with deciduous trees; rare.
50, 51. SJ 05, 26. AV, CF. (2) 2001. (52)

Inocybe oblectabilis (Britz.) Sacc.
Mycorrhizal with beech; uncommon.
51. Recorded only from Loggerheads Country Park (SJ 26) in 1985.

Inocybe petiginosa (Fr.) Gillet *Scurfy Fibrecap*
Mycorrhizal with beech; common.
50, 51. SH 85; SJ 05, 08, 16, 26, 34, 36. AV, CE, ER, GW, LO. (9) 1924. (48, 49, 52)

Inocybe posterula (Britz.) Sacc.
Mycorrhizal with conifers; uncommon.
51. Recorded only from Loggerheads Country Park (SJ 26) in 1972.

Inocybe praetervisa Quél.
Mycorrhizal with beech; uncommon.
50, 51. SH 97; SJ 26. AV, CE, ML. (3) 1979. (49, 52)

Inocybe proximella P. Karst.
Mycorrhizal with oak; uncommon.
50, 51. SH 85; SJ 26. WW. (2) 1924.

Inocybe pruinosa R. Heim
Mycorrhizal with willows in dunes; frequent.
51. Recorded only from Point of Ayr (SJ 18) in 1985. (52)

Inocybe pusio P. Karst.
Mycorrhizal with birch; uncommon.
51. Recorded only from Etna Country Park, Buckley (SJ 26) in 2002.

Inocybe rimosa (Bull.) P. Kumm. *Split Fibrecap*
Inocybe fastigiata (Schaeff.) Quél.
Mycorrhizal with broad-leaves; common.
50, 51. SH 85, 96; SJ 05–08, 16, 17, 23–26, 34, 36. AV, CN, ER, GW, LO, MA, MO, NF, PP. (25) 1910. (48, 49, 52)

Inocybe sindonia (Fr.) P. Karst.
Inocybe eutheles auctt.; *I. kuehneri* Stangl & Vesel.
Mycorrhizal with beech and oak; common.
50, 51. SH 87, 97; SJ 07, 15, 16, 24, 26, 34. AV, BO, CC, CE, ER, LO, MA, MF, ML. (15) 1880. (48, 49, 52)

Inocybe squamata J.E. Lange
Mycorrhizal with birch and oak; rare.
50, 51. SJ 34–36. ER, GW. (4) 1981. (49)

Inocybe subcarpta Kühn. & Bours.
Inocybe boltonii auct.
Mycorrhizal with conifers; frequent.
50. Recorded only from Coed Clwyd on Moel Famau (SJ 16) in 1977.

Inocybe whitei (Berk. & Br.) Sacc. *Blushing Fibrecap*
Inocybe pudica Kühn.
Mycorrhizal with conifers; uncommon.
50, 51. SH 76, 97; SJ 26. AV, LO, ML. (4) 1978. (49, 52)

Family Lycoperdaceae

Apioperdon pyriforme (Schaeff.) Vizzini *Stump Puffball*
Lycoperdon pyriforme Schaeff.
On stumps, logs and buried wood including roots; common – always on wood.
50, 51. SH 76, 85, 87, 97; SJ 05–08, 15–18, 23–27, 34–36, 43. AV, BO, CC, CE, DU, ER, GW, HC, LH, LO, MA, MF, ML, MO, NF, PA, PP, WW. (106) 1880. (48, 49, 52) **RDL/LC**

Bovista aestivalis (Bonord.) Demoulin *'Deceiving Bovist'*
In dune grassland; frequent.
51. Recorded only from dunes at Gronant (SJ 08) in 1990. (49, 52) **RDL/VU**

Bovista nigrescens Pers. *Brown Puffball*
In sheep pastures and other grasslands; common.
50, 51. SH 97; SJ 04, 06–08, 15–18, 23–26, 34–36, 43. AV, BO, ER, GW, HC, LO, ML, MO, NM, PA. (42) 1910. (48, 49, 52) **RDL/LC**

Bovista plumbea Pers. *Grey Puffball*
In turf in dunes, grassland and waste land; common.
50, 51. SH 88, 95, 97; SJ 06–08, 15, 17, 18, 24–27, 34, 35, 43. ER, HC, MA, MO, PA. (21) 1860. (48, 49, 52) **RDL/LC**

Calvatia gigantea (Batsch) Lloyd *Giant Puffball*
Langermannia gigantea (Bastch) Rostk.
In tall vegetation, especially among nettles, in well-manured ground, including woodland edges and gardens; common. Specimens over one metre in circumference have been measured in Flintshire.
50, 51. SH 95, 97; SJ 04, 06–08, 15–18, 23–27, 34–36, 43, 45, 54. BO, BW, ER, GW, LO, MA, ML, MO. (49) 1932. (48, 49, 52) **RDL/LC**

Species Accounts

Lycoperdon dermoxanthum Vittad. *Dwarf Puffball*
Bovista dermoxantha (Vittad.) de Toni; *Bovista pusilla* auct.
In sand dunes; common.
51. SJ 18, 27. PA. (2) 1978. (48, 52) **RDL/DD**

Lycoperdon echinatum Pers. *Spiny Puffball*
In beech litter on limestone; uncommon.
50, 51. SJ 05–07, 15, 16, 26, 34. ER, MA. (8) 1984. (49, 52) **RDL/LC**

Lycoperdon excipuliforme (Scop.) Pers. *Pestle Puffball*
Calvatia excipuliformis (Scop.) Perdeck; *Handkea excipuliformis* (Scop.) Kreisel
In open woodland and hedges; common.
50, 51. SH 85, 87, 95, 97; SJ 05–08, 13, 15–18, 24–27, 34–36, 43. AV, BP, CC, CE, ER, ML, MO, NF, NM, PA, PP, WW. (39) 1880. (48, 49, 52) **RDL/LC**

Lycoperdon lividum Pers. *Grassland Puffball*
Lycoperdon spadiceum Pers.
In sandy and calcareous grassland, especially in dunes; frequent.
50, 51. SH 97; SJ 06–08, 13, 15, 17, 18, 25–27, 35. BP, LO, HC, MO, PA. (26) 1953. (48, 49, 52) **RDL/LC**

Lycoperdon mammiforme Pers. *'Flaky Puffball'*
In beech litter on limestone; rare.
50, 51. SH 97; SJ 16, 26. AV, LO, MA, ML. (5) 1979. (49) **RDL/VU**

Lycoperdon molle Pers. *Soft Puffball*
In woodland litter; uncommon.
50, 51. SJ 08, 16, 17, 25, 26, 36. AV, BO, CN, GW, LO, MA, NF, WW. (12) 1953. (49, 52) **RDL/LC**

Lycoperdon nigrescens Pers. *Dusky Puffball*
Lycoperdon foetidum Bonord.
In woodland litter on acid soils; common.
50, 51. SH 87, 95, 97; SJ 05–08, 15–17, 24–26, 34–36. AV, BO, BP, CC, CE, CF, ER, GW, LH, LO, MF, NL, NM, PP, WW. (35) 1880. (48, 49, 52) **RDL/LC**

Lycoperdon perlatum Pers. *Common Puffball*
In woodland litter; common.
50, 51. SH 85, 87, 95, 97; SJ 05–08, 15–18, 23–27, 34–36, 43. AV, BO, BW, CC, CE, DU, ER, GW, HC, LH, LO, MA, MF, ML, MO, NF, NM, PA, PP, WW. (71) 1880. (48, 49, 52) **RDL/LC**

Lycoperdon pratense Pers. *Meadow Puffball*
Vascellum pratense (Pers.) Kreisel
In lawns and short turf, especially on basic soils; common.
50, 51. SH 77, 97; SJ 06–08, 15–17, 18, 23–27, 34–36, 43. AV, BP, ER, GW, HC, LO, MA, MF, ML, MO, NF, PA. (47) 1910. (48, 49, 52) **RDL/LC**

Lycoperdon umbrinum Pers. *Umber-brown Puffball*
On acid soil under conifers; uncommon,
50, 51. SJ 24, 26. CE. (2) 1910. (49) **RDL/VU**

Lycoperdon utriforme Bull. *Mosaic Puffball*
Calvatia utriformis (Bull.) Jaap; *Lycoperdon caelatum* Bull.; *Handkea utriformis* (Bull.) Kreisel
In meadows; common.
50, 51. SH 87, 97; SJ 06, 08, 13, 15-17, 25-27, 34, 36, 54. AV, BO, CC, GW, HC, ML, MO, NF. (22) 1880. (48, 49, 52) **RDL/LC**

Family Lyophyllaceae

Asterophora lycopodioides (Bull.) Ditmar *Powdery Piggyback*
Asterophora agaricoides Fr.
On decaying fruit bodies of *Russula nigricans*; uncommon.
51. Recorded only from Y Graig Nature Reserve, Tremeirchion (SJ 07) in 2001. (49, 52)

Asterophora parasitica (Bull.) Sing. *Silky Piggyback*
Nyctalis parasitica (Bull.) Fr.
On decaying fruit bodies of *Russula nigricans*; uncommon.
50, 51. SH 76, 87, 97; SJ 05, 07, 34. CC, ER, ML. (8) 1902. (48, 49, 52)

Calocybe carnea (Bull.) Donk *Pink Domecap*
Rugosomyces carneus (Bull.) M. Bon
In lawns and dune turf; uncommon.
50, 51. SH 97; SJ 07, 15, 17, 18, 23, 25, 26, 34, 35. AV, ER, ML, PA. (18) 1979. (48, 49, 52)

Calocybe gambosa (Fr.) Donk *St George's Mushroom*
Tricholoma gambosum (Fr.) P. Kumm.
On roadsides and edges of woods and fields; common.
50, 51. SH 96, 97; SJ 06-08, 15-18, 24-27, 34-36. AV, BO, BW, DU, ER, GW, LO, MA, ML, MO, PA. (55) 1910. (48, 49, 52) This species first appears around 23 April – St. George's Day.

Calocybe ionides (Bull. ex Pers.) Donk *Violet Domecap*
Rugosomyces ionides (Bull. ex Pers.) M. Bon
On calcareous sandy soil, with larches; rare.
50. Recorded only from Marford Quarry Nature Reserve (SJ 35) in 2000. (52)

Hypsizygus ulmarius (Bull.) Redhead *Elm Leech*
Lyophyllum ulmarium (Bull.) Kühn.
On fallen elm trunks; once frequent, now much rarer, with the decline of the host.
50, 51. SH 87; SJ 26, 34, 36. CC, ER, GW, WW. (4) 1880.

Leucocybe candicans (Pers.) Vizzini, P. Avarado, G. Moreno & Consiglio
Clitocybe candicans (Pers.) P. Kumm.
In woodland litter; frequent.
50, 51. SH 87; SJ 17, 26, 34. CC, CE, ER, LH, LO. (5) 1880.

Leucocybe connata (Schum.) Vizzini, Alvarado, G. Moreno & Consiglio *White Domecap*
Lyophyllum connatum (Schum.) Sing.
On road verges and at woodland margins; frequent.
50, 51. SH 95, 97; SJ 05, 08, 16, 17, 24-26, 35, 36. BP, CF, GW, LO, MF, MO, NM. (24) 1973. (48, 49)

Species Accounts

Leucocybe houghtonii (Phillips) Haluma & Pensakowski
Clitocybe houghtonii (Phillips) Dennis
In litter in beech woods on basic soils; uncommon.
50, 51. SJ 15, 26. CE, LO. (3) 1983.

Lyophyllum ambustum (Fr.) Singer
Tephrocybe ambusta (Fr.) Donk; *Collybia ambusta* (Fr.) Quél.
On bonfire sites in woodland; uncommon.
50. SH 87; SJ 24. CC, PP. (2) 1880. (49)

Lyophyllum anthracophilum (Lasch) M. Lange & Sivertsen
Tephrocybe anthracophila (Lasch) P.D. Orton
On bonfire sites in woods; frequent.
50, 51. SJ 17, 26. DU, MA, MO. (3) 1978. (49, 52)

Lyophyllum atratum (Fr.) Singer
Tephrocybe atrata (Fr.) Donk
On bonfire sites in woodland; uncommon.
50, 51. SH 85; SJ 18, 35. PA. (3) 1924. (48, 52)

Lyophyllum confusum (P.D. Orton) Gulden
Tephrocybe confusa (P.D. Orton) P.D. Orton
In conifer litter; uncommon.
50. Recorded only from Coed Clwyd, Moel Famau (SJ 16) in 1981. (49)

Lyophyllum decastes (Fr.) Sing. *Clustered Domecap*
At edges of paths in woodland; common.
50, 51. SH 87, 97; SJ 07, 15–18, 25–27, 35. AV, BO, CC, DU, ER, HC, LO, MF, NM, WW. (23) 1880. (49, 52)

Lyophyllum fumosum (Pers.) P.D. Orton
In woodland litter; uncommon.
50, 51. SH 87, 97; SJ 17, 26. CC, DU, LO. (4) 1880.

Lyophyllum inolens (Fr.) Kühner & Romagn.
Tephrocybe inolens (Fr.) M.M. Moser
In beech litter; rare.
50. Recorded only from Glan-yr-Afon, Marli (SH 97) in 1981.

Lyophyllum leucophaeatum (P. Karst.) P. Karst. *Smoky Domecap*
Lyophyllum gangraenosum (Fr.) Gulden
Under trees along roadside; frequent.
50, 51. SH 97; SJ 26, 35. AV, ML. (3) 1980. (49, 52)

Lyophyllum loricatum (Fr.) Kühn.
In woodland litter; rare.
50. Recorded only from Ceiriog Forest (SJ 13) in 1998. (49, 52)

Lyophyllum mephiticum (Fr.) Singer
Tephrocybe mephitica (Fr.) M.M. Moser
In needle litter; uncommon.
51. Recorded only from Nant-y-Ffrith (SJ 25) in 2003.

Lyophyllum putidum (Fr.) Singer
Tephrocybe putida (Fr.) M.M. Moser
In grassy woodland; uncommon.
50. Recorded only from Coed Coch (SH 87) in 1880.

Lyophyllum semitale (Fr.) Kalamees
In litter in base-rich woods; uncommon.
50, 51. SH 87; SJ 17. CC. (2) 1880.

Lyophyllum striipilea (Fr.) Kühner & Romagn
Tephrocybe striipilea (Fr.) Donk
In woodland litter; rare.
50. Recorded only from Coed Coch (SH 87) in 1870.

Ossicaulis lignatilis (Pers.) Redhead & Ginnes
Pleurotus lignatilis (Pers.) P. Kumm.; *Clitocybe lignatilis* (Pers.) P. Karst.
On rotten logs; uncommon.
51. Recorded only from Coed Nant Gain, Cilcain (SJ 16) in 1990.

Sagaranella tylicolor (Fr.) v. Hofst. et al.
Tephrocybe tylicolor (Fr.) M.M. Moser;
In woodland litter; frequent.
50, 51. SH 87; SJ 07, 16, 17, 24, 26, 35. AV, CC, LH, PP, WW. (7) 1880. (48, 52)

Sphagnurus paluster (Peck) Redhead & Hofst. *Sphagnum Greyleg*
Tephrocybe palustris (Peck) Donk; *Lyophyllum palustre* (Peck) Singer
Parasitic on *Sphagnum* in bogs; common.
50, 51. SJ 17, 25, 43. HM, LH, NM. (4) 1977. (48, 49, 52)

Tephrocybe murina (Batsch) M.M. Moser
On rotten branch; uncommon.
51. Recorded only from Hawarden (SJ 36) in 2019.

Tephrocybe rancida (Fr.) Donk *Rancid Greyleg*
Lyophyllum rancidum (Fr.) Singer
In woodland litter; frequent.
50, 51. SH 95, 97; SJ 05, 16, 24, 26, 35. AV, CF, MA, ML, MO, PP, WW. (10) 1975. (49, 52)

Family Macrocystidiaceae

Macrocystidia cucumis (Pers.) Joss. *Cucumber Cap*
On piles of small sticks and branches; frequent.
50, 51. SH 87, 97; SJ 06–08, 15, 16, 23–26, 34, 36. AV, BO, BP, CC, CE, ER, GW, LO, NF, PP, WW. (17) 1880. (48, 49, 52)

Species Accounts

Family Marasmiaceae

Crinipellis scabella (Alb. & Schwein.) Murrill *Hairy Parachute*
Crinipellis stipitaria (Fr.) Pat.
On dead grasses in turf and dunes; frequent.
50, 51. SH 97; SJ 08, 16, 18, 34, 35. BP, BW, PA. (7) 1978. (48, 49, 52)

Marasmius bulliardii Quél.
On beech leaf litter; un common.
50, 51. SJ 26. MA, WW. (2) 2001. (48) These are the first records for Wales. **PF**

Marasmius cohaerens (Pers.) Cooke & Quél.
In woodland litter and humus; frequent.
50, 51. SH 87, 97; SJ 07, 08, 15, 16, 26, 34, 36. AV, CC, CN, ER, GW, LO, MA. (11) 1880. (48, 49, 52)

Marasmius curreyi Berk. & Br.
Marasmius graminum auct.
In grass litter; frequent.
50, 51. SJ 16–18, 26, 34, 36. DU, ER, GW, LH, LO, MF, PA, WW. (10) 1972. (49, 52)

Marasmius epiphylloides (Rea) Sacc. & Trott. *'Ivy Parachute'*
On dead ivy leaves in litter; common.
50, 51. SH 97; SJ 16–18, 26, 27, 34. AV, BO, CN, ER, HC, LH, MA, WW. (14) 1977. (48, 49, 52)

Marasmius epiphyllus (Pers.) Fr. *Leaf Parachute*
In herbaceous woodland litter; common.
50, 51. SH 85, 87, 97; SJ 05–08, 15–18, 24–26, 34–36. AV, BO, BW, CC, CE, CF, CN, ER, GW, LH, LO, MA, NF, PP, WW. (38) 1880. (48, 49, 52)

Marasmius hudsonii (Pers.) Fr. *Holly Parachute*
In holly litter; frequent in the west, rare elsewhere.
50, 51. SH 87; SJ 16, 26. AV, CC, CE. (3) 1880.

Marasmius oreades (Bolt.) Fr. *Fairy Ring Champignon*
In lawns and grazed grassland, forming fairy rings; common.
50, 51. SH 77, 87, 96, 97; SJ 04–08, 15–18, 24, 26, 27, 34–36. AV, BO, BP, CC, CE, ER, GW, MA, MF, ML, MO, NM, PA, WW. (42) 1880. (48, 49, 52)

Marasmius rotula (Scop.) Fr. *Collared Parachute*
On twiggy litter in woods; common.
50, 51. SH 97; SJ 05–08, 15–17, 23, 24, 26, 34–36. AV, CE, CF, CN, DU, ER, GW, LH, LO, MA, MF, MO, WW. (40) 1910. (48, 49, 52)

Marasmius torquescens Quél.
Marasmius lupuletorum (Weinm.) Bres.
In beech litter; uncommon.
51. SJ 17, 26. CE. (2) 1984.

Marasmius wynnei Berk. & Br. *Pearly Parachute*
Marasmius globularis Fr.
In leaf litter; frequent.
50, 51. SH 87, 97; SJ 15–17, 26, 34, 36. AV, CC, CN, ER, GW, LH, LO, ML, WW. (15) 1858. (48, 49, 52) This species was named in honour of the Wynne family of Coed Coch, from where it was described.

Family Mycenaceae

Atheniella adonis (Bull.) Redhead et al. *Scarlet Bonnet*
Mycena adonis (Bull.) Gray; *M. coccinea* (Sow.) Quél.
On sticks in damp woodland; uncommon.
50, 51. SJ 17, 26, 34, 35. DU, ER, LO. (4) 1910. (48, 49, 52)

Atheniella flavoalba (Fr.) Redhead et al. *Ivory Bonnet*
Mycena flavoalba (Fr.) Quél.; *M. luteoalba* (Bolton) Gray
In unimproved grassland; common.
50, 51. SH 97; SJ 06–08, 16, 18, 23, 25, 26, 34–36. AV, BO, BP, CE, ER, GW, LO, MA, MF, ML, NF, NM, PA. (28) 1973. (48, 49, 52)

Delicatula integrella (Pers.) Fayod
On twigs; uncommon.
50, 51. SH 97; SJ 15, 17, 25, 34, 35. BO, ER, LH, NF. (6) 1910. (49, 52)

Hemimycena cucullata (Pers.) Sing.
Mycena gypsea (Fr.) Quél.
On twigs in litter; uncommon.
50, 51. SH 97; SJ 08, 17, 24–26, 35. AV, BP, NF, PP (7) 1978.

Hemimycena delectabilis (Peck) Singer
On decaying *Juncus* stems; uncommon.
50. Recorded only from Fenn's Moss (SJ 43) in 2008. (49, 52)

Hemimycena epichloe (Kühner) Singer
On dead grass; uncommon.
51. Recorded only from Rhydymwyn Nature Reserve (SJ 26) in 2007.

Hemimycena lactea (Pers.) Sing. *Milky Bonnet*
Mycena lactea (Pers.) P. Kumm.; *Hemimycena delicatella* (Peck) Sing.
On damp, fallen branches; uncommon.
50, 51. SJ 15–17, 24–26, 35. AV, LO, MF, NF. (8) 1910. (49, 52)

Hemimycena tortuosa (P.D. Orton) Redhead *Dewdrop Bonnet*
On damp, fallen branches; common.
50, 51. SH 97; SJ 05–08, 16, 17, 26. AV, BO, DU, LO, MA, ML, WW. (15) 1992. (48, 49, 52)

Mycena abramsii (Murrill) Murrill
On decaying wood; uncommon.
50. Recorded only from Nant Mill Country Park (SJ 24) in 1993. (52)

Species Accounts

Mycena acicula (Schaeff.) P. Kumm. *Orange Bonnet*
In moss on woodland floor; common.
50, 51. SH 87; SJ 05-08, 15-18, 24-27, 34-36. AV, CC, CE, CN, DU, ER, GW, LO, MA, NF, PP, WW. (33) 1880. (48, 49, 52)

Mycena adscendens (Lasch) Maas Geest. *Frosty Bonnet*
Mycena tenerrima (Berk.) Quél.
On damp sticks and bark; common.
50, 51. SH 87, 97; SJ 05, 08, 15, 17, 25, 26, 34, 35. AV, CC, CE, ER, LH, LO, ML, WW. (17) 1880. (48, 49, 52)

Mycena aetites (Fr.) Quél. *Drab Bonnet*
In leaf litter; frequent.
50, 51. SJ 07, 16, 26, 34, 35. AV, ER, MF, WW. (8) 1976. (48, 49, 52)

Mycena amicta (Fr.) Quél.
In leaf litter; frequent.
50, 51. SJ 16, 17, 25, 26, 34. AV, ER, LO, MF, NF, NM, WW. (10) 1910. (48, 49, 52)

Mycena arcangeliana Bres. *Angel's Bonnet*
Mycena oortiana Hora
On sticks and small, fallen branches; common.
50, 51. SH 95, 97; SJ 05-08, 15-17, 24-26, 34-36. AV, BO, BP, CE, CN, ER, LO, MA, ML, NF, PP, WW. (30) 1977. (48, 49, 52)

Mycena aurantiomarginata (Fr.) Quél.
In needle litter; uncommon.
50, 51. SH 97; SJ 24. BO. (2) 1910.

Mycena bulbosa (Cejp) Kühn.
On bases of *Juncus* stems, in centre of clumps; frequent.
50, 51. SH 85; SJ 36. GW. (2) 1972. (48, 49, 52)

Mycena capillaripes Peck *Pinkedge Bonnet*
Mycena rubromarginata auct.
On conifer litter; frequent.
50, 51. SJ 16, 17, 25, 26, 34, 36. ER, GW, LH, MF, MO, NF, WW. (10) 1976. (48)

Mycena capillaris (Schum.) P. Kumm. *Beechleaf Bonnet*
On fallen beech leaves; frequent.
50, 51. SH 97; SJ 06, 16-18, 26, 34, 36. BO, CN, ER, GW, LH, LO, MA, MO, WW. (11) 1910. (48, 49, 52)

Mycena chlorantha (Fr.) P. Kumm.
On sand in marram dunes; uncommon.
51. Recorded only from Point of Ayr (SJ 18) in 1979.

Mycena cinerella (P. Karst.) P. Karst. *Mealy Bonnet*
In short grassland; frequent.
50, 51. SH 95, 97; SJ 06, 08, 15, 16, 23-26, 34, 35. AV, BO, ER, LO, MF, ML, NM, WW. (18) 1972. (49, 52)

Mycena clavularis (Batsch) Sacc.
On the bark of living trees; uncommon.
50, 51. SJ 07, 16-18, 34. AV, ER, LO. (6) 1978. (48, 49, 52)

Mycena corynephora Maas Geest.
On mossy bark of living trees; uncommon.
50, 51. SH 76, 87; SJ 02, 17, 18, 22, 26. LH, WW. (7) 1993. (48, 49)

Mycena diosma Krieglst. & Schwöbel
In beech litter; uncommon.
50. SJ 26, 34. ER, MA. (2) 1999.

Mycena epipterygia (Scop.) Gray *Yellowleg Bonnet*
On decaying bracken stems; common.
50, 51. SH 85, 87, 95, 97; SJ 05, 07, 13, 15-18, 24-26, 35, 36, 43. AV, BO, CC, CE, CF, DU, GW, LH, LO, MF, ML, NF, NM, PP, WW. (39) 1880. (48, 49, 52)

Mycena epipterygioides A. Pears.
On conifer stumps; frequent.
50, 51. SJ 15-17. LH, MF. (4) 1977.

Mycena filopes (Bull.) P. Kumm. *Iodine Bonnet*
Mycena amygdalina auct.
On litter; common.
50, 51. SH 85, 97; SJ 05, 07, 08, 15-18, 23, 24, 26, 34-36. AV, BP, CE, CF, CN, DU, ER, GW, LH, LO, MA, ML, PA, PP, WW. (27) 1910. (48, 49, 52)

Mycena flavescens Velen.
In limestone grassland; uncommon.
51. Recorded only from Craig Fawr (SJ 08) in 1983. (48)

Mycena galericulata (Scop.) Gray *Common Bonnet*
On stumps and logs; common.
50, 51. SH 85, 87, 95-97; SJ 05-08, 13, 15-18, 24-27, 34-36, 43. AV, BO, BW, CC, CE, CF, CN, DU, GW, HC, LH, LO, MA, MF, ML, MO, NF, PA, PP, WW. (74) 1862. (48, 49, 52)

Mycena galopus (Pers.) P. Kumm. var. **galopus** *Milking Bonnet*
In leaf litter; common.
50, 51. SH 85, 87, 95, 97; SJ 05-08, 13, 15-18, 23-27, 34-36. AV, BO, BP, CC, CE, CF, CN, ER, GW, HC, LH, LO, MA, MF, ML, MO, NF, PP, WW. (62) 1880. (48, 49, 52)

Mycena galopus var. **candida** J.E. Lange *White Milking Bonnet*
In leaf litter; common.
50, 51. SH 85; SJ 16, 17, 26, 34. AV, ER, LH, MA, MO. (9) 1924. (48, 49, 52)

Mycena galopus var. **nigra** Rea *Black Milking Bonnet*
Mycena leucogala (Cooke) Sacc.
In leaf litter, especially in acid woodland; common.
50, 51. SH 85, 97; SJ 05, 07, 16-18, 24-27, 34-36, 43. AV, CE, ER, GW, HC, HM, LO, MF, NF, NM, PP, WW. (23) 1910. (48, 49, 52)

Mycena haematopus (Pers.) P. Kumm. *Burgundydrop Bonnet*
On fallen trunks and branches; common.
50, 51. SH 85, 97; SJ 05–08, 15–18, 24–26, 34, 36. AV, BO, CE, CN, DU, ER, GW, LH, LO, MA, MF, ML, NF, WW. (31) 1862. (48, 49, 52)

Mycena inclinata (Fr.) Quél. *Clustered Bonnet*
On fallen trunks, branches and stumps of oak; common.
50, 51. SH 85, 87, 97; SJ 05–08, 15–18, 23–26, 34–36. AV, BO, BP, BW, CC, CE, ER, GW, LH, LO, MA, ML, NF, PP, WW. (40) 1880. (48, 49, 52)

Mycena kuehneriana A.H. Smith
In beech leaf litter; rare.
50. Recorded only from Loggerheads Country Park (SJ 16) in 2001. (49)

Mycena leptocephala (Pers.) Gillet *Nitrous Bonnet*
Mycena chlorinella (J. Lange) Sing.
In woodland litter; common.
50, 51. SH 95, 97; SJ 05–08, 15–18, 23–27, 34–36. AV, BO, BP, CE, CF, CN, DU, ER, HC, LO, MA, MF, ML, MO, PA, PP, WW. (38) 1910. (48, 49, 52)

Mycena maculata P. Karst.
On oak stumps; rare.
51. Recorded only from Celyn Wood (SJ 26) in 1984.

Mycena megaspora Kauffm.
Mycena uracea A. Pears.
On peat and *Sphagnum* in moorland; frequent.
50, 51. SJ 16, 25. HM, MF. (2) 1978. (49, 52)

Mycena meliigena (Berk. & Cooke) Sacc.
Mycena corticola auct.
On the trunks of living trees; uncommon.
50, 51. SJ 06, 07, 16–18, 24–26, 34, 36. DU, GW, LO, NF, PP. (10) 1910.

Mycena metata (Fr.) P. Kumm.
Mycena phyllogena (Pers.) Sing.
In leaf litter; frequent.
50, 51. SJ 06, 07, 16–18, 24–26, 34–36. AV, ER, GW, LH, MF, NF, PP, WW. (17) 1910. (48, 49, 52)

Mycena mucor (Batsch) Gillet
On oak leaf litter; frequent.
51. SJ 16, 17, 26. LH, MO, WW. (5) 1975.

Mycena olivaceomarginata (Massee) Massee *Brownedge Bonnet*
Mycena avenacea auct.
In unimproved grassland; frequent.
50, 51. SH 97; SJ 08, 15, 16, 18, 24, 26, 34, 35. BO, CE, ER, LO, MA, MF, MO, PA, WW. (17) 1973. (49, 52)

Mycena pearsoniana Sing.
In alder litter; uncommon.
50. Recorded only from Loggerheads Country Park (SJ 16) in 1985. (49, 52)

Mycena pelianthina (Fr.) Quél. *Blackedge Bonnet*
In beech litter; frequent.
50, 51. SH 97; SJ 16, 26, 36. GW, LO. (4) 1976. (49)

Mycena polyadelpha (Lasch) Kühn.
On fallen oak leaves; frequent.
50, 51. SJ 08, 16, 17, 24, 34. CN, ER, LH, MA. (7) 1977.

Mycena polygramma (Bull.) Gray *Grooved Bonnet*
On stumps and fallen trunks; common.
50, 51. SH 85, 97; SJ 05, 07, 08, 16, 17, 24–26, 34. AV, BP, ER, LO, MA, ML, NF, PP, WW. (22) 1910. (48, 49, 52)

Mycena pseudocorticola Kühn.
On mossy bark of living trees; frequent.
50, 51. SJ 06, 07, 24, 34. ER. (4) 1977. (48, 49, 52)

Mycena pura (Pers.) P. Kumm. *Lilac Bonnet*
In leaf litter, especially under beech; common.
50, 51. SH 85, 87, 97; SJ 05–08, 15–18, 23–27, 34–36. AV, BO, BP, BW, CC, CE, CN, ER, GW, LH, LO, MA, MF, ML, MO, NF, NM, PA, PP, WW. (67) 1974. (48, 49, 52)

Mycena rosea (Bull.) Gramberg *Rosy Bonnet*
Mycena rosea var. *rosea* Gillet
In woodland litter; frequent.
50, 51. SH 97; SJ 07, 15, 24, 26, 34, 35. CE, ER, MA, PP, WW. (12) 1993. (48, 52)

Mycena rosella (Fr.) P. Kumm. *Pink Bonnet*
In needle litter; rare.
50. SH 87; SJ 15. CC. (2) 1880.

Mycena sanguinolenta (Alb. & Schwein.) P. Kumm. *Bleeding Bonnet*
In leaf litter, especially on acid soils; common.
50, 51. SH 85, 87, 95; SJ 05–07, 13, 15–17, 23–26, 34–36. AV, CC, CE, CF, DU, ER, GW, LH, LO, MF, NF, NM, PP, WW. (32) 1880. (48, 49, 52)

Mycena smithiana Kühner
On fallen oak leaves; rare.
51. Recorded only from Rhydymwyn Nature Reserve (SJ 26) in 2007. (49)

Mycena stipata Maas Geest. & Schwöbel
Mycena alcalina auct.
On stumps and fallen trunks and branches; common.
50, 51. SH 85, 87, 97; SJ 05, 07, 15–17, 23–26, 34, 36. AV, BP, CC, CE, CN, ER, GW, LH, LO, MA, MF, ML, NF, PP, WW. (29) 1880. (48, 49, 52)

Species Accounts

Mycena stylobates (Pers.) P. Kumm. *Bulbous Bonnet*
On leaf litter; frequent.
50, 51. SJ 05, 06, 15, 17, 26, 34, 36. AV, CF, ER, GW, LH, MA, WW. (12) 1910. (48, 49, 52)

Mycena vitilis (Fr.) Quél. *Snapping Bonnet*
In woodland litter; common.
50, 51. SH 87, 95, 97; SJ 05–08, 15–18, 23–26, 34–36, 43. AV, BO, BP, CE, CN, ER, MA, ML, MO, NF, PP, WW. (43) 1910. (48, 49, 52)

Mycena vulgaris (Pers.) P. Kumm.
In conifer litter; frequent.
50, 51. SH 85, 87; SJ 16, 17, 24–26, 35. CC, CE, LO, MF, NM. (9) 1880. (49, 52)

Phloeomana hiemalis (Osbeck) Redhead
Mycena hiemalis (Osbeck) Quél.
On mossy trunks of living trees; uncommon.
50, 51. SH 97; SJ 16, 17, 26, 34. AV, CE, CN, ER, LH, MF, ML. (10) 1976. (48, 49, 52)

Phloeomana olida Bres. *Rancid Bonnet*
Mycena olida Bres.
On the bark of dead and living trees; uncommon.
50, 51. SJ 07, 16, 17, 23, 25, 26, 34, 36, 37. AV, CN, ER, GW, LO, NF, WW. (15) 1977. (49, 52)

Phloeomana speirea (Fr.) Redhead *Bark Bonnet*
Mycena speirea (Fr.) Gillet
On mossy dead wood; frequent.
50, 51. SJ 05–08, 15–18, 24, 26, 34, 36. AV, BP, DU, ER, GW, LO, MF, PP, WW. (18) 1978. (48, 49, 52)

Resinomyces saccharifera (Berk. & Br.) Redhead
Mycena pudica Hora
On *Carex* litter; uncommon.
51. Recorded only from Hawarden Woods (SJ 36) in 1972. (48, 49, 52)

Roridomyces rorida (Fr.) Rexer *Dripping Bonnet*
Mycena rorida (Fr.) Quél.
On dead bramble stems; frequent.
50, 51. SH 85; SJ 08, 15–18, 23, 26. DU, LH, WW. (10) 1924. (48, 49, 52)

Family Nidulariaceae

Crucibulum laeve (Huds.) Kambly *Common Bird's Nest*
Crucibulum vulgare Tul. & C. Tul.
On dead herbaceous stems on the ground; common.
50, 51. SH 87; SJ 07, 08, 15, 16, 18, 26, 36. CC, GW, LO, PA. (9) 1880. (48, 49, 52)

Cyathus olla (Batsch) Pers. *Field Bird's Nest*
On woody debris on soil in arable fields and gardens; common.
50, 51. SH 87; SJ 06, 07, 18, 26, 35. CC, LO, MO, PA. (7) 1880. (48, 52)

Cyathus stercoreus (Schwein.) de Toni *'Dung Bird's Nest'*
On rabbit pellets among marram in dunes; rare.
51. Recorded only from Point of Ayr (SJ 18) since 1986. This was the second British record of a species rare in Europe. (48, 52)

Cyathus striatus (Huds.) Pers. *Fluted Bird's Nest*
On buried sticks in woodland litter; common.
50, 51. SH 85, 87, 97; SJ 15, 16, 23, 26, 34. CC, CE, ER, LO, ML. (9) 1880. (48, 49, 52)

Mycocalia denudata (Fr. & Nordholm) J.T. Palmer *'Common Mycocalia'*
On dead leaves in clumps of rushes on moorland; uncommon.
50. Recorded only from Foel Fenlli (SJ 16) in 1957. (49)

Mycocalia minutissima (J.T. Palmer) J.T. Palmer *'Tiny Mycocalia'*
On dead leaves in clumps of rushes in damp moorland; uncommon.
50. SH 93, 95. (2) 1957.

Nidularia deformis (Willd.) Fr. *'Pea-shaped Bird's Nest'*
Nidularia farcta (Roth) Fr.
On small sticks in damp woodland; uncommon.
50. Recorded only from Eryrys (SJ 25) in 2000.

Family Omphalotaceae

Gymnopus androsaceus (L.) Antonin & Noordel. *Horsehair Parachute*
Marasmius androsaceus (L.) Fr.; *Setulipes androsaceus* (L.) Antonin
On needle and heather litter in moorland and conifer forest; common.
50, 51. SH 85, 87, 95, 97; SJ 05, 07, 13, 15–17, 23, 25, 26, 34, 43. AV, CC, CE, CF, ER, HM, LH, LO, MA, MF, ML, WW. (27) 1880. (48, 49, 52)

Gymnopus aquosus (Bull.) Antonin & Noordel.
Collybia aquosa (Bull.) P. Kumm.; *C. dryophila* var. *aquosa* (Bull.) Quél.
In woodland litter; rare.
50. SH 85, 87. CC. (2) 1880. (49)

Gymnopus brassicolens (Romagn.) Antonin & Noordel. *Cabbage Parachute*
Micromphale brassicolens (Romagn.) P.D. Orton
On small sticks and fallen branches in deep litter; uncommon.
51. Recorded only from Bodelwyddan (SH 97) in 1996.

Gymnopus confluens (Pers.) Antonin, Halling & Noordel. *Clustered Toughshank*
Collybia confluens (Pers.) P. Kumm.
In woodland litter; common.
50, 51. SH 85, 87, 97; SJ 05–08, 13, 15–17, 24–26, 34–36. AV, CC, CE, CN, ER, GW, LH, LO, MA, ML, MO, NF, PP, WW. (47) 1880. (48, 49, 52)

Gymnopus dryophilus (Bull.) Murrill *Russet Toughshank*
Collybia dryophila (Bull.) P. Kumm.
In woodland litter; common.
50, 51. SH 85, 87, 95–97; SJ 05–08, 15–18, 24–26, 34–36, 43. AV, BO, CC, CE, ER, GW, LH, LO, MA, MF, ML, MO, NF, PA, PP, WW. (54) 1880. (48, 49, 52)

Gymnopus erythropus (Pers.) Antonin, Halling & Noordel. *Redleg Toughshank*
Collybia erythropus (Pers.) P. Kumm.; *C. kuehneriana* Sing.
In woodland litter; common.
50, 51. SH 85, 87, 96, 97; SJ 05–08, 15–17, 23, 24, 26, 34–36. AV, CC, CE, ER, GW, LH, LO, MA, ML, MO, PP, WW. (27) 1880. (49, 52)

Gymnopus fuscopurpureus (Pers.) Antonin, Halling & Noordel.
Collybia fuscopurpurea (Pers.) P. Kumm.
In beech litter; rare.
50, 51. SH 97; SJ 26, 27, 34, 36. ER, GW, MA, ML. (5) 1974.

Gymnopus fusipes (Bull.) Gray *Spindle Toughshank*
Collybia fusipes (Bull.) Quél.
At base of trunks of standing deciduous trees, especially oaks; common.
50, 51. SH 97; SJ 06, 15, 17, 23, 24, 26, 34, 36. ER, GW, LH, LO, ML, PP, WW. (14) 1910. (49)

Gymnopus impudicus (Fr.) Antonin, Halling & Noordel.
Micromphale impudicum (Fr.) P.D. Orton; *Collybia impudica* (Fr.) Sing.
In woodland litter; rare.
50. Recorded only from Glan-yr-Afon, Marli (SH 97) in 1981.

Gymnopus ocior (Pers.) Antonin & Noordel.
Collybia ocior (Pers.) Vilgalys & O.K. Mill.; *C. succinea* (Fr.) Quél.
In woodland litter; uncommon.
50. Recorded only from Coed Coch (SH 87) in 1880. (48, 49)

Gymnopus peronatus (Bolt.) Antonin, Halling & Noordel. *Wood Woollyfoot*
Collybia peronata (Bolt.) P. Kumm.
In woodland litter; very common.
50, 51. SH 76, 85, 87, 97; SJ 05–08, 13, 15–18, 23–26, 34–36. AV, BO, BO, CC, CE, CF, CN, DU, ER, GW, HM, LH, LO, MA, MF, ML, MO, NF, NM, PP, WW. (71) 1880. (48, 49, 52)

Marasmiellus candidus (Bolt.) Sing.
Marasmius candidus (Bolt.) Fr.; *Marasmiellus albus-corticis* (Secr.) Sing.
On stumps; uncommon.
50, 51. SH 97; SJ 05, 07, 16, 17, 26, 34. AV, CF, ER, LO, WW. (9) 1981.

Marasmiellus foetidus (Sow.) Antonin, Halling & Noordel. *Foetid Parachute*
Micromphale foetidum (Sow.) Sing.
In leaf litter; frequent.
50, 51. SH 97; SJ 15–17. BO, LH, MF. (4) 1977. (49, 52)

Marasmiellus perforans (Hoffm.) Antonin, Halling & Noordel. *Stinking Parachute*
Micromphale perforans (Hoffm.) Gray
In beech leaf litter; frequent.
51. SJ 17, 25, 26. LH, NF, WW. (3) 1976. (48, 49)

Marasmiellus ramealis (Bull.) Sing. *Twig Parachute*
Marasmius ramealis (Bull.) Fr.
On small dead branches, especially of brambles; common.
50, 51. SH 76, 85, 97; SJ 05, 15–18, 24, 26, 34, 36. AV, CE, CN, ER, GW, LH, LO, MA, MF, ML, MO, NF, PA, PP, WW. (34) 1910. (48, 49, 52)

Marasmiellus vaillantii (Pers.) Sing. *Goblet Parachute*
Marasmiellus languidus (Lasch) Sing.; *Marasmius calopus* auct.
In grass litter in woodland; frequent.
50, 51. SH 85, 87, 96; SJ 07, 08, 15–17, 24, 26, 34, 35. CC, CE, ER, LH, LO, MA, MF, PP. (18) 1859. (48, 49, 52)

Mycetinis alliaceus (Jacq.) A. Wilson & Desjardin *Garlic Parachute*
Marasmius alliaceus (Jacq.) Fr.
In litter in beechwoods; now rare.
50. Recorded only from Erddig (SJ 34) in 1910. This species is more or less confined to the south of England at the present day but was once more widespread.

Mycetinis scorodonius (Fr.) Wilson & Desjardin
Marasmius scorodonius (Fr.) Fr.
In woodland litter, especially of beech; uncommon, becoming rare.
50, 51. SJ 15, 17, 26, 36. CE, GW, LH, MO. (5) 1975. (49) The strong smell of garlic is distinctive and makes these recent records significant.

Rhodocollybia butyracea (Bull.) Lennox var. **butyracea** *Butter Cap*
Collybia butyracea (Bull.) P. Kumm. var. *butyracea*
In woodland leaf litter; common.
50, 51. SH 76, 87, 95–97; SJ 05–08, 15–18, 23–27, 34–36, 43, 54. AV, BO, BP, CC, CE, CN, ER, GW, HC, LH, LO, MA, MF, ML, MO, NF, PP, WW. (54) 1880. (48, 49, 52)

Rhodocollybia butyracea var. **asema** (Fr.) Antonin, Halling & Noordel.
Collybia butyracea var. *asema* (Fr.) Quél.
In woodland litter; common.
50, 51. SH 97; SJ 05, 13, 16, 17, 26, 34–36. CE, ER, GW, LO, MA, WW. (12) 1998.

Rhodocollybia distorta (Fr.) Singer
Collybia distorta (Fr.) Quél.
In woodland litter; rare.
50, 51. SH 85; SJ 26, 34, 36. GW, MO. (4) 1924. (48, 49, 52)

Rhodocollybia maculata (Alb. & Schwein.) Sing. *Spotted Toughshank*
Collybia maculata (Alb. & Schwein.) P. Kumm.
In leaf litter, especially under conifers; common.
50, 51. SH 85, 95–97; SJ 05, 07, 08, 13, 15–17, 23–26, 34, 36, 43. AV, BO, CE, CF, ER, GW, LH, LO, MA, MF, ML, MO, NF, NM, WW. (47) 1910. (48, 49, 52)

Species Accounts

Family Physalacriaceae

Armillaria mellea *sensu lato Honey Fungus*
Parasitic on trees and shrubs of many species; common. This is an aggregate of several closely related species which have different ecological preferences and may be distinguished on macroscopic characters. All records for the aggregate are given first followed by three segregate species.
50, 51. SH 85, 87, 96, 97; SJ 05–08, 15–18, 23–27, 34–37. AV, BO, BP, CC, CE, CF, CN, DU, ER, GW, HC, LH, LO, MA, ML, MO, NF, NM, PP, WW. (90) 1880. (48, 49, 52)

Armillaria lutea Gillet *Bulbous Honey Fungus*
Armillaria gallica Marxm. & Romagn.; *A. bulbosa* (Barla) Kile & Watling
Parasitic on a wide range of woody plants; the commonest member of the group.
50, 51. SH 87, 96, 97; SJ 05–08, 15–18, 24–27, 34, 35. AV, BO, BP, CE, CN, ER, HC, LO, MA, ML, MO, NF, PP, WW. (53) 1978. (48, 49, 52)

Armillaria mellea (Vahl) P. Kumm. *sensu stricto Honey Fungus*
Associated with oaks in ancient woodland, rarely in other habitats; frequent.
50, 51. SH 97; SJ 05–07, 15–18, 24–27, 34, 35. AV, BO, CE, DU, ER, HC, NF, PP, WW. (23) 1978. (48, 49, 52)

Armillaria ostoyae (Romagn.) Herink *Dark Honey Fungus*
Armillaria obscura auct.; *A. polymyces* auct.
Parasitic on conifers, often fruiting high up on dead standing trunks; common.
50, 51. SJ 05, 16, 24–26, 34. AV, CF, ER, MO, NF. (7) 1978. (48)

Desarmillaria tabescens (Scop.) R.A. Koch & Aime *Ringless Honey Fungus*
Armillaria tabescens (Scop.) Emel
On dead wood, possibly non-parasitic; uncommon.
51. SJ 25, 26. MO, NM. (2) 1978. Easily recognised by the absence of a ring on the stem.

Calyptella capula (Holmsk.) Quél.
On dead herbaceous and grass stems, small twigs, etc.; common.
50, 51. SH 85; SJ 07, 16, 17, 18, 26, 34, 36. AV, DU, ER, GW, LH, LO, PA, WW. (16) 1972. (48, 49, 52)

Cylindrobasidium laeve (Pers.) Chamuris
Cylindrobasidium evolvens (Fr.) Jülich
On wood and bark of fallen branches, especially sycamore; common.
50, 51. SH 85, 87; SJ 05, 15–17, 26, 34, 36. CC, CE, ER, GW, LH, LO. (13) 1880. (48, 49, 52)

Flammulina velutipes (Curt.) Sing. *Velvet Shank*
On dead wood, especially standing trunks, during the winter months; common.
50, 51. SH 87, 97; SJ 02, 05, 07, 15–18, 23–27, 34–36. BO, CC, CE, CN, DU, ER, GW, HC, LH, LO, MA, ML, MO, NF, PP, WW. (48) 1880. (48, 49, 52)

Hymenopellis radicata (Relh.) R.H. Petersen *Rooting Shank*
Xerula radicata (Relh.) Dorfelt; *Oudemansiella radicata* (Relh.) Sing.;
Collybia radicata (Relh.) Quél.
In woodland leaf litter; common.
50, 51. SH 87, 95, 97; SJ 05–07, 15–17, 23–27, 34–36. AV, BO, CC, CE, ER, GW, HC, LH, LO, MA, ML, PP, WW. (36) 1880. (48, 49, 52)

Mucidula mucida (Schrad.) Pat. *Porcelain Fungus*
Oudemansiella mucida (Schrad.) Höhn.
On dead trunks and branches of beech, both standing and fallen; common.
50, 51. SH 85, 95, 97; SJ 05, 16, 17, 24, 26, 36. AV, CF, GW, LO, MA, PP. (12) 1972. (48, 49, 52)

Mycenella bryophila (Vogl.) Sing.
Mycena bryophila Vogl.
In terrestrial woodland mosses; uncommon.
51. SJ 17, 26. AV. (2) 1978. (52)

Rhizomarasmius setosus (Sow.) Antonin & A. Urb
Marasmius setosus (Sow.) Noordel.; *M.recubans* auct.; *M. saccharinus* auct.
In beech litter; uncommon.
50, 51. SH 87, 97; SJ 07, 26. AV, BO, CC. (4) 1880. (48, 49, 52)

Rhizomarasmius undatus (Berk.) Antonin & Noordel. *'Bracken Parachute'*
Marasmius undatus (Berk.) Fr.; *M. chordalis* Fr.
On dead bracken stems; frequent.
50, 51. SJ 07, 08, 16, 17, 25–27, 34. ER, HC, LH, MA, MF, WW. (13) 1910. (49, 52)

Strobilurus esculentus (Wulf.) Sing. *Sprucecone Cap*
Pseudohiatula esculenta (Wulf.) Sing.
On buried spruce cones in spring; common.
50, 51. SH 97; SJ 05, 16, 17, 26. AV, CF, ML. (5) 1978. (48, 49)

Strobilurus stephanocystis (Hora) Sing.
Pseudohiatula stephanocystis Hora
On buried pine cones in spring; uncommon.
50. Recorded only from Pant-yr-Ochain, Gresford (SJ 35) in 2005. (52)

Strobilurus tenacellus (Pers.) Sing. *Pinecone Cap*
Pseudohiatula tenacella (Pers.) Métrod
On buried pine cones in spring; common.
50, 51. SJ 05, 15–17, 26. AV, CE, CF, CN, WW. (10) 1985. (48, 49, 52)

Xerula pudens (Pers.) Sing.
Oudemansiella longipes (P. Kumm.) M.M. Moser
In woodland leaf litter; uncommon.
50, 51. SJ 06, 07, 26, 34. GW. (4) 1910.

Family Pleurotaceae

Hohenbuehelia culmicola Bon
On dead culms of *Leymus* grass; rare.
51. Recorded only from Point of Ayr (SJ 18) in 1978. This was the first British record; the fungus is now known from sites in Scotland and Yorkshire.

Hohenbuehelia petaloides (Bull.) Schulzer
Hohenbuehelia geogenia (DC.) Singer
On sawdust heaps; rare.
50. SH 85; SJ 17, 26. MA. (3) 1924. (49)

Hohenbuehelia reniformis (G. Mey.) Sing.
On twigs on the forest floor; rare.
50. Recorded only from Erddig (SJ 34) in 1910. (52)

Hohenbuehelia tremula (Schaeff.) Thorn & G.L. Barron
On soil and sawdust; rare.
50. Recorded only from Coed Coch (SH 87) in 1880. (49)

Pleurotus cornucopiae (Paul.) Roll. *Branching Oyster*
On fallen trunks; frequent.
50, 51. SH 97; SJ 04, 06–08, 15–17, 24, 26, 34–36. AV, CE, CN, DU, ER, GW, LO, MF, ML, MO, WW. (29) 1972. (49, 52) An unusual specimen was found on waste wood in the boot of a car. The fruit bodies were coralloid, small and devoid of pigment. This growth abnormality has been well documented for fungi growing in darkness.

Pleurotus dryinus (Pers.) P. Kumm. *Veiled Oyster*
On fallen trunks and stumps; uncommon.
50, 51. SH 97; SJ 26, 34. BO, CE, ER, LO, ML, WW. (6) 1974. (48, 49, 52)

Pleurotus ostreatus (Jacq.) P. Kumm. *Oyster Mushroom*
On dead parts of standing tree trunks, also on fallen trunks; common.
50, 51. SH 97; SJ 05–08, 15–17, 24–27, 34, 36. AV, BO, CE, CF, CN, ER, GW, HC, HM, LO, MA, PP, WW. (27) 1958. (48, 49, 52)

Pleurotus pulmonarius (Fr.) Quél. *Pale Oyster*
On fallen trunks; uncommon.
51. SJ 08, 26. BP, CE, MO, WW. (5) 1985.

Resupinatus applicatus (Batsch) Gray *Smoked Oysterling*
On rotten branches of broad-leaves; common.
50, 51. SJ 06–08, 16–18, 24, 26, 34, 35, 54. CN, DU, LO, PA, WW. (13) 1910. (48, 49, 52)

Resupinatus griseopallidus (Desm.) Knudsen & Elbone
Arrhenia griseopallida (Desm.) Watling; *Omphalina griseopallida* (Desm.) Quél.; *Phaeotellus griseopallidus* (Desm.) Kühn. & Lamoure.
In mossy grassland; rare.
50, 51. SJ 16, 26, 35. BW, LO. (3) 1982.

Resupinatus poriaeformis (Pers.) Thorn, Moncalvo & Redhead
Stigmatolemma poriiforme (Pers.) W.B. Cooke; *S. urceolatum* (Fr.) Donk
On dead, fallen wood; rare.
51. Recorded only from Celyn Wood (SH 26) in 1984. (48, 49)

Family Pluteaceae

Melanoleuca cinerifolia (Bon) Bon *'Dune Cavalier'*
In sand at the front of marram dunes; frequent.
51. Recorded only from Point of Ayr (SJ 18) since 1998. (48, 52)

Melanoleuca cognata (Fr.) Konrad & Maubl. *Spring Cavalier*
In woodland litter; frequent.
50, 51. SH 96; SJ 06, 07, 15, 16, 18, 25, 26, 35. AV, MA, MF, NF, NM, PA, WW. (13) 1976. (49, 52)

Melanoleuca grammopodia (Bull.) Pat.
In grassy places in woods; frequent.
50, 51. SH 87, 97; SJ 06–08, 16, 17, 25, 26, 35. BO, CC, MA, MF, MO. (14) 1880. (49, 52)

Melanoleuca langei (Boekhout) Bon
In woodland litter; uncommon.
50, 51. SH 87, 97; SJ 07, 24, 35. BO, CC. (5) 1880. (49, 52)

Melanoleuca melaleuca (Pers.) Murrill
In short turf in sandy soil; uncommon.
50. Recorded only from Marford Quarry Nature Reserve (SJ 35) in 1997. (48, 52)

Melanoleuca polioleuca (Fr.) Konrad & Maubl. *Common Cavalier*
In grass at edge of woods and roads; common.
50, 51. SH 87, 95–97; SJ 06–08, 15–18, 24–26, 34–36, 43. AV, BO, CC, CN, DU, ER, GW, LH, LO, MA, MF, ML, MO, NF, PA, WW. (32) 1880. (48, 49, 52)

Melanoleuca strictipes (P. Karst.) Jul. Schaeff.
In hedgerows at woodland margins; uncommon.
51. Recorded only from Mold (SJ 26) in 1978.

Pluteus atromarginatus (Sing.) Kühn.
On rotting conifer wood; uncommon.
51. SJ 26. LO, WW. (2) 1977.

Pluteus aurantiorugosus (Trog) Sacc.
On rotting elm stumps; rare.
50, 51. SJ 07, 34. ER. 1910.

Pluteus cervinus (Schaeff.) Kühn. *Deer Shield*
On fallen logs and stumps of deciduous trees; common.
50, 51. SH 76, 85, 87, 96–98; 05–08, 15–18, 23–27, 34–36, 43. AV, BO, CC, CE, CF, DU, ER, GW, HC, LH, LO, MA, MF, ML, MO, NF, PP, WW. (66) 1880. (48, 49, 52)

Pluteus chrysophaeus (Schaeff.) Quél. *Yellow Shield*
Pluteus luteovirens Rea
On small fallen branches; uncommon.
50, 51. SH 97; SJ 24, 26, 34. CE, ER, ML, PP, WW. (5) 1979. (49, 52)

Pluteus cinereofuscus J.E. Lange
On hardwood stumps; uncommon.
50, 51. SJ 15, 26. AV, WW. (3) 1976. (49, 52)

Pluteus ephebeus (Fr.) Gillet
On sawdust and wood chips; uncommon.
51. SJ 26, 36. GW, WW. (2) 1981.

Pluteus griseoluridus P.D. Orton
On rotten wood; uncommon.
51. Recorded only from Bodelwyddan (SH 97) in 2001. (48, 49, 52)

Pluteus hispidulus (Fr.) Gillet
On mossy beech stumps; uncommon.
51. Recorded only from Hawarden Woods (SJ 26) in 1972.

Pluteus leoninus (Schaeff.) P. Kumm. *Lion Shield*
On beech stumps; uncommon.
50, 51. SJ 24, 25, 35. NF, PP. (3) 1986.

Pluteus luctuosus Boud.
On elm stump; uncommon.
50. Recorded only from Pen Parc Llwyd, Henllan (SJ 06) in 2004.

Pluteus nanus (Pers.) P. Kumm. *Dwarf Shield*
On stumps; uncommon.
50, 51. SJ 16, 17, 34. ER, LH, LO. (5) 1910. (49, 52)

Pluteus pellitus (Pers.) P. Kumm. *Ghost Shield*
On fallen beech trunk; uncommon.
50. Recorded only from Glan-yr-Afon, Marli (SH 97) in 1980. (49)

Pluteus petasatus (Fr.) Gillet
On sawdust piles; uncommon.
51. SH 97; SJ 08. BO. (2) 1984.

Pluteus plautus (Weinm.) Gillet *Satin Shield*
On stumps and branches; uncommon.
51. SJ 16, 26. CN (2) 1989. (48, 52)

Pluteus podospileus Sacc.
On fallen beech branches; uncommon.
50, 51. SJ 23, 26, 34, 36. AV, ER, GW. (4) 1993. (49, 52)

Pluteus romellii (Britz.) Sacc. *Goldleaf Shield*
Pluteus lutescens auct.
On fallen deciduous wood; common.
50, 51. SH 85, 97; SJ 06, 15–17, 26, 34, 35. ER, LO, MA, ML, MO, WW. (13) 1924. (48, 49, 52)

Pluteus salicinus (Pers.) P. Kumm. *Willow Shield*
On hardwood stumps and logs; common.
50, 51. SH 97; SJ 06, 08, 15–17, 25, 26, 34, 35. AV, BO, CE, CN, ER, LO, ML, MO, NF, WW. (23) 1910. (48, 49)

Pluteus thomsonii (Berk. & Br.) Dennis *Veiled Shield*
On beech stumps and branches; frequent.
51. SJ 16, 26, 36. AV, CE, GW, LO, WW. (5) 1976. (52)

Pluteus umbrosus (Pers.) P. Kumm. *Velvet Shield*
On hardwood stumps; frequent.
50, 51. SH 87, 97; SJ 16, 25, 26, 34, 35. ER, LO, ML, NF, WW. (7) 1979. (49)

Volvopluteus gloiocephalus (DC.) Vizzini, Contu & Justo *Stubble Rosegill*
Volvariella gloiocephala (DC.) Boekhout & Enderla; *V. speciosa* (Fr.) Sing.
In woodland clearings and on rich, often mulched, soil in gardens; frequent.
50, 51. SH 87, 97; SJ 08, 17, 26. AV, BP, CC, ML, WW. (7) 1902. (48, 49, 52)

Family Porotheleaceae

Henningsomyces candidus (Pers.) Kuntze
Solenia candida Pers.
On rotten wood, especially beech; common.
50, 51. SH 97; SJ 34. BO, ER. (2) 1977.

Hydropus trichoderma (Joss.) Sing.
On decayed wood of deciduous trees; very rare.
50. Recorded only from Glan-yr-Afon, Marli (SH 97) in 1980. This is the second British and first Welsh record.

Megacollybia platyphylla (Pers.) Kotl. & Pouzar *Whitelaced Shank*
Tricholomopsis platyphylla (Pers.) Sing.
In woodland litter, attached to buried wood; common.
50, 51. SH 85, 95–97; SJ 07, 15–17, 25, 26, 34–36. AV, BO, CE, ER, GW, ML, NF, NM, WW. (20) 1910. (48, 49, 52)

Family Psathyrellaceae

Coprinellus angulatus (Peck) Redhead, Vilgalys & Hopple
Coprinus angulatus Peck
On burnt ground, especially bonfire sites in woodland; frequent.
51. SJ 08, 26. BP, MO. (2) 1977. (48, 49, 52)

Coprinellus congregatus (Bull.) P. Karst.
Coprinus congregatus (Bull.) Fr.
On dung; rare.
51. Recorded only from Point of Ayr (SJ 18) in 2001. (52)

Coprinellus deliquescens (Fr.) P. Karst.
Coprinus silvaticus Peck
In broad-leaved woodland litter; common.
50, 51. SJ 05, 07, 08, 16, 17, 24–27, 34, 35. AV, CE, DU, ER, HC, LO, MO, NF. (21) 1975. (52)

Coprinellus disseminatus (Pers.) J.E. Lange *Fairy Inkcap*
Coprinus disseminatus (Pers.) Gray
On stumps of hardwoods; common.
50, 51. SH 97; SJ 05–07, 15–17, 24–27, 34, 36. AV, BO, CE, CN, DU, ER, GW, HC, LH, LO, MA, ML, MO, WW. (37) 1910. (48, 49, 52)

Coprinellus domesticus (Bolton) Vilgalys, Hopple & Johnson *Firebug Inkcap*
Coprinus domesticus (Bolton) Gray
On logs of hardwood trees; frequent.
50, 51. SH 97; SJ 05, 08, 15–18, 25–27, 34, 35. AV, BW, DU, ER, LH, LO, ML, MO, WW. (25) 1974. Often with the aerial brown mycelium – ozonium.

Coprinellus ellisii (P.D. Orton) Redhead, Vilgalys & Moncalvo
Coprinus ellisii P.D. Orton
On stumps of deciduous trees; uncommon.
50, 51. SH 97; SJ 26, 34, 35. BO, ER, MO. (5) 1977. (49)

Coprinellus heptemerus (M. Lange & A.H. Smith) Vilgalys, Hopple & Johnson
Coprinus heptemerus M. Lange & A.H. Smith
On dung of cow and rabbit; common.
50, 51. SJ 16, 25, 26. HM, MA, MO. (3) 1977. (52)

Coprinellus hetersetulosus (Watling) Vilgalys, Hopple & Johnson
Coprinus heterosetulosus Watling
On cow dung; frequent.
51. Recorded only from Hope Mountain (SJ 25) in 1978.

Coprinellus hiascens (Fr.) Redhead, Vilgalys & Moncalvo
Coprinus hiascens (Fr.) Quél.
In basic grassland; uncommon.
50, 51. SJ 08, 34. (2) 1982. (49)

Coprinellus impatiens (Fr.) J.E. Lange
Coprinus impatiens (Fr.) Quél.
In beech leaf litter; uncommon.
50. Recorded only from Maes Mynan (SJ 17) in 1994. (52)

Coprinellus micaceus (Bull.) Vilgalys, Hopple & Johnson *Glistening Inkcap*
Coprinus micaceus (Bull.) Fr.
On stumps and logs; common.
50, 51. SH 87, 88, 97; SJ 05–08, 15–18, 23–27, 34–36. AV, BO, BP, CC, CE, CN, DU, ER, GW, HC, LH, LO, MA, MF, ML, MO, NF, PA, PP, WW. (72) 1880. (48, 49, 52)

Coprinellus radians (Desm.) Vilgalys, Hopple & Johnson
Coprinus radians (Desm.) Fr.
On rotten logs; uncommon.
50, 51. SH 97; SJ 07, 17, 26. BO, DU, MA, MO. (5) 1862. (48, 49, 52)

Coprinellus truncorum (Scop.) Redhead, Vilgalys & Moncalvo
Coprinus truncorum (Scop.) Fr.
On stumps; uncommon.
50, 51. SJ 17, 26, 34. ER, MO. (3) 1977. (49)

Coprinellus xanthothrix (Romagn.) Vilgalys, Hopple & Johnson
Coprinus xanthothrix Romagn.
In woodland litter and on woodwork of an old, but still in use, estate car; rare.
51. SJ 16, 26. CN, MO. 1973. (48)

Coprinopsis acuminata (Romagn.) Redhead, Vilgalys & Moncalvo
Humpback Inkcap
Coprinus acuminatus (Romagn.) P.D. Orton
On bare soil near stumps in broad-leaved woodland; frequent.
50, 51. SH 97; SJ 17, 26, 36. DU, ML, MO. (5) 1977. (48, 49)

Coprinopsis atramentaria (Bull.) Redhead, Vilgalys & Moncalvo *Common Inkcap*
Coprinus atramentarius (Bull.) Fr.
On or near buried wood and stumps; common. **Poisonous** if eaten with alcohol.
50, 51. SH 85, 88, 97; SJ 05–08, 16–18, 24–27, 34–36, 43, 44. AV, BO, CE, CN, DU, ER, GW, LH, LO, ML, MO, NF, PP, WW. (51) 1910. (48, 49, 52)

Coprinopsis cinerea (Schaeff.) Redhead, Vilgalys & Moncalvo
Coprinus cinereus (Schaeff.) Gray
On dung and stable heaps; common.
50, 51. SJ 07, 16, 17, 26, 27, 34, 43. AV, ER, LH, MO, WW. (8) 1910. (48, 52)

Coprinopsis cinereofloccosus (P.D. Orton) Redhead, Vilgalys & Moncalvo
Coprinus cinereofloccosus P.D. Orton
On woodland soil; uncommon.
51. Recorded only from Rhydymwyn Nature Reserve (SJ 26) in 2008.

Coprinopsis cordispora (T. Gibbs) P.M. Kirk
Coprinus cordisporus T. Gibbs
On sheep dung; uncommon.
51. Recorded only from Gwespyr (SJ 18) in 1990. (48, 49, 52)

Coprinopsis cortinata (J.E. Lange) Gminder
Coprinus cortinatus J.E. Lange
In basic woodland soil; uncommon.
50, 51. SH 97; SJ 26. ML, MO. (2) 1986. (48, 49, 52)

Coprinopsis echinospora (Buller) Redhead, Vilgalys & Moncalvo
Coprinus echinosporus Buller
On small sticks in woodland litter; rare.
50. Recorded only from Nant-y-Belan, Wynnstay (SJ 34) in 1982.

Coprinopsis ephemeroides (DC.) G. Moreno
Coprinus ephemeroides (DC.) Fr.
On sheep dung; common.
51. SJ 08, 16. AV, BP. (2) 1990. (48)

Coprinopsis erythrocephala (Lév.) Redhead, Vilgalys & Moncalvo
Coprinus erythrocephalus (Lév.) Fr.
On sandy soil and compost; rare.
50, 51. SJ 26, 34. AV. (2) 1997.

Coprinopsis friesii (Quél.) P. Karst.
Coprinus friesii Quél.
In leaf litter; frequent.
50, 51. SJ 16, 17, 24, 26. BW, LH, LO. (5) 1977. (49, 52)

Species Accounts

Coprinopsis jonesii (Peck) Redhead, Vilgalys & Moncalvo *Bonfire Inkcap*
Coprinus jonesii Peck; *Coprinus lagopides* sensu auct.
On bonfire sites; frequent.
50, 51. SJ 17, 25, 34. DU, ER. (3) 1978. (48, 49)

Coprinopsis laanii (Kits van Wav.) Redhead, Vilgalys & Moncalvo
Coprinus laanii Kits van Wav.
On the ends of logs in woodland, often with thick algal film; uncommon.
50, 51. SJ 26, 35. CE. (2) 1995.

Coprinopsis lagopus (Fr.) Redhead, Vilgalys & Moncalvo *Hare'sfoot Inkcap*
Coprinus lagopus (Fr.) Fr.
On soil and leaf litter in woods; common.
50, 51. SH 85, 97; SJ 05–08, 15–17, 24, 26, 34, 35, 43. AV, CF, CN, ER, LH, LO, MA, MF, ML, MO, PP, WW. (30) 1910. (48, 49, 52)

Coprinopsis marcescibilis (Berk. & Br.) Örstadius & E. Larss.
Psathyrella marcescibilis (Berk. & Br.) Sing.
In woodland soil; frequent.
50, 51. SJ 08, 15, 17, 25, 26, 34, 35. WW. (11) 1982.

Coprinopsis nivea (Pers.) Redhead, Vilgalys & Moncalvo *Snowy Inkcap*
Coprinus niveus (Pers.) Fr.
On cow dung and dung heaps; frequent.
50, 51. SJ 08, 25, 34. BP, ER, HM. (6) 1977. (48, 49, 52)

Coprinopsis picacea (Bull.) Redhead, Vilgalys & Moncalvo *Magpie Inkcap*
Coprinus picaceus (Bull.) Gray
In beech leaf litter on base-rich soils; uncommon.
50, 51. SH 97; SJ 05, 15, 16, 26, 34. AV, ER, MA, MF, ML. (12) 1977.

Coprinopsis radiata (Bolton) Redhead, Vilgalys & Moncalvo
Coprinus radiatus (Bolton) Gray
On rotting straw; uncommon.
50, 51. SJ 17, 34. ER, LH. (2) 1910. (48, 49, 52)

Coprinopsis romagnesianus (Sing.) Redhead, Vilgalys & Moncalvo
Coprinus romagnesianus Sing.
On birch stumps; rare.
50. Recorded only from Glan-yr-Afon, Marli (SH 97) in 1980. (49, 52)

Coprinopsis semitalis (P.D. Orton) Redhead, Vilgalys & Moncalvo
Coprinus semitalis P.D. Orton
On bare limestone soil; uncommon.
51. Recorded only from Craig Fawr (SJ 08) in 1983. (48, 49)

Coprinopsis stercorea (Fr.) Redhead, Vilgalys & Moncalvo
Coprinus stercoreus Fr.
On cow and rabbit dung; common.
50, 51. SJ 26, 34. ER, MO. (2) 1976. (48, 49)

Lacrymaria lachrymabunda (Bull.) Pay. *Weeping Widow*
Lacrymaria velutina (Pers.) P. Kumm.
In garden beds, grassland and clearings in woods; common.
50, 51. SH 87, 97; SJ 05–07, 15–18, 24–27, 34–36. AV, BO, CC, CE, DU, ER, GW, LH, ML, MO, NF, WW. (42) 1863. (49, 52)

Lacrymaria pyrotricha (Holmsk.) Konrad & Maubl.
In woodland soil; rare.
50, 51. SJ 24, 26, 27, 34. CE, ER, HC. (4) 1910. (49)

Parasola auricoma (Pat.) Redhead, Vilgalys & Hopple
Coprinus auricomus Pat.
In broad-leaved woodland litter; frequent.
50, 51. SH 97; SJ 06, 08, 34, 35. ML. (6) 1910. (49)

Parasola conopilus (Fr.) Örstadius & E. Larss. *Conical Brittlestem*
Psathyrella conopilus (Fr.) A. Pears. & Dennis; *Psathyrella conopilea* auct.
At edges of paths in woods; common.
50, 51. SH 95, 97; SJ 05, 07, 08, 15–18, 24, 26, 34–36. AV, BO, BP, CE, DU, ER, GW, LH, LO, MA, WW. (25) 1910. (49, 52)

Parasola leiocephala (P.D. Orton) Redhead, Vilgalys & Hopple
Coprinus leiocephalus P.D. Orton
In soil and litter at edges of paths in woods; frequent.
50, 51. SJ 06, 07, 13, 16, 17, 25, 26, 34, 35. AV, DU, ER, LH, LO, MA, MO, WW. (15) 1976. (52)

Parasola plicatilis (Fr.) Redhead, Vilgalys & Hopple *Pleated Inkcap*
Coprinus plicatilis (Fr.) Fr.
In short grassland, including playing fields and lawns; common.
50, 51. SH 85, 87, 95, 97; SJ 04–08, 15–18, 23–27, 34–36. AV, BO, CC, CE, CF, ER, GW, HC, LH, LO, MA, MF, ML, MO, NF, PA, WW. (59) 1880. (48, 49, 52)

Psathyrella ammophila (Dur. & Lév.) P.D. Orton *Dune Brittlestem*
In marram dunes; common.
51. SJ 08, 18. PA. (2) 1978. (48, 52)

Psathyrella artemisiae (Pers.) Konrad & Maubl. *Petticoat Brittlestem*
Psathyrella squamosa sensu auct.
In lawns; frequent.
50, 51. SJ 26, 34. ER, MO. (2) 1978. (48, 49, 52)

Psathyrella bipellis (Quél.) A.H. Sm.
At edges of paths in open woodland; rare.
51. Recorded only from Celyn Woods (SJ 26) in 1995.

Psathyrella candolleana (Fr.) Maire *Pale Brittlestem*
In woodland litter; common.
50, 51. SH 76, 85, 96, 97; SJ 06–08, 15–17, 23–26, 34–36. AV, BO, CC, CE, CN, ER, GW, LH, LO, MA, MF, ML, MO, NF, PP, WW. (39) 1910. (48, 49, 52)

Species Accounts

Psathyrella caput-medusae (Fr.) Konrad & Maubl. *Medusa Brittlestem*
On conifers stumps; rare.
50. Recorded only from World's End (SJ 24) in 1910. This was the last Welsh record. **RDL/VU**

Psathyrella corrugis (Pers.) Konrad & Maubl. *Red Edge Brittlestem*
Psathyrella gracilis auct.
In woodland litter; common.
50, 51. SH 87, 97; SJ 05–08, 15–18, 24, 26, 34–36, 43. AV, BO, BP, CC, CE, CF, CN, DU, ER, GW, LH, LO, MF, MO, PP, WW. (40) 1880. (48, 49, 52)

Psathyrella fatua (Fr.) Konrad & Maubl.
On soil in woods; rare.
50. Recorded only from Coed Coch (SH 87) in 1880.

Psathyrella flexispora P.D. Orton
In marram dunes; rare.
51. Recorded only from Point of Ayr (SJ 18) in 1979. This remains the only Welsh record.

Psathyrella friesii Kits van Wav.
Psathyrella fibrillosa (Pers.) Sing.
In grassy clearings in woodland; frequent.
50. SH 85, 87. CC. (3) 1880. (48, 49)

Psathyrella gordonii (Berk. & Br.) A. Pears. & Dennis
In woodland soil; rare and poorly understood.
50. Recorded only from Coed Coch (SH 87) in 1880.

Psathyrella gossypina (Bull.) A. Pears. & Dennis
In woodland soil; frequent.
50. SH 87; SJ 35. CC. (2) 1880. (48)

Psathyrella hirta Peck
Psathyrella coprobia (J.E. Lange) A.H. Sm.
On horse dung; uncommon.
51. Recorded only from Mold (SJ 26) in 1978. (48)

Psathyrella leucotephra (Berk. & Br.) P.D. Orton
In woodland litter; uncommon.
50, 51. SJ 06, 08. (2) 1990. (52)

Psathyrella microrrhiza (Lasch) Konrad & Maubl. *Rootlet Brittlestem*
In woodland soil; frequent.
50, 51. SJ 05, 08, 15, 16, 18, 24–26, 34, 36. AV, ER, LO, PA, PP, WW. (16) 1982. (49, 52)

Psathyrella multipedata (Peck) A.H. Smith *Clustered Brittlestem*
At base of stumps and in grass; frequent.
50, 51. SJ 05, 26, 35, 36. CE, WW. (7) 1983. (49, 52)

Psathyrella obtusata (Pers.) A.H. Smith
In woodland soil; frequent.
50, 51. SJ 07, 16, 17, 26, 34. AV, ER, LH, LO, MA, MF, WW. (10) 1972. (48, 52)

Psathyrella olympiana A.H. Sm. f. **amstelodamensis** Kits van Wav.
On decayed wood; very rare.
51. Recorded only from Rhydymwyn Nature Reserve (SJ 26) in 2007. New to Wales.

Psathyrella pennata (Fr.) Konrad & Maubl. *'Bonfire Brittlestem'*
On bonfire sites in woodland; frequent.
51. SJ 17, 26. DU, MO, WW. (4) 1977.

Psathyrella piluliformis (Bull.) P.D. Orton *Common Stump Brittlestem*
Psathyrella hydrophilum (Bull.) Maire
On stumps; common.
50, 51. SH 85, 87, 97; SJ 05-08, 15-18, 23-27, 34-36, 43. AV, BO, CC, CE, CN, DU, ER, GW, LH, LO, MF, ML, NF, PP, WW. (43) 1880. (48, 49, 52)

Psathyrella potteri A.A. Smith
Psathyrella prona f. *cana* Kits van Wav.; *Psathyrella atomata* auct.
In woodland soil; frequent.
50, 51. SJ 06, 16, 26, 34, 35. AV, ER, WW. (7) 1910. (49, 52)

Psathyrella prona (Fr.) Gillet f. **prona**
Psathyrella subatomata J.E. Lange
In woodland soil; rare.
51. SJ 07, 16. AV. (2) 1990.

Psathyrella pseudogracilis (Romagn.) M.M.Moser
In dune grassland; rare.
51. Recorded only from Point of Ayr (SJ 18) in 1995. New to Wales.

Psathyrella pygmaea (Bull.) Singer
On decayed wood; uncommon.
51. Recorded only from Rhydymwyn Nature Reserve (SJ 26) in 2007. New to North Wales.

Psathyrella sarcocephala (Fr.) Sing.
In litter in base-rich woods; frequent.
50, 51. SH 97; SJ 07, 16, 24, 26, 34. AV, BO, ER, PP. (8) 1910. (49)

Psathyrella spadicea (Schaeff.) Sing. *Chestnut Brittlestem*
In woodland litter; frequent.
50, 51. SH 87, 97; SJ 06, 07, 15, 26, 34-36. BO, CC, CE, ER, GW, ML, WW. (12) 1880. (49, 52)

Psathyrella spadiceogrisea (Schaeff.) G. Bertrand *Spring Brittlestem*
In woodland soil; common.
50, 51. SH 97; SJ 15, 17, 26. BO, CE, DU, LH, ML, MO. (8) 1976. (48, 49, 52)

Psathyrella sphagnicola (Maire) J. Favre *'Bog Britttlestem'*
In *Sphagnum* bogs; uncommon.
51. Recorded only from Llyn Helyg (SJ 25) in 1978. (48)

Psathyrella tephrophylla (Romagn.) Bon
In base-rich woodland soil; rare.
51. Recorded only from Rhydymwyn Nature Reserve (SJ 26) in 2012. (52)

Species Accounts

Family Pseudoclitocybaceae

Pseudoclitocybe cyathiformis (Bull.) Sing. *Goblet*
Cantharellula cyathiformis (Bull.) Sing.
In woodland litter; frequent.
50, 51. SH 97; SJ 05, 07, 15, 16, 24–27, 34–36. AV, CE, CF, ER, GW, MA, ML, ML, WW. (19) 1976. (48, 49, 52)

Family Pterulaceae

Radulomyces confluens (Fr.) M.P. Christ.
On rotten beech and ash wood; frequent.
50. SJ 36, 34. ER, MA. (2) 1977.

Family Schizophyllaceae

Fistulina hepatica (Schaeff.) With. *Beefsteak Fungus*
On oak and sweet chestnut trees, producing a reddish wood stain and frequently leading to 'stag-headed' trees; common.
50, 51. SH 77, 85, 87, 97; SJ 05–08, 16, 17, 23, 26, 34, 36. BO, BW, CC, CN, ER, GW, LH, ML, MO, WW. (25) 1880. (48, 49, 52)

Schizophyllum commune (Fr.) Fr. *Split Gill*
On fallen beech trunks; uncommon but increasing its range northwards.
50, 51. SH 97; SJ 07, 16, 36. BO, GW, ML. (5) 1974. (49, 52)

Family Strophariaceae

Agrocybe arvalis (Fr.) Sing.
In wood chips used as a garden mulch, increasing.
51. Recorded only from Rhuddlan (SJ 8) in 1996. (48, 52)

Agrocybe molesta (Lasch) Sing. *Bearded Fieldcap*
Agrocybe dura (Bolton) Sing.
In lawns and gardens; frequent.
51. SJ 17, 25–27. CE, MO, NM, WW. (6) 1978. (52)

Agrocybe paludosa (J.E. Lange) Kühn. & Romagn.
In wet grassland; uncommon.
50, 51. SJ 07, 18, 25, 34, 35. ER, PA. (5) 1974. (48, 49, 52)

Agrocybe pediades (Fr.) Fayod *Common Fieldcap*
incl. *Agrocybe semi-orbicularis* (Bull.) Fayod; *A. pusilla* (Fr.) Watling
In grassland; frequent.
50, 51. SH 97; SJ 06–08, 15–18, 25, 34, 35. AV, BO, NM, PA. (20) 1981. (48, 49, 52)

Agrocybe praecox (Pers.) Fayod *Spring Fieldcap*
In woodland soil in spring; common.
50, 51. SJ 07, 08, 16, 17, 25, 26, 34, 35. ER, MO, WW. (15) 1910. (48, 49, 52)

Agrocybe putaminum (Maire) Singer
On wood chippings and bark used as a mulch on shrubberies; increasing.
50, 51. SJ 08, 16, 17, 26, 36. LO, MO. (5) 1992.

Agrocybe rivulosa Nauta
On a pile of sycamore wood chippings; rare.
51. Recorded only from Hawarden (SJ 36) in 2015.

Cyclocybe erebia (Fr.) Vizzini & Mathey *Dark Fieldcap*
Agrocybe erebia (Fr.) Sing.
In woodland soil; common.
50, 51. SH 87, 97; SJ 05, 15–17, 24–26, 34–36. AV, CC, DU, ER, GW, LO, ML, MO, WW. (18) 1880. (48, 49, 52)

Deconica coprophila (Bull.) P. Karst.
Psilocybe coprophila (Bull.) P. Kumm.
On herbivore dung, espically of sheep; common.
50, 51. SJ 25, 34, 36. ER, GW, NF. (5) 1978. (48, 49, 52)

Deconica crobula (Fr.) Romagn.
Psilocybe crobula (Fr.) Singer
On woody debris; uncommon.
51. Recorded only from Nant-y-Frith (SJ 25) in 1910.

Deconica horizontalis (Bull.) Noordel.
Melanotus horizontalis (Bull.) P.D. Orton
On dead wood, often worked; uncommon.
50, 51. SH 77; SJ 16, 17. CC, LH. (3) 1994.

Deconica inquilinus (Fr.) Romagn.
Psilocybe inquilinus (Fr.) Bres.
On soil among grass; uncommon.
50, 51. SH 97; SJ 07, 17, 35. BO. (5) 1996.

Deconica merdaria (Fr.) Noordel.
Psilocybe merdaria (Fr.) Ricken
On old herbivore dung; frequent.
50, 51. SJ 16, 18. CC, PA. (3) 1976. (48, 49, 52)

Deconica montana (Pers.) P.D. Orton *Mountain Brownie*
Psilocybe montana (Pers.: Fr.) P. Kumm.
On soil with mosses; frequent.
50, 51. SJ 97; SJ 16, 35. BO, CC, MQ. (3) 1976. (49)

Deconica phillipsii (Berk. & Broome) Noordel.
Melanotus phillipsii (Berk. & Broome) Singer
On herbaceous remains; uncommon.
50, 51. SJ 06, 17, 26, 34. DU, ER, LO, WW. (6) 1985. (52)

Species Accounts

Hemipholiota destruens (Boud.) Quél.
Pholiota populnea (Pers.) Kuyp. & Tjall. Beuk; *P. destruens* (Boud.) Gillet
On standing poplar trunks; rare.
51. Recorded only from Drury (SJ 36) in 1978.

Hypholoma capnoides (Fr.) P. Kumm. *Conifer Tuft*
On conifer stumps; frequent.
50, 51. SH 85, 87, 97; SJ 05, 13, 16, 24–26, 34, 36. AV, BO, CC, CE, CF, CN, ER, GW, LO, MF, NF, PP. (14) 1974. (48, 49, 52)

Hypholoma elongatum (Pers.) Ricken *Sphagnum Brownie*
In *Sphagnum* bogs; frequent.
50, 51. SJ 16, 17, 25. HM, LH, MF. (3) 1978. (48, 49, 52)

Hypholoma epixanthum (Fr.) Quél.
On stumps; rare.
50. Recorded only from Erddig (SJ 34) in 1910.

Hypholoma ericaeoides P.D. Orton
On wet soil under alders; uncommon.
50. Recorded only from Glan-yr-Afon, Marli (SH 97) in 1979. (49)

Hypholoma ericaeum (Pers.) Kühn.
On bare peat in moorland; frequent.
50, 51. SJ 08, 16, 25, 43. HM, MF, NM. (5) 1978. (49)

Hypholoma fasciculare (Huds.) P. Kumm. *Sulphur Tuft*
On stumps and fallen logs, throughout the year; common.
50, 51. SH 76, 77, 85, 87, 95–97; SJ 05–08, 13, 15–18, 23–27, 34–36, 43. AV, BO, BW, CC, CE, CF, CN, DU, ER, GW, HC, HM, LH, LO, MA, MF, ML, MO, NF, PA, PP, WW. (105) 1880. (48, 49, 52)

Hypholoma lateritium (Schaeff.) P. Kumm. *Brick Tuft*
Hypholoma sublateritium (Fr.) Quél.
On stumps; frequent.
50, 51. SH 76, 85, 87, 95, 97; SJ 05, 07, 08, 15, 17, 26, 27, 34. CC, ER, HC, LO, WW. (17) 1880. (48, 49)

Hypholoma marginatum (Pers.) J. Schroet. *Snakeskin Brownie*
Hypholoma dispersum (Fr.) Quél.
On litter at edge of paths in conifer plantations; common.
50, 51. SH 95; SJ 05, 07, 16, 24–26, 34. CF, ER, MA, MF, NF, NM, WW. (15) 1910. (48, 49, 52)

Hypholoma myosotis (Fr.) M. Lange *Olive Brownie*
Pholiota myosotis (Fr.) Sing.
In wet, peaty soil; uncommon.
50, 51. SH 87; SJ 17, 25. CC, LH, NM. (3) 1880. (49, 52)

Hypholoma polytrichi (Fr.) Ricken
In clumps of *Polytrichum commune* in damp, acid woods; uncommon.
50, 51. SH 85; SJ 16, 25. HM, MF. (3) 1924. (49)

Hypholoma radicosum J.E. Lange *Rooting Brownie*
On stumps; rare.
50. SH 87; SJ 34. CC, ER. (2) 1880. Not seen in North East Wales since 1910. (48, 49)

Hypholoma subericaeum (Fr.) Kühn.
On bare peat at edge of lake; uncommon.
51. Recorded only from Hope Mountain (SJ 25) in 1978. (49, 52)

Hypholoma udum (Pers.) Kühn. *Peat Brownie*
On bare peat; frequent.
50, 51. SH 76; SJ 16, 25. HM, MF. (3) 1978. (49, 52)

Kuhneromyces mutabilis (Schaeff.) Sing. & A.H. Sm. *Sheathed Woodtuft*
Pholiota mutabilis (Schaeff.) P. Kumm.; *Galerina mutabilis* (Schaeff.) P.D. Orton
On stumps; common.
50, 51. SH 85, 87, 96, 97; SJ 05–07, 15–18, 24–27, 34–36, 54. AV, BO, CC, CE, CF, CN, ER, GW, HC, LO, MA, ML, MO, NF, PA, PP, WW. (43) 1880. (48, 49, 52)

Leratiomyces ceres (Cooke & Massee) Spooner & Bridge
In mixed woodland litter; rare.
50. Recorded only from Bodnant Garden (SH 77) in 2017.

Leratiomyces squamosus (Pers.) Bridge, Spooner, Beaver & Park
Stropharia squamosa (Pers.) Quél.; *Psilocybe squamosa* (Pers.) P.D. Orton
In woodland litter and on wood chips used in gardens; uncommon.
50, 51. SJ 23, 26, 34, 36. MO. (4) 1910.

Panaeolina foenisecii (Pers.) R. Maire *Brown Mottlegill*
Panaeolus foenisecii (Pers.) Kühn.
In rich grassland and lawns; common.
50, 51. SH 87, 97; SJ 06, 08, 16, 17, 24–26, 34–36. AV, BO, CC, CE, ER, GW, HC, LH, LO, MA, MO, NF, WW. (28) 1880. (48, 49, 52)

Panaeolus acuminatus (Schaeff.) Quél. *Dewdrop Mottlegill*
incl. *Panaeolus rickenii* Hora
In grassland on nitrogen-rich soils; common.
50, 51. SH 95, 97; SJ 05–08, 15–18, 23–27, 34–36. AV, BO, BP, CE, CF, DU, ER, GW, HC, HM, LO, MA, MO, NF, PA, WW. (43) 1973. (48, 49, 52)

Panaeolus cinctulus (Bolton) Sacc. *Banded Mottlegill*
Panaeolus subbalteatus (Berk. & Br.) Sacc.
On dung and in nitrogen-rich grassland; frequent.
50, 51. SJ 25, 26, 34. ER, HM, WW. (5) 1977. (48, 49, 52)

Panaeolus fimicola (Pers.) Gillet *Turf Mottlegill*
Panaeolus ater (Lange) Bon
In rich grassland; frequent.
50, 51. SH 97; SJ 07, 15, 18, 24–27, 34–36. AV, ER, GW, HC, ML, PA. (17) 1977. (48, 49, 52)

Panaeolus papilionaceus (Bull.) Quél. var. **papilionaceus** *Petticoat Mottlegill*
incl. *Panaeolus sphinctrinus* (Fr.) Quél.
In rich grassland; frequent.
50, 51. SH 87, 97; SJ 06, 08, 15-17, 23-27, 34-36, 44. AV, BO, DU, ER, GW, HC, HM, LO, MO, NF, NM, WW. (35) 1910. (48, 49, 52)

Panaeolus papilionaceus var. **parvisporus** Ew. Gerhardt *'Bell Mottlegill'*
Panaeolus campanulatus (Bull.) Quél.
In rich grassland and lawns; common.
50, 51. SH 85, 87, 96, 97; SJ 05, 07, 08, 15-18, 23-26, 34-36. AV, BO, BW, DU, ER, GW, LO, MF, MO, NF, PA, WW. (36) 1910. (48, 49)

Panaeolus semiovatus (Sow.) S. Lundell *Egghead Mottlegill.*
Panaeolus fimiputris auct.
On and near cattle and sheep dung; common.
50, 51. SH 87; SJ 07, 08, 15-18, 23-26, 35. AV, BO, CC, CE, HM, MA, MO, NF. (26) 1880. (48, 49, 52)

Pholiota adiposa (Fr.) P. Kumm.
On soil close to tree bases; uncommon.
50, 51. SJ 25-27, 34. ER, HC, LO. (5) 1973.

Pholiota alnicola (Fr.) Sing. *Alder Scalycap*
On dead wood, especially of alder; common.
50, 51. SH 77, 85, 87, 97; SJ 26, 34. CC, ER, ML. (6) 1880. (48, 49, 52)

Pholiota aurivella (Batsch.) P. Kumm. *Golden Scalycap*
Pholiota cerifera (P. Karst.) P. Karst.
On dead parts of trunks of living trees, and on fallen wood; uncommon.
50, 51. SH 96; SJ 05, 06, 08, 17, 34-36. CF, ER. (8) 1978. (48)

Pholiota connisans (Fr.) M.M. Moser
On dead wood of alder and willow; rare.
50. Recorded only from Coed Coch (SH 87) in 1880.

Pholiota flammans (Batsch.) P. Kumm. *Flaming Scalycap*
On conifer stumps and fallen trunks; uncommon but increasing.
50, 51. SH 97; SJ 05, 06, 13, 15, 16, 34, 43. BO, CF, CN, ER, MF. (13) 1974. (48, 49, 52)

Pholiota gummosa (Lasch.) Sing. *Stick Scalycap*
Pholiota ochrochlora (Fr.) P.D. Orton
On buried wood in woodland soils; frequent.
50, 51. SJ 05, 06, 08, 16-18, 25-27, 34-36. AV, CE, ER, GW, MO, NF, WW. (19) 1976. (48, 49, 52)

Pholiota highlandensis (Peck) Quadr. *Bonfire Scalycap*
Pholiota carbonaria (Fr.) Sing.
On bonfire sites in woods; uncommon.
50, 51. SH 97; SJ 17, 26, 34, 35. DU, ER, MO. (6) 1977. (48, 49, 52)

Pholiota lenta (Pers.) Sing.
On buried wood; uncommon.
50, 51. SJ 26, 34. ER, LO. (3) 1978.

Pholiota mixta (Fr.) Kuyp. & Tjall. Beuk.
On mixed leaf litter; rare.
51. Recorded only from Point of Ayr (SJ 18) in 1972.

Pholiota scamba (Fr.) M.M. Moser
On buried wood in woodland soil; uncommon.
50, 51. SJ 24, 25, 34. NF. (3) 1910. (48, 49)

Pholiota spumosa (Fr.) Sing.
On conifer wood; uncommon.
51. Recorded only from Cwm Woods (SJ 07) in 1998.

Pholiota squarrosa (Weigel) P. Kumm. *Shaggy Scalycap*
Parasitic on tree trunks, fruiting at ground level; common.
50, 51. SH 77, 97; SJ 05–07, 15, 16, 24–26, 34–36. AV, BO, CE, CF, ER, GW, ML, MO, NF, NM. (23) 1910. (48, 49, 52)

Pholiota tuberculosa (Schaeff.) P. Kumm.
On decayed wood; rare.
50. SH 87; SJ 34. CC, ER. (2) 1880. (48)

Protostropharia semiglobata (Batsch) Redhead et al. *Dung Roundhead*
Stropharia semiglobata (Batsch) Quél.
On cow and sheep dung; common.
50, 51. SH 87, 95–97; SJ 04–08, 13, 15–18, 23–27, 34–36, 43. AV, BO, CC, CE, CF, ER, GW, HC, HM, LO, MF, MO, NF, WW. (58) 1880. (48, 49, 52)

Psilocybe cyanescens Wakef. *Blueleg Brownie*
In grass, especially on sand; uncommon.
50, 51. SH 97; SJ 08, 18, 34. BO, ER, PA. (4) 1978. (48, 49, 52)

Psilocybe semilanceata (Fr.) P. Kumm. *Liberty Cap*
In grassland, especially playing fields; common.
50, 51. SH 85, 87, 95, 97; SJ 05–07, 13, 16, 17, 23, 25, 26, 34–36. AV, BO, CC, ER, GW, LO, MF, NM. (25) 1880.

Psilocybe subviscida (Peck) Kaufmann var. **velata** Noordel. & Verduin
In grass litter; rare.
50, 51. SH 85; SJ 26. MO. (2) 1924. (48, 49, 52)

Stropharia aeruginosa (Curt.) Quél. *'Verdigris Roundhead'*
In rich soil in gardens and woodland edges; common.
50, 51. SH 85, 87, 95, 97; SJ 06–08, 15–18, 24–27, 34–36. AV, BO, BP, CC, CE, ER, GW, LH, LO, MF, ML, MO, NF, PA, WW. (49) 1880. (48, 49, 52)

Stropharia caerulea Kreisel *Blue Roundhead*
Stropharia cyanea auct.
Among nettles at woodland edge; common.
50, 51. SH 97; SJ 05–08, 15, 18, 25, 26, 34–36. AV, BO, ER, GW, MO, PA, WW. (20) 1985. (48, 49, 52)

Species Accounts

Stropharia coronilla (Bull.) Quél. *Garland Roundhead*
In short grassland and dune meadows; common.
50, 51. SH 97; SJ 06, 16–18, 23, 25, 26, 35, 36. AV, LO, MA, MO, PA. (15) 1978. (48, 52)

Stropharia inuncta (Fr.) Quél. *Smoky Roundhead*
In grassy soil; uncommon.
50, 51. SJ 15, 24–26, 34, 35. CE, ER. (9) 1910. (49, 52)

Stropharia melanosperma (Bull.) Gillet
On woodland soil and, recently, on wood chips used as a garden mulch; rare.
50. SH 87; SJ 06, 24. CC. (3) 1880. This species was considered to be rare, or even extinct, but a few recent records from England, on wood chips, suggest a new lease of life. It was found at Foxhall (SJ 06) on wood chips in 1996.

Stropharia pseudocyanea (Desm.) Morg. *Peppery Roundhead*
Stropharia albocyanea (Fr.) Quél.
In leaf litter in woodland; frequent.
50, 51. SJ 15, 17, 18, 23–26. AV, DU, HC, PA. (10) 1910. (48, 49, 52)

Family Tricholomataceae

Aspropaxillus giganteus (Sibth.) Kuhner & Maire *Giant Funnel*
Leucopaxillus giganteus (Sibth.) Sing.
In woodland litter; common.
50, 51. SH 87, 95, 97; SJ 05, 13, 16, 17, 24–26, 34–36. AV, BO, CC, CE, DU, ER, GW, LO, MA, NF, PP, WW. (22) 1880. (49, 52)

Collybia amanitae (Batsch) Kreisel *Piggyback Shanklet*
Collybia cirrhata (Pers.) Quél.
On decaying agarics in woodland litter; frequent.
50, 51. SJ 16, 26, 34. CE, MA, MF. (5) 1984. (48, 49, 52)

Collybia cookei (Bres.) J.D. Arnold *Splitpea Shanklet*
On decaying agarics in woodland litter; frequent.
50, 51. SH 96; SJ 13, 15, 17, 24, 34, 35. ER, PP. (7) 1985. (48, 49, 52)

Collybia tuberosa (Bull.) P. Kumm. *Lentil Shanklet*
On decaying fungi, especially *Russula* and *Lactarius* spp.; frequent.
50, 51. SH 87, 95; SJ 05, 07, 17, 26, 34, 35. CC, ER, MA. (9) 1880. (49)

Dermoloma cuneifolium (Fr.) Bon *Crazed Cap*
In unimproved grassland; frequent.
50, 51. SH 85, 97; SJ 04–06, 08, 17, 23, 25, 26, 34, 35. CF, ER, HC, MA, ML. (18) 1910. (48, 49, 52)

Dermoloma pseudocuneifolium Bon
In limestone turf; uncommon.
51. Recorded only from Y Graig Nature Reserve, Tremeirchion (SJ 07) in 2000.

Leucopaxillus paradoxus (Constantin & L.M. Dufour) Boursier
In limestone woodland litter; rare.
50, 51. SH 97; SJ 17. ML. 1978.

Pogonoloma spinulosum (Kühn. & Romagn.) Sanchez et al. *Aromatic Meadowcap*
Porpoloma spinulosum (Kühn. & Romagn.) Sing.
In unimproved grassland; rare.
51. Recorded only from Ffrwd Quarry Nature Reserve (SJ 25) in 2001. **RDL/VU**

Pseudobaeospora dichroa Bas
On base-rich woodland soil; all four British species are rare.
51. Recorded only from Rhydymwyn Nature Reserve (SJ 26) in 2007. Only six previous British collections of the genus are recorded.

Tricholoma acerbum (Bull.) Quél. *Bitter Knight*
Mycorrhizal with oak and beech; uncommon.
50, 51. SH 85, 87; SJ 08, 26. BP, CC, LO. (4) 1985. (49, 52)

Tricholoma album (Schaeff.) P. Kumm. *White Knight*
Mycorrhizal with birch; frequent.
50, 51. SH 76, 85, 87, 97; SJ 16, 25, 26. CC, CN, LO, MA, ML. (10) 1880. (48, 49, 52)

Tricholoma atrosquamosum (Chev.) Sacc. var. **squarrulosum** (Bres.) M. Chr. & Noordel. *Dark Scaled Knight*
Tricholoma squarrulosum Bres.
Mycorrhizal with conifers; uncommon.
51. Recorded only from Nercwys Mountain (SJ 25) since 1981.

Tricholoma cingulatum (Almfert) Jacobasch *Girdled Knight*
Mycorrhizal with willows; uncommon.
50, 51. SH 87; SJ 08, 15–17, 25, 26, 35, 36. AV, BW, CC. (12) 1880. (48, 49, 52)

Tricholoma columbetta *Blue Spot Knight*
Mycorrhizal with beech; uncommon.
50, 51. SH 76, 87; SJ 15, 26, 34. CC, ER, LO. (5) 1880. (48, 49)

Tricholoma equestre (L.) P. Kumm. *Yellow Knight*
Tricholoma flavovirens (Pers.) S. Lundell & Nannf.
Mycorrhizal with pine; rare.
50. Recorded only from Coed Coch (SH 87) in 1880. (48, 49)

Tricholoma focale (Fr.) Ricken *Booted Knight*
Mycorrhizal with pine; probably extinct in England and Wales, rare in Scotland.
50. Recorded only from Coed Coch (SH 87) in 1880.

Tricholoma fulvum (Bull.) Bigeard & H. Guill. *Birch Knight*
Tricholoma flavobrunneum (Fr.) P. Kumm.
Mycorrhizal with birch; common.
50, 51. SH 76, 96, 97; SJ 15–17, 24–27, 36, 43. AV, GO, LH, LO, MF, NF, PP, WW. (16) 1978. (48, 49, 52)

Tricholoma imbricatum (Fr.) P. Kumm. *Matt Knight*
Mycorrhizal with pine; rare.
50, 51. SJ 07, 16, 17, 26. LO, MF. (4) 1981. (49)

Tricholoma inamoenum (Fr.) Gillet
Mycorrhizal with a variety of trees; rare.
50, 51. SH 87; SJ 26. CC, LO. (2) 1880.

Tricholoma inocybeoides A. Pears.
Mycorrhizal with birch; uncommon.
50. Recorded only from Nant Mill Country Park (SJ 24) in 1991.

Tricholoma lascivum (Fr.) Gillet *Aromatic Knight*
Mycorrhizal with beech on limestone; frequent.
50, 51. SJ 17, 23, 26. LO. (3) 1972. (49, 52)

Tricholoma orirubens Quél.
Mycorrhizal with beech, on limestone; uncommon.
50, 51. SJ 08, 16, 26. AV, BP, LO, MA. (5) 1978.

Tricholoma pessundatum (Fr.) Quél.
Mycorrhizal with pine; rare.
50. Recorded only from Coed Coch (SH 87) in 1880.

Tricholoma portentosum (Fr.) Quél.
Mycorrhizal with conifers; uncommon.
50. Recorded only from Coed Coch (SH 87) in 1880. There is an old record from v.c. 49.

Tricholoma psammopus (Kalchbr.) Quél. *Larch Knight*
Mycorrhizal with larch; frequent.
50, 51. SJ 23, 26, 35. LO. (3) 1978. (48, 49)

Tricholoma saponaceum (Fr.) P. Kumm. var. **saponaceum** *Soapy Knight*
Mycorrhizal with oak and beech; common.
50, 51. SH 77, 85, 87; SJ 07, 15–17, 26, 34–36. AV, CC, CE, ER, GW, LO, MA. (15) 1880. (48, 49, 52)

Tricholoma saponaceum var. **squamosum** (Cooke) Rea
Mycorrhizal with various trees on acid soil; rare.
50, 51. SJ 05, 15, 16. AV. (3) 2000. (49)

Tricholoma scalpturatum (Fr.) Quél. *Yellowing Knight*
Tricholoma argyraceum auct.
Mycorrhizal with broad-leaves, especially beech; frequent.
50, 51. SH 85; SJ 05, 07, 16, 17, 24–26, 34, 36. AV, CF, ER, GW, LO, MA, MF, WW. (18) 1910. (48, 49, 52)

Tricholoma sciodes (Pers.) C. Martín
Mycorrhizal with beech and birch; frequent.
50, 51. SJ 05, 16, 26, 34, 36. AV, CE, CF, ER, GW, LO, MA. (8) 1977. (48, 49, 52)

Tricholoma sejunctum (Sow.) Quél. *Deceiving Knight*
Mycorrhizal with conifers; frequent.
50, 51. SH 97; SJ 26. LO, ML. (2) 1978. (49, 52)

Tricholoma stiparophyllum (S. Lundell) P. Karst.
Mycorrhizal with birch; frequent, but confused with *T. album*.
51. Recorded only from Rhydymwyn Nature Reserve (SJ 26) in 2007. (49, 52)

Tricholoma sulphurescens Bres.
Tricholoma resplendens auct.
Mycorrhizal with deciduous trees on acid soil; rare.
50. SH 85, 87; SJ 24, 34. CC. (4) 1880. (49) Not seen in North Wales since 1924.

Tricholoma sulphureum (Bull.) P. Kumm. var. **sulphureum** *Sulphur Knight*
Mycorrhizal with oak and beech; frequent.
50, 51. SH 87, 97; SJ 07, 15, 16, 24, 26, 36. AV, CC, GW, LO, MA, ML, WW. (12) 1880. (48, 49, 52)

Tricholoma sulphureum var. **hemisulphureum** Kühner
Mycorrhizal with *Helianthemum nummularium*; uncommon.
50. Recorded only from Bryn Alyn (SJ 25) in 2003. (49)

Tricholoma terreum (Schaeff.) P. Kumm. Grey Knight
incl. *Tricholoma myomyces* (Pers.) J.E. Lange
Mycorrhizal with pines; common.
50, 51. SH 77, 85, 87, 97; SJ 16, 17, 24–27, 34–36. AV, CC, CE, CN, ER, GW, LO, MA, ML, NM, PP. (26) 1880. (48, 49, 52)

Tricholoma ustale (Fr.) P. Kumm. *Burnt Knight*
Mycorrhizal with beech; common.
50, 51. SH 77, 85, 96; SJ 05, 06, 15, 24, 26, 34, 36. CE, CF, ER, GW, LO, MA, MO, MF. (19) 1924. (48, 49, 52)

Tricholoma ustaloides Romagn.
Tricholoma albobrunneum auct.
Mycorrhizal with beech; frequent.
50, 51. SH 85, 97; SJ 16, 17, 23, 24, 26, 34–36. AV, CE, ER, GW, LO, MF, ML, PP, WW. (13) 1910. (48, 49, 52)

Tricholoma vaccinum (Schaeff.) P. Kumm. *Scaly Knight*
Mycorrhizal with pine; rare.
50. SH 87; SJ 34. CC, ER. (2) 1880. (49) Last seen in our area in 1910.

Tricholoma virgatum (Fr.) P. Kumm. *Ashen Knight*
Mycorrhizal with broad-leaves; frequent.
50, 51. SH 85; SJ 05, 08, 16, 23, 26, 34. AV, CE, ER, LO, MA, MO. (11) 1910. (48, 49, 52)

Family Tubariaceae

Flammulaster carpophilus (Fr.) Earle var. **subincarnatus** (Joss. & Kühn.) Vellinga
On fallen beech mast; uncommon.
51. Recorded only from Loggerheads Coiuntry Park (SJ 26) in 1977.

Flammulaster granulosus (J.E. Lange) Watling
On soil in deciduous woodland; uncommon.
50, 51. SJ 16, 26, 34, 35. AV, CE, WW. (7) 1983. (48, 49, 52)

Species Accounts

Flammulaster limulatus (Fr.) Watling
On fallen beech wood; rare.
50. Recorded only from Glan-yr-Afon, Marli (SH 97) in 1980. (49, 52)

Flammulaster muricatus (Fr.) Watling
On fallen branch; uncommon.
51. Recorded only from Rhydymwyn Nature Reserve (SJ 26) in 2007.

Phaeocollybia lugubris (Fr.) R. Heim *Russet Rootshank*
In conifer litter; rare.
50. Reported only from Coed Coch (SH 87) in 1880; now confined to Scotland.

Phaeomarasmius erinaceus (Fr.) Kühn.
On sticks in woodland litter; frequent.
50, 51. SJ 15, 17, 18, 25, 35. DU, NF, PA. (6) 1910. (48, 49, 52)

Tubaria conspersa (Pers.) Fayod *Felted Twiglet*
On small sticks in woodland litter; frequent.
50, 51. SH 97; SJ 08, 17, 26, 35. AV, BO. (5) 1998. (48, 49, 52)

Tubaria dispersa (Pers.) Sing.
Tubaria autochthona (Berk. & Br.) Sacc.
On hawthorn fruits under hedges; uncommon.
50, 51. SJ 04, 07, 15, 16, 18, 26, 34. AV, ER, LO, MA, MO, PA. (11) 1977. (52)

Tubaria furfuracea (Pers.) Gillet *Scurfy Twiglet*
On twigs under hedges and in woodland litter; common.
50, 51. SH 77, 85, 97; SJ 05–08, 15–18, 24–27, 34–36. AV, BP, CE, CF, DU, ER, GW, LH, LO, MA, MF, ML, MO, NF, NM, PA, WW. (57) 1910. (48, 49, 52)

Tubaria hiemalis Bon *Winter Twiglet*
On twigs and litter in hedges in winter; common.
50, 51. SJ 15, 16, 26. MA, MO. (5) 1994. (48, 52) Usually included in the last species but differs in microscopic characters and its appearance only in winter.

Family Typhulaceae

Panellus mitis (Pers.) Sing. *Elastic Oysterling*
On ends of rotten conifer logs; frequent.
50, 51. SH 87; SJ 13, 16, 26. CC, CE, LO, MF. (6) 1880. (48, 49, 52)

Panellus stipticus (Bull.) P. Karst. *Bitter Oysterling*
On fallen trunks and branches; common.
50, 51. SH 85; SJ 16, 24, 26, 36. CE, GW, MF, MO. (6) 1910. (48, 49, 52)

Sarcomyxa serotina (Pers.) P. Karst. *Olive Oysterling*
Panellus serotinus (Pers.) Kühn.
On fallen trunks, especially of beech; frequent.
50, 51. SH 95, 97; SJ 05, 16, 17, 25, 26, 34. CE, CN, ER, LO, ML. (10) 1977. (48, 49, 52)

Tricholomopsis decora (Fr.) Sing. *Prunes and Custard*
On rotting conifer stumps and fallen trunks; rare south of Scotland.
51. Recorded only from Cwm Woods (SJ 07) in 1995.

Tricholomopsis rutilans (Schaeff.) Sing. *Plums and custard*
On rotting conifer stumps and trunks; common.
50, 51. SH 77, 87, 95, 97; SJ 05-08, 13, 15-17, 23-27, 34, 36, 43. AV, BO, CC, CE, CF, ER, GW, HC, LH, LO, MA, MF, ML, NF, NM, WW. (49) 1880. (48, 49, 52)

Typhula contorta (Holmsk.) Olariaga
Macrotyphula fistulosa var. *contorta* (Holmsk.) Nannf. & L. Holm; *Clavariadelphus fistulosus* var. *contortus* (Holmsk.) Corner
On attached, dead birch twigs, often at head height; uncommon.
50, 51. SJ 15, 24, 26, 35. LO. (4) 1978.

Typhula erythropus (Pers.) Fr. *Redleg Club*
On leaf litter in damp woods; common.
50, 51. SH 97; SJ 07, 08, 15-17, 25-27, 34, 43. AV, BO, BW, DU, ER, HC, LH, LO, MA, NF, WW. (23) 1910. (48, 49, 52)

Typhula fistulosa (Holmsk.) Olariaga *Pipe Club*
Macrotyphula fistulosa (Holmsk.) R.H. Petersen var. *fistulosa*; *Clavariadelphus fistulosus* (Holmsk.) Corner var. *fistulosus*
In deep woodland litter; uncommon.
50, 51. SJ 17, 24, 26, 27, 35, 36. AV, GW, LH, LO, PP. (9) 1985. (48, 49, 52)

Typhula juncea (Holmsk.) Olariaga
Macrotyphula juncea (Fr.) Berthier; *Clavariadelphus junceus* (Fr.) Corner
In deep woodland litter; uncommon.
50, 51. SJ 07, 15. (2) 1994. (48, 49, 52)

Typhula micans (Pers.) Berthier
Pistillaria micans (Pers.) Fr.
On fallen leaves; uncommon.
50, 51. SJ 26, 34. ER, MO. (2) 1910.

Typhula phacorrhiza (Reichardt) Fr.
In leaf litter in damp woodland; uncommon.
50. SJ 15, 34. ER. (2) 1979. (49, 52)

Typhula quisquiliaris (Fr.) Corner *Bracken Club*
On dead bracken stems; frequent.
50, 51. SH 85, 87; SJ 16, 17, 25, 26, 34. AV, CC, CE, ER, LH, MF, MO, NF. (11) 1880. (48, 49)

Typhula setipes (Grev.) Berthier
Pistillaria setipes Grev.; *Typhula gyrans* (Batsch) Fr.; *T. ovata* auct.; *T. sclerotioides* auct.; *T. pusilla* auct.
On leaf litter and herbaceous stems in marshy places; common.
50, 51. SH 97; SJ 07, 17, 18, 24, 26, 34. BO, ER, LH, MO, PA, WW. (9) 1910. (49)

Species Accounts

Family Volvariaceae

Volvariella murinella (Quél.) Courtec.
In unimproved grassland; rare.
50. Recorded only from Glan-yr-Afon, Marli (SH 97) in 1979.

Order AMYLOCORTICIALES
Family Plicaturaceae

Plicaturopsis crispa (Pers.) D.A. Reid
Plicatura crispa (Pers.) Rea
On dead hazel branches; rare.
50. Recorded only from Hafod Wood (SH 85) in 1924.

Order ATHELIALES
Family Atheliaceae

Athelia arachnoidea (Berk.) Jülich
Corticium centrifugum auct.
On lichens on tree trunks and on rotten wood; common.
50, 51. SH 85, 87; SJ 05, 15, 26, 34. CC, CE, ER. (7) 1880. (52)

Athelia epiphylla Pers.
Parasitic on lichens on trees; common.
50, 51. SH 85; SJ 24, 26, 34. ER, LO, WW. (5) 1924.

Byssocorticium pulchrum (S. Lundell) M.P. Christ.
On rotten wood; rare.
50. Recorded only from Aberduna Nature Reserve, Maeshafn (SJ 26) in 2001. (48)

Byssocorticium terrestre (DC.) Bondartsev & Sing.
On soil and rotten wood; rare.
50. Recorded only from Erddig (SJ 34) in 1910.

Order AURICULARIALES
Family Auriculariaceae

Auricularia auricula-judae (Bull.) Wettst. *Jelly Ear*
Hirneola auricula-judae (Bull.) Berk.
On dying parts of trunks and attached branches, especially of elder, but on many other species of trees, including barberry, beech and laburnum; common.
50, 51. SH 77, 87, 97; SJ 05–08, 15–18, 23–27, 34–36. AV, BO, CC, CE, CN, DU, ER, GW, HC, LH, LO, MA, ML, MO, NF, PA, PP, WW. (97) 1880. (48, 49, 52)

Auricularia mesenterica (Dicks.) Pers. *Tripe Fungus*
On fallen trunks, especially of elm; frequent.
50, 51. SJ 06, 07, 15, 23, 36. GW. (5) 1976. (49, 52)

Family Exidiaceae

Eichleriella deglubens (Berk. & Broome) D.A. Reid
On rotten wood of deciduous trees; uncommon.
50, 51. SJ 06, 16, 26. CN, MA, WW. (4) 1993.

Exidia glandulosa (Bull.) Fr. *Witches' Butter*
Exidia truncata Fr.
On dead branches, both fallen and attached, especially of oak; common.
50, 51. SJ 08, 15-17, 23-25, 34-36. CN, DU, ER, GW, PP, WW. (21) 1978. (48, 49, 52) Typically top-shaped fruit bodies distinguish this from *E. plana*, which is often confused with it; the latter has brain-like and flattened fruit bodies.

Exidia nigricans (With.) P. Roberts
Exidia plana (F.H. Wigg.) Donk; *E. glandulosa* in part
On dead attached and fallen branches, especially of beech and sycamore; common.
50, 51. SH 76, 85; SJ 07, 08, 16-18, 25, 26, 34-36. AV, BW, CE, ER, GW, LH, LO, MO, NM, WW. (29) 1924.

Exidia recisa (Ditmar) Fr.
On dead attached twigs of willows, usually above streams; uncommon.
50, 51. SJ 17, 26, 27, 34. AV, ER, WW. (5) 1978. (48)

Exidia saccharina (Alb. & Schwein.) Fr.
On dead pine branches; rare.
50, 51. SJ 15, 25, 26. CE, LO, NM. (4) 1984. (48)

Exidia thuretiana (Lév.) Fr. *White Brain*
Exidia albida auct.
On fallen branches, especially of beech and sycamore; common.
50, 51. SH 85, 87, 95, 97; SJ 05-08, 16, 17, 24-26, 34-36. AV, BO, CC, CE, CF, DU, ER, GW, LO, MA, MF, MO, NF, PP, WW. (36) 1880. (48, 49, 52)

Exidiopsis calcea (Pers.) K. Wells
On rotten wood; uncommon.
50, 51. SJ 07, 24, 34. (4) 1910.

Exidiopsis effusa (Sacc.) Møll.
On rotten wood; uncommon.
51. Recorded only from Celyn Woods (SJ 26) in 1990.

Guepinia helvelloides (DC.) Fr. *Salmon Salad*
Tremiscus helvelloides (DC.) Donk
On woodland soil; rare.
50, 51. SJ 05, 17. CF, DU. (2) 1993.

Myxarium nucleatum Wallr. *Crystal Brain*
Exidia nucleata (Schwein.) Burt
On fallen sticks and small branches, especially nof beech and sycamore; common.
50, 51. SH 97; SJ 05-07, 15-18, 24-27, 34-36. BO, CE, ER, GW, LH, LO, MA, NF, NM, PA, PP, WW. (38) 1976. (48, 49, 52)

Pseudohydnum gelatinosum (Scop.) P. Karst. *Jelly Tooth*
On rotten stumps and trunks of conifers; frequent.
50, 51. SH 85; SJ 05, 17, 24, 26. CF, LH, LO, MO, PP, WW. (9) 1924. (48, 49, 52)

Order BOLETALES
Family Boletaceae

Boletus edulis Bull. *Penny Bun*
Mycorrhizal with broad-leaves, especially beech; frequent.
50, 51. SH 76, 87, 95–97; SJ 05–07, 16–18, 24–27, 35, 36. BO, CC, CF, GW, LH, LO, MA, MF, ML, NM, WW. (31) 1880. (48, 49, 52)

Boletus reticulatus Schaeff. *Summer Bolete*
Boletus aestivalis Fr.
Mycorrhizal with beech and oak; uncommon.
50, 51. SH 87, 96; SJ 26, 36. CC, GW, LO. (4) 1976. (49)

Buchwaldoboletus lignicolor (Kallenb.) Pilát *Wood Bolete*
Pulveroboletus lignicola (Kallenb.) E.A. Dick & Snell; *Boletus lignicola* Kallenb.
On spruce stumps, associated with the polypore *Phaeolus schweinitzii*; rare, but increasing.
50, 51. SH 76, 97; SJ 17. BO, LH. (3) 1979. (49)

Butyriboletus appendiculatus (Schaeff.) Arora & Frank *Oak Bolete*
Boletus appendiculatus Schaeff.
Mycorrhizal with broad-leaves, especially oak; uncommon.
50, 51. SJ 23, 26. LH. 1972. (49)

Caloboletus calopus (Pers.) Vizzini *Bitter Beech Bolete*
Boletus calopus Pers.
Mycorrhizal with beech; uncommon.
50, 51. SH 87, 95; SJ 17, 26. CC, LO. (4) 1880. (48, 49, 52)

Caloboletus radicans (Pers.) Vizzini *Rooting Bolete*
Boletus radicans Pers.; *Boletus albidus* Roques
Mycorrhizal with oak; uncommon.
50, 51. SJ 17, 24, 36. GW. (3) 1994. (49, 52)

Chalciporus piperatus (Bull.) Bat. *Peppery Bolete*
Boletus piperatus Bull.
Mycorrhizal with birch on acid soils; frequent.
50, 51. SH 85, 95–97; SJ 05, 13, 16, 17, 24–26, 36. CE, CF, GW, MF, PP. (20) 1910. (48, 49, 52)

Cyanoboletus pulverulentus (Opat.) Vizzini *Inkstain Bolete*
Boletus pulverulentus Opat.
Mycorrhizal with isolated oaks, especially in parkland; uncommon.
50, 51. SH 97; SJ 15, 16, 34, 36. CN, ER, GW, ML. (5) 1979. (48, 49, 52)

Hemileccinum impolitum (Fr.) Šutara *Iodine Bolete*
Boletus impolitus Fr.
Mycorrhizal with oak; uncommon.
50, 51. SH 97; SJ 34. BO, ML. (3) 1910. (49)

Hortiboletus engelii (Hlaváček) Biketova & Wasser
Xerocomellus engelii (Hlaváček) Šutara; *Xerocomus declivitatum* (C. Martín) Klofac;
Boletus declivitatum (C. Martín) Watling; *Boletus communis* auct.
Mycorrhizal with oak; probably common but confused with the previous two species.
50, 51. SJ 05, 16, 25, 26, 35, 36. AV, CF, GW, MO, NF. (6) 2003.

Hortiboletus rubellus (Krombh.) Wu et al. *Ruby Bolete*
Xerocomellus rubellus (Krombh.) Šutara; *Xerocomus rubellus* (Krombh.) Quél.;
Boletus rubellus Krombh.; *Boletus versicolor* Rostk.
Mycorrhizal with oak; rare.
50, 51. SH 97; SJ 23, 26. BO, WW. (3) 1979. (52)

Imleria badia (Fr.) Vizzini *Bay Bolete*
Boletus badius (Fr.) Fr.; *Xerocomus badius* (Fr.) E.J. Gilb.
Mycorrhizal with conifers; common.
50, 51. SH 76, 95-97; SJ 05, 07, 16, 17, 23-26, 34, 36, 43. AV, BO, CE, CF, ER, GW, LH, LO, MF, ML, NF, NM, PP, WW. (42) 1973. (48, 49, 52)

Leccinellum crocipodium (Letell.) Bresinsky & Manfr. Binder *Saffron Bolete*
Leccinum crocipodium (Letell.) Watling
Mycorrhizal with oak; uncommon.
51. SJ 26, 27. HC. (2) 1984.

Leccinellum pseudoscabrum (Kallemb.) Bresinsky & Manfr. Binder *Hazel Bolete*
Leccinum pseudoscaber (Kallemb.) Šutara; *L. carpini* (R. Schulz.) D.A. Reid
Mycorrhizal with hazel and hornbeam; uncommon.
50, 51. SJ 15, 17, 26, 34. DU, ER, MA. (4) 1979. (52)

Leccinum aurantiacum (Bull.) Gray *Orange Oak Bolete*
Leccinum quercinum (Pilát) Pilát & Dermeck
Mycorrhizal with oak; uncommon.
51. Recorded only from Celyn Wood (SJ 26) in 1984.

Leccinum roseofractum Watling *Blushing Bolete*
Mycorrhizal with birch; uncommon.
50, 51. SJ 16, 24, 36. AV. (3) 1983. (48, 49, 52)

Leccinum roseotinctum Watling
Mycorrhizal with birch; uncommon.
50, 51. SJ 24, 26. LO. (2) 1998.

Lecccinum scabrum (Bull.) Gray *Brown Birch Bolete*
Mycorrhizal with birch; common.
50, 51. SH 76, 77, 85, 96, 97; SJ 05, 16, 17, 23-27, 34-36, 43. AV, BO, CE, CN, ER, GW, LH, LO, MA, MF, ML, NF. (38) 1924. (48, 49, 52)

Leccinum variicolor Watling *Mottled Bolete*
Mycorrhizal with birch; the commonest species of the genus with birch in Britain.
50, 51. SH 85, 96; SJ 05-07, 16, 17, 24, 25, 35, 43. AV, CE, CF, CN, NF, PP. (18) 1983. (48, 49, 52)

Species Accounts

Leccinum versipelle (Fr. & Hök) Snell *Orange Birch Bolete*
Boletus testaceoscaber Secr.
Mycorrhizal with birch; frequent.
50, 51. SH 76, 77; SJ 05, 17, 25, 36. CF, NF. (7) 1976. (48, 49, 52)

Neoboletus luridiformis (Rostk.) Gelardi *Scarletina Bolete*
Boletus luridiformis Rostk.; *B. erythropus* auct.
Mycorrhizal with oak and, rarely, with *Helianthemum*; frequent.
50, 51. SH 76, 97; SJ 05, 13, 17, 25–27, 36. AV, BO, CE, CF, GW, HC, LH, LO, ML, MO, NM. (17) 1974. (48, 49, 52)

Porphyrellus porphyrosporus (Fr. & Hök) E.-J. Gilb. *Dusky Bolete*
P. pseudoscaber (Secr.) Sing.; *Tylopilus porphyrosporus* (Fr. & Hök) Smith & Thiers
Mycorrhizal with oak; uncommon.
50, 51. SH 85; SJ 05, 16, 24–26, 36. AV, CE, GW, WW. (9) 1910. (48, 49)

Pseudoboletus parasiticus (Bull.) Šutara *Parasitic Bolete*
Boletus parasiticus Bull.; *Xerocomus parasiticus* (Bull.) Quél.
Appearing to be parasitic on *Scleroderma citrinum* but probably co-mycorrhizal with oak; frequent.
50, 51. SJ 35, 36, 43. GW. (4) 1978. (48, 49)

Strobilomyces strobilaceus (Scop.) Berk. *Old Man of the Woods*
Strobilomyces floccopus (Vahl) P. Karst.
Mycorrhizal with beech; rare.
50. Recorded only from Nant Mill Country Park (SJ 24) in 1997.

Suillellus luridus (Schaeff.) Murrill *Lurid Bolete*
Boletus luridus Schaeff.
Mycorrhizal with oak and, in the absence of trees, *Helianthemum*; frequent.
50, 51. SH 97; SJ 07, 15, 18, 23, 25, 26, 34, 35. BO, ER, LO, MA, NM, WW. (14) 1974. (49, 52)

Suillellus queletii (Schulzer) Vizzini *Deceiving Bolete*
Boletus queletii Schulzer
Mycorrhizal with beech and evergreen oak; uncommon.
50, 51. SH 96; SJ 08, 16, 26, 34. BW, ER, LO, MO. (6) 1977. (49)

Tylopilus felleus (Bull.) P. Karst. *Bitter Bolete*
Mycorrhizal with beech; uncommon.
50. SH 85, 97; SJ 05, 24, 26. CE, ML, PP. (5) 1979. (49, 52)

Xerocomellus chrysenteron (Bull.) Šutara *Red Cracking Bolete*
Xerocomus chrysenteron (Bull.) Quél.; *Boletus chrysenteron* Bull.
Mycorrhizal with beech and conifers; common.
50, 51. SH 76, 77, 87, 95–97; SJ 04, 05, 07, 08, 16–18, 23–27, 34, 36. AV, BO, CC, CE, CN, ER, HC, LO, MA, MF, ML, MO, NM, PP, WW. (51) 1880. (48, 49, 52) Some of the older records may refer to the next two species and *X. ripariellus*.

Xerocomellus cisalpinus (Sinmonini, Ladurner & Peintner) Šutara
Xerocomus cisalpinus Simonini, Ladurner & Peintner; *Boletus cisalpinus* (Simonini, Ladurner & Peintner) Watling & A. Hills
Mycorrhizal with oak, beech and conifers; probably common.
50, 51. SJ 05, 26, 35, 36. CF, GW, LO, MO. (6) 2002. This newly described species has probably been misidentified as X. *chrysenteron*. This species is clearly better placed in *Xerocomellus* but the combination has not yet been published.

Xerocomellus porosporus (Imler ex G. Moreno) Šutara *Sepia Bolete*
Xerocomus porosporus (Imler ex G. Moreno) Contu; *Boletus porosporus* Imler ex G. Moreno
Mycorrhizal with oak; frequent.
50, 51. SH 95; SJ 05, 15, 16, 26, 27, 34, 36. AV, ER, HC, MF, WW. (10) 1977. (49)

Xerocomellus pruinatus (Fr. & Hök) Šutara *Matt Bolete*
Xerocomus pruinatus (Fr. & Hök) Quél.; *Boletus pruinatus* Fr. & Hök
Mycorrhizal with oak; common.
50, 51. SH 85; SJ 06, 16, 23, 24, 26, 34, 36. AV, CE, ER, GW, MO, PP, WW. (12) 1910. (49, 52)

Xerocomellus ripariellus (Redeuilh) Šutara
Xerocomus ripariellus Redeuilh; *Boletus ripariellus* (Redeuilh) Watling & A. Hills
Mycorrhizal with alder and willow; probably frequent.
50, 51. SJ 17, 25. LH. (2) 1978. (48, 49) Probably misidentified as B. *chrysenteron* in the past.

Xerocomus ferrugineus (Schaeff.) Bon
Boletus ferrugineus Schaeff.; *B. spadiceus* Fr.; *B. lanatus* Rostk.
Mycorrhizal with beech and birch; frequent.
51. SJ 16, 25, 26, 36. AV, CE, NF. (4) 1981. (48, 49, 52)

Xerocomus subtomentosus (L.) Quél. *Suede Bolete*
Boletus subtomentosus L.
Mycorrhizal with oak and birch; common.
50, 51. SH 87, 97; SJ 05, 07, 08, 17, 25, 26, 34, 36. BO, CC, CE, CF, ER, GW, LH, ML, NF, WW. (14) 1880. (48, 49, 52)

Family Coniophoraceae

Coniophora puteana (Schum.) P. Karst. *Wet Rot*
On damp rotten wood in woods and buildings; common.
50, 51. SJ 08, 18, 24, 26, 34. AV, CE, ER, MO, PP, WW. (7) 1910. (49, 52)

Leucogyrophana mollusca (Fr.) Pouzar
On decayed conifer wood; uncommon.
50. Recorded only from World's End (SJ 24) in 1910.

Serpula himantioides (Fr.) P. Karst.
On rotten conifer wood; frequent.
51. SJ 26, 35. (2) 1972.

Species Accounts

Serpula lacrymans (Wulf.) J. Schröt. *Dry Rot*
On wood in buildings, never found in nature in Britain; common.
50, 51. SJ 08, 16, 26, 27, 34–36. MO. (11) 1973. This fungus was probably imported with tropical hardwoods and is very much at home in centrally heated houses!

Family Gomphidiaceae

Chroogomphus rutilus (Schaeff.) O.K. Mill. *Copper Spike*
Gomphidius viscidus auct.
Mycorrhizal with pines; frequent.
50, 51. SH 97; SJ 15, 16, 24. AV, PP. (5) 1902. (48, 49, 52)

Gomphidius glutinosus (Schaeff.) Fr. *Slimy Spike*
Mycorrhizal with conifers; frequent.
50, 51. SH 87, 97; SJ 05, 16, 26. CC, LO, MF. (5) 1869. (49, 52)

Gomphidius maculatus Fr.
Gomphidius gracilis Berk. & Br.
Mycorrhizal with larch; rare.
50, 51. SJ 15, 16, 24, 26, 35. LO. (5) 1910. (49, 52)

Gomphidius roseus (Fr.) Fr. *Rosy Spike*
Mycorrhizal with pine, always in association with *Suillus bovinus*; uncommon.
50, 51. SH 87; SJ 07, 26. CC, LO. (3) 1880. (48, 49, 52)

Family Gyroporaceae

Gyroporus castaneus (Bull.) Quél. *Chestnut Bolete*
Mycorrhizal with oak; uncommon.
51. SH 97; SJ 17. BO, LH. (2) 1962. (49)

Gyroporus cyanescens (Bull.) Quél.
Mycorrhizal with birch; rare.
50. Recorded only from Fairy Glen (SH 85) in 1924. (49)

Family Hygrophoropsidaceae

Hygrophoropsis aurantiaca (Wulf.) Maire var. **aurantiaca** *False Chanterelle*
In litter under conifers; common.
50, 51. SH 77, 87, 95, 97; SJ 05–08, 15–18, 24–27, 34–36, 43. BO, CC, CE, CF, ER, GW, HC, HM, MF, ML, MO, NM, PA, PP, WW. (39) 1880. (48, 49, 52)

Hygrophoropsis aurantiaca var. **pallida** (Cooke) Kühn. & Romagn.
In litter under conifers and on moorland; uncommon.
50, 51. SH 87, 96; SJ 07, 16, 25, 26. CC, HM, MF, MO, NM, WW. (8) 1880. (48, 49)

Family Melanogastraceae

Melanogaster broomieanus Berk. *'Bath Truffle'*
On or just under soil surface, associated with broad-leaved trees; uncommon.
50, 51. SJ 17, 26, 36. GW, MA. (3) (49) **PF**

Family Paxillaceae

Paxillus ammoniavirescens Dessi & Contu *Greening Rollrim*
Paxillus validus C. Hahn
Mycorrhizal with *Eucalyptus and Salix*; frequent.
50, 51. SJ 26, 35. MO. (2) 2004.

Paxillus involutus (Batsch) Fr. *Brown Rollrim*
Mycorrhizal with birch; common.
50, 51. SH 76, 77, 85, 95, 97; SJ 05–07, 13, 15–18, 23–27, 34–36, 43.
AV, BO, CC, CE, CF, CN, ER, GW, LH, LO, MF, ML, MO, NF, NM, PP, WW. (72) 1880. (48, 49, 52) This has been shown recently to be a species-complex, with several taxa in the British Isles, distinguished largely by chemical reactions.

Paxillus olivellus P.-A. Moreau, J.E. Chaumetton, H. Gryta & P. Jangeat
Paxillus rubicundulus P.D. Orton; *P. filamentosus* auct.
Mycorrhizal with alder; uncommon.
50, 51. SJ 08, 35. (2) 1998. (48, 49, 52)

Family Rhizopogonaceae

Rhizopogon roseolus (Corda) Th. Fr. *'Rosy False Truffle'*
In soil under pines; uncommon.
51. Recorded only from Crown Wood, Whitford (SJ 17) in 1993.

Family Sclerodermataceae

Scleroderma areolatum Ehrenb. *Leopard Earthball*
In soil in oak woodland; common.
50, 51. SH 85, 97; SJ 05, 07, 08, 16–18, 24, 26, 27, 34–36. ER, GW, LH, LO, MF, ML, PP, WW. (19) 1981. (48, 49, 52)

Scleroderma bovista Fr. *Potato Earthball*
In sandy soil and dry woodlands and gardens; common.
50, 51. SJ 07, 08, 16–18, 24, 26, 34. CN, ER, LO, MO, PA, WW. (14) 1954. (48, 49, 52)

Scleroderma cepa Pers. *'Onion Earthball'*
On sandy and marl soils in woodland and open habitats; uncommon.
50, 51. SJ 05, 17, 34. DU, ER. (3) 1984.

Scleroderma citrinum Pers. *Common Earthball*
Scleroderma aurantium auct.; *S. vulgare* Hornem.
Mycorrhizal with oak and birch in acid woodlands; common.
50, 51. SH 76, 77, 85, 97; SJ 05–08, 16, 17, 23, 24, 26, 34–36, 43, 53. CE, CF, CN, ER, GW, LH, MA, MF, ML, MO, PP, WW. (35) 1910. (48, 49, 52)

Species Accounts

Scleroderma verrucosum (Bull.) Pers. *Scaly Earthball*
In rich woodland soil; common.
50, 51. SH 77, 85, 87, 97; SJ 05, 07, 08, 15–17, 23–26, 34–36, 43. AV, BO, CC, ER, GW, MO, NM, PP, WW. (34) 1880. (48, 49, 52)

Family Suillaceae

Suillus bovinus (L.) Roussel *Bovine Bolete*
Mycorrhizal with pine; frequent.
50, 51. SH 87, 95, 97; SJ 07, 16, 25, 26. CC, LO, MF, NF. (8) 1880. (48, 49, 52)

Suillus collinitus (Fr.) Kuntze
Suillus fluryi Huijsm.
Mycorrhizal with pine on limestone; uncommon.
50. SH 97; SH 26. MA. (2) 1999. (52)

Suillus granulatus (L.) Roussel *Weeping Bolete*
Mycorrhizal with pine; frequent.
50, 51. SH 97; SJ 17, 24–26, 35. NM, WW. (7) 1978. (48, 49, 52)

Suillus grevillei (Klotsch) Sing. *Larch Bolete*
Boletus elegans Schum.
Mycorrhizal with larch; common.
50, 51. SH 85, 87; SJ 05, 07, 13, 15–17, 23–26, 34, 35. AV, CC, CE, CF, ER, HM, LO, MA, MF, NM, WW. (36) 1880. (48, 49, 52)

Suillus luteus (L.) Roussel *Slippery Jack*
Mycorrhizal with pine; common.
50, 51. SH 87, 95, 97; SJ 05, 07, 16, 25, 26, 34–36. CC, CE, ER, GW, MF, NM, WW. (17) 1880. (48, 49, 52)

Suillus variegatus (Sow.) Richon & Roze *Velvet Bolete*
Mycorrhizal with pine; frequent.
50, 51. SH 95, 97; SJ 16, 25–27, 35. AV, HC, MF, NF, NM (9) 1976. (48, 49, 52)

Suillus viscidus (L.) Roussel *Sticky Bolete*
Suillus aeruginascens Snell; *S. laricinus* (Berk.) Kuntze
Mycorrhizal with larch; uncommon.
50, 51. SH 85, 87, 97; SJ 15, 16, 18, 24, 26, 35. AV, CC, LO, PA, WW. (12) 1880. (49, 52)

Family Tapinellaceae

Tapinella atrotomentosa (Batsch) Šutara *Velvet Rollrim*
On pine and spruce stumps; uncommon.
50. Recorded only from Llyn Syberi (SH 76) in 1980. (49, 52)

Tapinella panuoides (Fr.) E. Gilb. *Oyster Rollrim*
Paxillus panuoides (Fr.) Fr.
On beech logs; uncommon.
50, 51. SH 85; SJ 26. LO. (2) 1924. (48, 49, 52)

Order CANTHARELLALES
Family Botryobasidiaceae

Botryobasidium aureum Parmasto
On rotten logs, especially of beech; common.
50, 51. SH 97; SJ 06, 07, 15, 16, 26, 34, 36. AV, BO, CE, CN, ER, GW, LO, WW. (11) 1986. (48, 49, 52) Recorded only as the *Haplotrichum* anamorph.

Botryobasidium candicans J. Erikss.
On rotten wood; common.
51. Recorded only from Rhydymwyn Nature Reserve (SJ 26) in 2007.

Botryobasidium conspersum J. Erikss.
On rotten wood; common.
50. Recorded only from the Legacy mine site near Ruabon (SJ 24) in 1994. (48, 49)

Family Cantharellaceae

Cantharellus amethysteus (Quél.) Sacc.
In woodland soil under beech or oak; rare.
50. Recorded only from Mostyn Uchaf near Llansannan (SH 96) in 1999. (49) **RDL/VU**

Cantharellus cibarius Fr. *Chanterelle*
In woodland litter that is not too acid, probably mycorrhizal with beech and pine; frequent in the north and west, disappearing in the south and east, as it is over much of Continental Europe.
50, 51. SH 76, 77, 85, 87, 95-97; SJ 05, 06, 08, 14, 16, 17, 24-26, 34. AV, BO, CC, ER, HM, LH, LO, ML, NF, NM, PP. (26) 1880. (48, 49, 52)

Craterellus cornucopioides (L.) Pers. *Horn of Plenty*
In leaf litter under beech; frequent.
50, 51. SH 85; SJ 26, 34. ER, WW. (4) 1924. (48, 49, 52) **RDL/LC**

Craterellus lutescens (Pers.) Fr. *Golden Chanterelle*
Cantharellus aurora (Batsch) Kuyp.; *C. lutescens* auct.
In damp mossy woodland soil; uncommon, except in the west.
50. Recorded only from Coed Hafod, near Betws-y-Coed (SH 85) in 1924. (49) **RDL/VU**

Craterellus sinuosus (L.) Fr. *Sinuous Chanterelle*
Pseudocraterellus undulatus (Pers.) Rauschert; *P. sinuosus* (Fr.) D.A. Reid; *C. sinuosus* (Fr.) Fr.
In woodland litter; uncommon.
50, 51. SH 85, 87; SJ 16, 17, 26. AV, CC, CE, LH. (5) 1880. **RDL/LC**

Craterellus tubaeformis (Fr.) Quél. *Trumpet Chanterelle*
Cantharellus tubaeformis (Bull.) Fr.; *C. infundibuliformis* (Scop.) Fr.
In mossy woodland soil; frequent in the west, rare elsewhere.
50. SH 85, 95; SJ 05. (4) 1924. (48, 49, 52) **RDL/LC**

Species Accounts

Family Clavulinaceae

Clavulina cinerea (Bull.) J. Schröt. *Grey Coral*
In woodland litter; common.
50, 51. SH 85, 96, 97; SJ 05-08, 15-17, 24-26, 34, 35, 43. AV, BP, BW, CE, CF, ER, LO, MA, MF, ML, MO, NF, PP, WW. (38) 1910. (48, 49, 52)

Clavulina coralloides (L.) J. Schröt. *Crested Coral*
Clavulina cristata (Holmsk.) J. Schröt.
In woodland litter; common.
50, 51. SH 77, 85, 87, 97; SJ 05, 16, 17, 23-27, 34. AV, BO, CC, CE, CF, ER, HC, LH, MA, MF, ML, MO, NF, PP, WW. (33) 1880. (48, 49, 52)

Clavulina rugosa (Bull.) J. Schröt. *Wrinkled Club*
In woodland litter; common.
50, 51. SH 77, 87, 96, 97; SJ 05, 07, 15, 16, 24, 26, 34. AV, CC, CE, CF, ER, LO, MF, ML, PP (23) 1880. (48, 49, 52)

Family Hydnaceae

Hydnum repandum L. *Wood Hedgehog*
In woodland litter; common.
50, 51. SH 76, 85, 87, 95-97; SJ 05, 06, 13, 15, 17, 18, 24, 26, 34, 36. CC, CF, ER, GW, LO, MA. (24) 1880. (48, 49, 52)

Hydnum rufescens Pers. *Terracotta Hedgehog*
In woodland litter; frequent, usually on more acidic soil than the last species.
50, 51. SH 85; SJ 05, 16, 25, 26. AV, LO, NM. (7) 1924. (48, 49, 52)

Order CORTICIALES
Family Corticiaceae

Dendrothele acerina (Pers.) P.A. Lemke
On the bark of living maples; common.
50. Recorded only from Coed Coch (SH 87) in 1880.

Vuilleminia comedens (Nees) Maire *Waxy Crust*
On attached branches of living trees and shrubs, especially oak; common.
50, 51. SH 77, 85, 87; SJ 15, 16, 23, 26, 27, 34-36. CC, CE, ER, GW, LO, MO, WW. (23) 1880. (48, 49)

Vuilleminia coryli Boidin, Lanq & Gilles
On attached branches of *Corylus*; frequent.
50, 51. SJ 06, 25-27, 36. GW, LO, WW. (7) 2014.

Order GEASTRALES
Family Geastraceae

Geastrum coronatum Pers. *'Crowned Earthstar'*
In leaf litter on limestone soils; rare.
50, 51. SH 97; SJ 16, 17, 34. AV. (4) 1879. (52) **RDL/VU**

Geastrum fimbriatum Fr. *Sessile Earthstar*
Geastrum sessile (Sow.) Pouzar
On basic soil in broad-leaved woodland; frequent.
50, 51. SJ 08, 15–17, 26, 34–36. ER, LO, MA. (11) 1910. (49) **RDL/LC**

Geastrum fornicatum (Huds.) Hook. *Arched Earthstar*
On soil in woodland and hedges; rare, except in the extreme south of the British Isles.
50. SJ 15, 24. (2) 1992. These were the first records for North Wales but the species was found in Anglesey (v.c. 52) in 2002, suggesting that climate change may be affecting the distribution of this and several other southern fungi. **RDL/NT**

Geastrum marginatum Vittad. *Tiny Earthstar*
Geastrum minimum Schwein.
On sandy soil over limestone in open woodland; rare.
50. Recorded only from Nant Gwyn (SJ 06) in 1997. This is the first Welsh and fifth British record. **RDL/VU**

Geastrum pectinatum Pers. *'Beaked Earthstar'*
In woodland litter; uncommon.
51. SJ 16, 25. AV. (2) 1952. (48, 52) **RDL/VU**

Geastrum quadrifidum Pers.
On limestone soil under beech; uncommon.
50. Recorded only from Henllan (SJ 06) in 2014. **RDL/VU**

Geastrum rufescens Pers. *'Rosy Earthstar'*
Geastrum vulgatum Vittad.
In deciduous woodland litter on base-rich soils; frequent.
50, 51. SH 87; SJ 08, 15, 26, 34. AV, ER, LO. (7) 1880. (52) **RDL/LC**

Geastrum schmidelii Vittad. *Dwarf Earthstar*
Geastrum nanum Pers.
On calcareous soil in woodland; uncommon, except in sand dunes.
50, 51. SJ 15, 16. AV. (2) 1953. This species has not been found in the dunes at Point of Ayr but is not uncommon in several Welsh sand dune systems. **RDL/VU**

Geastrum striatum DC. *Striate Earthstar*
In woodland litter; uncommon.
51. Recorded only from Dyserth (SJ 08) in 1962. (48, 49, 52) **RDL/LC**

Geastrum triplex Jungh. *Collared Earthstar*
In woodland litter, especially in beech woodland on limestone; fairly common.
50, 51. SH 97; SJ 07, 08, 15–18, 24–26, 34–36. AV, CN, DU, ER, GW, HM, LO, MA, ML, WW. (30) (48, 49, 52) **RDL/LC**

Sphaerobolus stellatus Tode *Shooting Star*
On dead herbaceous stems, rotten wood and dung; common.
50, 51. SH 85, 87, 97; SJ 07, 16, 17, 24, 26, 34, 36. BO, CC, DU, ER, GW, LO. (11) 1880. (48, 49, 52)

Species Accounts

Order GLOEOPHYLLALES
Family Gloeophyllaceae

Gloeophyllum sepiarium (Wulf.) P. Karst. *Conifer Mazegill*
On logs and stumps of conifers; common.
50, 51. SH 95; SJ 05, 06, 13, 17. CF, LH. (5) 1996. (48, 49, 52)

Neolentinus lepideus (Fr.) Redhead & Ginns *Scaly Sawgill*
Lentinus lepideus (Fr.) Fr.
On pine stumps and worked softwood; uncommon.
51. SJ 08, 36. GW. (2) 1972.

Order GOMPHALES
Family Ramariaceae

Phaeoclavulina abietina (Pers.) Giachini
Ramaria abietina (Pers.) Quél.; *R. ochraceovirens* (Jungh.) Donk
On soil and litter under conifers; frequent.
50, 51. SJ 16, 24, 26. AV, MF, PP, WW. (4) 1983. (49, 52)

Phaeoclavulina decurrens (Pers.) Giachini *Ochre Coral*
Ramaria decurrens (Pers.) R.H. Petersen
On soil under conifers; rare.
50. Recorded only from Coed Coch (SH 87) in 1880; this is the only Welsh record.

Phaeoclavulina eumorpha (P. Karst.) Giachini
Ramaria eumorpha (P. Karst.) Corner
On woodland soil; uncommon.
51. Recorded only from Hawarden (SJ 36) in 2019.

Ramaria botrytis (Pers.) Ricken *Rosso Coral*
In litter under beech; rare.
51. Recorded only from Loggerheads Country Park (SJ 26) in 1973. (52)

Ramaria flava (Schaeff.) Quél.
In litter under beech; uncommon.
51. Recorded only from Loggerheads Country Park (SJ 26) in 1983.

Ramaria stricta (Pers.) Quél. *Upright Coral*
On buried dead wood and roots of hardwoods; common.
50, 51. SH 87, 97; SJ 08, 18, 26, 34, 35. BO, CC, CE, ER, ML, WW. (13) 1880. (48, 49, 52)

Order HYMENOCHAETALES
Family Hymenochaetaceae

Coltricia perennis (L.) Murrill *Tiger's Eye*
On acid, sandy or peaty soil under heath or conifers; frequent.
50, 51. SH 85; SJ 07, 34. ER. (3) 1910. (48, 49, 52)

Fuscoporia ferrea (Pers.) G. Cunn. *Cinnamon Porecrust*
Phellinus ferreus (Pers.) Bourdot & Galzin
On living and moribund trunks and branches of broad-leaved trees and shrubs; common.
50, 51. SJ 08, 15, 23, 26, 34, 35. AV, ER. (8) 1982. (48, 49, 52)

Fuscoporia ferruginosa (Schrad.) G. Cunn. *Rusty Porecrust*
Phellinus ferruginosus (Schrad.) Bourdot & Galzin
On dead trunks and branches of broad-leaved trees; common.
50, 51. SJ 15, 16, 26, 34, 35. AV, ER, LO. (5) (48, 49, 52)

Hymenochaete rubiginosa (Dicks.) Lév. *Oak Curtain Crust*
On dry, decayed stumps and fallen branches of oak; common.
50, 51. SH 85; SJ 05–07, 15–17, 24–26, 34–36. BW, CE, CF, CN, ER, GW, LH, LO, MO, NF, PP, WW. (35) 1910. (48, 49, 52)

Hymenochaetopsis corrugata (Fr.) S.H. Ite & Jiao Yang *Glue Crust*
Hymenochaete corrugata (Fr.) Lév.
On branches of hazel, often 'glueing' touching areas together; common.
50, 51. SJ 06, 25, 26, 36. AV, GW, LO, NF, WW. (9) 1910. (48, 49, 52)

Hymenochaetopsis tabacina (Sowerby) S.H. Ite & Jiao Yang
Hymenochaete tabacina (Sowerby) Lév.
On attached branches of *Salix*; frequent.
51. Recorded only from Wepre Woods (SJ 26) in 2014.

Inonotus hispidus (Bull.) P. Karst. *Shaggy Bracket*
Parasitic on ash trunks; common.
50, 51. SH 97; SJ 06, 07, 14–16, 24, 26, 34. AV, CN, ER, ML, MO. (15) 1910. (49, 52)

Mensularia radiata (Sow.) Lazzaro-Ibiza *Alder Bracket*
Inonotus radiatus (Sow.) P. Karst.
Parasitic on alder trunks; common.
50, 51. SH 76, 77, 85, 97; SJ 06, 08, 15–17, 23, 26, 34–36. CE, ER, GW, LH, LO, ML, WW. (19) (48, 49, 52)

Phellinopsis conchatus (Pers.) Y.C. Dai
Phellinus conchatus (Pers.) Quél.
On dead elm trunk; uncommon.
51. Reported from Caergwrle (SJ 35) by Green (1902). (48)

Phellinus igniarius (L.) Quél. *Willow Bracket*
Parasitic on trunks of willows and sallows; uncommon.
50, 51. SJ 08, 16, 26, 34. CE. (4) 1910. (49)

Phellinus laevigatus (Fr.) Bourdot & Galzin
On dead birch trunk; rare.
51. Recorded only from Bodelwyddan (SH 97) in 2001.

Species Accounts

Phellinus pomaceus (Pers.) Maire *Cushion Bracket*
Phellinus tuberculosus (Baumg.) Niemalä
On dead and dying blackthorn trunks; uncommon.
50, 51. SJ 06, 07, 24–26, 34. ER. (6) 1910. (48, 49, 52)

Pseudoinonotus dryadeus (Pers.) T. Wagner & M. Fisch. *Oak Bracket*
Inonotus dryadeus (Pers.) Murrill
Parasitic on oak trunks; uncommon.
50, 51. SH 97; SJ 06, 08, 15, 18, 23, 26, 34. BO, ER, MO. (10) (49, 52)

Family Podoscyphaceae

Cotylidia pannosa (Sow.) D.A. Reid *Woolly Rosette*
On soil in deciduous woodland; rare.
50. Recorded only from Coed Coch (SH 87) in 1880. **RDL/EN**

Cotylidia undulata (Fr.) P. Karst. *Stalked Rosette*
On soil in deciduous woodland; rare.
50. Recorded only from Colwyn Bay (SH 87) in 1891. **RDL/VU**

Loreleia postii (Fr.) Redhead, Moncalvo, Vilgalys & Lutzoni
Omphalina postii (Fr.) P. Karst.
On thalli of *Marchantia polymorpha*; rare.
51. Recorded only from Y Graig Nature Reserve (SJ 07) in 2000.

Muscinupta laevis (Fr.) Redhead et al.
Cyphellostereum laeve (Fr.) D.A. Reid
On living woodland mosses, especially *Mnium hornum*; uncommon.
50, 51. SH 85; SJ 16, 26. CN, LO, MF. (4) 1924.

Rickenella fibula (Bull.) Raith. *Orange Mosscap*
Mycena fibula (Bull.) P. Kumm.; *Omphalina fibula* (Bull.) Quél.
In short turf and terrestrial moss sheets; common.
50, 51. SH 77, 85, 87, 97; SJ 05–08, 15–18, 23–27, 34–36. AV, BO, CC, CE, ER, GW, LH, LO, MA, MF, ML, NM, PA, WW. (45) 1880. (48, 49, 52)

Rickenella swartzii (Fr.) Kuyp. *Collared Mosscap*
Rickenella setipes auct.; *Mycena swartzii* (Fr.) A.H. Smith
In short turf, especially lawns; common.
50, 51. SH 77, 95, 97; SJ 07, 08, 15–18, 23, 24, 26, 34–36. AV, BO, CE, ER, GW, LO, MA, PP, WW. (22) 1973. (48, 49, 52)

Family Schizoporaceae

Basidioradulum radula (Fr.) Nobles *Toothed Crust*
Radulum orbiculare Fr.
On rotten wood and fallen bark; uncommon.
50, 51. SH 87; SJ 08, 24–26, 34. AV, BP, CC, ER, NF. (7) 1880. (49, 52)

Hyphodontia alutacea (Fr.) J. Erikss.
On rotten beech wood; uncommon.
50. Recorded only from Coed Coch (SH 87) in 1880. (48)

Hyphodontia arguta (Fr.) J. Erikss.
On rotten beech wood; uncommon.
51. Recorded only from Bodelwyddan (SH 97) in 1863. (48)

Hyphodontia aspera (Fr.) J. Erikss.
On rotten wood; rare.
51. Recorded only from Celyn Woods (SJ 26) in 1986.

Hyphodontia barbajovis (Bull.) J. Erikss.
On rotten wood; uncommon.
50, 51. SJ 05, 08, 26. CE. (3) 1994. (48, 49)

Hyphodontia pallidula (Bres.) J. Erikss.
On decaying conifer wood; common.
50. Recorded only from Fairy Glen (SH 85) in 1924. (49, 52)

Hyphodontia rimosissima (Peck) Gilb.
On rotten wood; rare.
50. Recorded only from Cilygroeslwyd Nature Reserve (SJ 15) in 1985.

Hyphodontia sambuci (Pers.) J. Erikss. *Elder Whitewash*
Lyomyces sambuci (Pers.) P. Karst.
On living and dead wood of trees and shrubs, especially elder; common.
50, 51. SH 85, 87, 97; SJ 05–08, 15–17, 24–27, 34–36. AV, CC, CE, CN, ER, GW, LO, MA, MO, NM, WW. (40) 1880. (48, 49, 52)

Oxyporus obducens (Pers.) Donk
On rotten wood of deciduous trees, especially elm; rare.
51. SH 97; SJ 26. AV, BO. (2) 1869. (52) Not found since 1869 (from Bodelwyddan) until 2007, when it was refound at Rhydymwyn Nature Reserve.

Oxyporus populinus (Schumach.) Donk *Poplar Bracket*
Parasitic on trunks of old deciduous trees, especially sycamore; uncommon.
50, 51. SJ 15, 16. CN. (2) 1993. (48, 52)

Schizopora paradoxa (Schrad.) Donk *Common Porecrust*
Irpex obliquus (Schrad.) Fr.; *Poria vaporaria* auct.
On rotten wood; common.
50, 51. SH 85, 87, 97; SJ 05–08, 15–18, 24–26, 34, 36. BP, CC, CE, DU, ER, GW, LH, LO, MA, ML, NF, NM, PA, PP, WW. (34) 1880. (48, 49, 52)

Trichaptum abietinum (Pers.) Ryvarden *Purplepore Bracket*
Hirschioporus abietinus (Pers.) Donk
On dead trunks of conifers, both fallen and standing, especially pine; common.
50, 51. SH 76, 85, 87, 95, 97; SJ 02, 07, 15–17, 24–26, 34–36. AV, CE, CN, ER, GW, LH, LO, NM, WW. (29) (48, 49, 52)

Species Accounts

Order PHALLALES
Family Clathraceae

Clathrus archeri (Berk.) Dring *Devil's Fingers*
In flower beds and in litter in woodland; rare – an establishing alien from New Zealand!
50. Recorded only from Penley churchyard (SJ 43) in 2010. This is the most northerly record for this species.

Family Hysterangiaceae

Hysterangium nephriticum Berk.
In beech litter; rare.
50. Recorded only from Big Covert, Maeshafn (SJ 26) in 2001. This is the first record from Wales.

Family Phallaceae

Mutinus caninus (Huds.) Fr. *Dog Stinkhorn*
On soil near stumps and on rotten wood; common.
50, 51. SH 85, 87, 95, 97; SJ 05, 07, 15-17, 23, 24, 26, 34-36, 54. AV, BO, CC, CF, ER, GE, LO, ML, MO, PP, WW. (30) 1880. (48, 49, 52)

Phallus impudicus L. *Stinkhorn*
In woodland soil, usually near rotten wood; common.
50, 51. SH 76, 77, 85, 87, 97; SJ 05-08, 13, 15-17, 23-26, 34, 36. AV, BO, BW, CC, CE, CF, CN, ER, GW, HC, LH, LO, MA, MF, ML, MO, NF, NM, PP, WW. (81) 1880. (48, 49, 52)

Order POLYPORALES
Family Fomitopsidaceae

Daedalea quercina (L.) Pers. *Oak Mazegill*
On oak stumps and trunks; common.
50, 51. SH 76, 77, 85, 87, 97; SJ 05, 15-18, 23-26, 34. BO, CC, CE, CN, ER, LO, MA, ML, NF, PP, WW. (28) 1880. (48, 49, 52)

Fomitopsis betulina (Bull.) B.K. Cui, M.L. Han & Y.C. Dai *Birch Polypore*
Piptoporus betulinus (Bull.) P. Karst.
Parasitic on trunks of birches, causing them to snap off at head height; common.
50, 51. SH 76, 85, 97; SJ 05-08, 13, 15-17, 23-27, 34-36, 43, 44. AV, BO, BW, CE, CF, CN, ER, GW, HC, LH, LO, MA, MF, ML, MO, NF, NM, WW. (69) 1910. (48, 49, 52) **RDL/LC**

Postia caesia (Schrad.) P. Karst. *Conifer Bluing Bracket*
Tyromyces caesius (Schrad.) Murrill
On rotten conifer wood; uncommon.
50, 51. SH 95, 97; SJ 05, 06, 08, 15, 17, 24-26, 35, 36, 43. CE, CF, GW, LO, ML, NF, NM, PP, WW. (26) 1910. (48, 49, 52)

Postia fragilis (Fr.) Jülich
On rotten conifer wood; rare.
50. Recorded only from World's End (SJ 24) in 1910. (49, 52)

Postia leucomallella (Murrill) Jülich
Tyromyces gloeocystidiatus Kotl. & Pouzar
On very rotten conifer wood; frequent.
50, 51. SH 87; SJ 26. CC, MO. (2) 1880. (52)

Postia ptychogaster (F. Ludw.) Vesterh.
Tyromyces ptychogaster (F. Ludw.) Donk
On rotten conifer logs; uncommon.
50, 51. SH 95; SJ 05, 16, 17, 24, 26, 34, 36. AV, CF, ER, GW, LH. (9) 1910. (49, 52) Recorded only in our area as the anamorph – *Ptychogaster albus* Corda.

Postia stiptica (Pers.) Jülich *Bitter Bracket*
Tyromyces stipticus (Pers.) Kotl. & Pouzar
On rotten conifer wood; common.
50, 51. SH 97; SJ 05, 15, 16, 25, 26, 34, 36, 43. BO, CE, CF, ER, GW, LO, MA, MF, NM, WW. (14) (48, 49, 52)

Postia subcaesia (A. David) Jülich *Bluing Bracket*
Tyromyces subcaesius A. David
On rotten hardwood, especially of sycamore; common.
50, 51. SJ 05, 06, 15, 16, 24, 26, 27, 35, 36. AV, CF, CN, GW, LO, MA, MF, PP, WW. (16) (48, 49, 52)

Postia tephroleuca (Fr.) Jülich *Greyling Bracket*
Tyromyces lacteus (Fr.) Murrill
On dead wood of deciduous trees, especially beech; common.
50, 51. SH 97; SJ 06, 15, 16, 18, 24, 26, 35, 36. AV, CE, GW, MA, MF, WW. (13) (49)

Family Ganodermataceae

Ganoderma applanatum (Pers.) Pat. *Artist's Bracket*
Parasitic on trunks of beech; common.
50, 51. SH 77, 87, 88, 97; SJ 06, 08, 15, 17, 18, 24–26, 34–36. CE, ER, GW, MA, ML, PP, WW. (22) (48, 49, 52)

Ganoderma australe (Fr.) Pat. *Southern Bracket*
Ganoderma adspersum (Schulzer) Donk; *G. europaeum* Steyart
Parasitic on beech and, less commonly, on a wide range of tree trunks; common.
50, 51. SH 76, 77, 97; SJ 06, 07, 15–18, 23–26, 34–36, 43. AV, BO, CE, CN, DU, ER, LO, MA, MO, NF, PP, WW. (34) (48, 49, 52)

Ganoderma lucidum (Curtis.) P. Karst. *Lacquered Bracket*
On base of stumps of oak, possibly parasitic; uncommon.
50. SJ 24, 35. PP. (2) 1993.

Ganoderma resinaceum Boud.
Parasitic on beech trunks; uncommon.
50. Recorded only from Erddig (SJ 34) in 1910. (49, 52)

Species Accounts

Family Hapalopilaceae

Bjerkandera adusta (Willd.) P. Karst. *Smoky Bracket*
On dead wood of mainly deciduous trees, especially beech; common.
50, 51. SH 77, 85, 97; SJ 05, 07, 15–18, 24–27, 34, 35. AV, BO, CE, CN, ER, GW, LO, MA, MF, MO, NF, PP, WW. (37) 1910. (48, 49, 52)

Bjerkandera fumosa (Pers.) P. Karst. *Big Smoky Bracket*
On rotten trunks and stumps of hardwoods, especially sycamore; frequent.
50, 51. SJ 16, 26, 24, 34. CN, ER, PP, WW. (4) 1910.

Ceriporia excelsa (S. Lundell) Parmasto
On rotten trunks of deciduous trees; uncommon.
50. Recorded only from Erddig (SJ 34) in 1910. (48) Not seen in North Wales since 1913.

Ceriporia metamorphosa (Fuckel) Ryvarden & Gilb.
On rotten oak wood; rare.
50. SH 87; SJ 34. CC, ER. 1880. Not seen in North Wales since 1910.

Ceriporia purpurea (Fr.) Donk
On fallen beech trunks; rare.
51. Recorded only from Loggerheads Country Park (SJ 26) in 1988.

Ceriporia reticulata (Hoffm.) Domański
On rotten wood; common.
51. SH 97; SJ 26, AV, BO. (2) 1988. (48, 49, 52)

Ceriporia viridans (Berk. & Br.) Donk
On rotten wood; common.
50, 51. SH 97; SJ 26, 34. AV, BO, ER. (3) 1910. (48, 49)

Hapalopilus nidulans (Fr.) P. Karst. *Cinnamon Bracket*
On rotten wood of deciduous trees; uncommon.
50, 51. SJ 26, 34. ER, LO. (3) 1848. (48, 49, 52)

Leptoporus mollis (Pers.) Quél.
On dead pine wood; probably extinct.
50. Only reported from World's End (SJ 24) in 1910. **RDL/RE**. Not seen in the United Kingdom since 1957.

Spongipellis spumeus (Sow.) Pat.
On living and dead wood of deciduous trees; uncommon.
51. SJ 25, 26. NF, WW. (2) 1986.

Family Hyphodermataceae

Brevicellium olivascens (Bres.) K.H. Larss. & Hjortstam
Grandinia granulosa auct.
On rotten wood; common.
50, 51. SH 87; SJ 16, 34. CC, CN, ER. (3) 1880. (48, 49, 52)

Bulbillomyces farinosus (Bres.) Jülich
On wet wood in situations liable to flood; common.
50, 51. SH 85, 87; SJ 07, 08. CC. (4) 1880. (49) Only collected as the anamorph *Aegerita candida* Pers. which resembles minute white polystyrene beads and is dispersed by water.

Hyphoderma setigerum (Fr.) Donk
On rotten deciduous wood; common.
50, 51. SH 85; SJ 24, 26. MA, WW. (4) 1924. (48, 49)

Hypochnicium polonense (Bres.) Å. Strid
On very wet, rotten wood; uncommon.
51. Recorded only from Rhydymwyn Nature Reserve (SJ 26) in 2007. New to North Wales.

Hypochnicium punctulatum (Cooke) J. Erikss.
On rotten wood; common.
50. Recorded only from Coed Coch (SH 87) in 1880. (49)

Peniophorella praetermissa (P. Karst.) K.H. Larss.
Hyphoderma praetermissum (P. Karst.) J. Erikss. & Å. Strid
On rotten wood; common.
50, 51. SJ 16, 18, 26. AV, MA, PA. (3) 1995. (48, 49)

Subulicystidium longisporum (Pat.) Parmasto
On rotten sycamore wood; frequent.
51. Recorded only from Pengwern College (SJ 07) in 2007. (48, 49, 52)

Family Meripilaceae

Abortiporus biennis (Bull.) Sing. *Blushing Rosette*
Heteroporus biennis (Bull.) Lázara Ibiza
On soil, but growing on buried wood or dead roots; frequent.
50, 51. SJ 05, 16, 17, 27. AV. (4) 1987. (48, 52)

Antrodia albida (Fr.) Donk
On fallen dead branches of deciduous trees and shrubs; common.
50. Recorded only from Erddig (SJ 34) in 1910. (48, 49)

Antrodia xantha (Fr.) Ryvarden
On rotten conifer wood; uncommon.
50. Recorded only from Coed Coch (SH 87) in 1880.

Grifola frondosa (Dicks.) Gray *Hen of the Woods*
On roots of deciduous trees, especially oak; uncommon.
50, 51. SJ 06, 23, 24, 26, 34–36, 43. CE, ER, GW, MA, PP. (12) 1979. (48, 49, 52)

Meripilus giganteus (Pers.) P. Karst. *Giant Polypore*
Parasitic on roots of beech; common.
50, 51. SH 76, 77, 87, 98; SJ 05–08, 15, 16, 18, 23, 24, 26, 27, 34–36. AV, ER, GW, HC, MF, ML, MO, PP, WW. (29) 1910. (48, 49, 52)

Species Accounts

Neoantrodia serialis (Fr.) Audet
Antrodia serialis (Fr.) Donk
On rotten wood and decayed house door; uncommon.
51. SJ 16, 26, 36. CN, GW, MO. (3) 1981.

Physisporinus sanguinolentus (Alb. & Schwein.) Pilát *Bleeding Porecrust*
Rigidoporus sanguinolentus (Alb. & Schwein.) Donk
On damp, rotten wood; common.
50, 51. SH 85, 97; SJ 05, 16, 17, 23–26, 34. AV, BO, CF, DU, ER, LO, MA, ML, WW. (16) (48, 49, 52)

Physisporinus vitreus (Pers.) P. Karst.
Rigidoporus vitreus (Pers.) Donk
On very rotten wood; uncommon.
50, 51. SJ 05, 25, 26, 35. AV, CE, CF, NF. (6) 1986. (49, 52)

Rigidoporus ulmarius (Sow.) Imazeki *'Elm Bracket'*
Parasitic on trunks of elms; once common now rare with the decline of the host.
50, 51. SJ 16, 17, 23, 26, 27, 35. AV, DU, HC, MO. (8) 1796. Last seen in our area in 2007. (52)

Family Meruliaceae

Byssomerulius corium (Pers.) Parmasto *Netted Crust*
Merulius corium (Pers.) Fr.
On rotten wood; common.
50, 51. SH 87, 97; SJ 05, 15–17, 24–27, 34–36. AV, BO, CC, CE, ER, GW, LH, LO, MF, NM, WW. (25) 1880. (48, 49, 52)

Merulius tremellosus Schrad. *Jelly Rot*
Phlebia tremellosa (Schrad.) Burds. & Nakasone
On rotten wood; common.
50, 51. SH 77, 97; SJ 05, 06, 15, 23–27, 34–36. CE, ER, GW, LO, NF, NM, PP, WW. (18) 1979. (48, 49, 52)

Mycoacia uda (Fr.) Donk
On rotten wood, especially of sycamore; common.
50, 51. SH 85, 87, 97; SJ 05, 06, 08, 16, 26, 34. BO, CC, CN, ER, LO, WW. (11) 1880. (48, 49)

Phlebia livida (Pers.) Bres.
On rotten wood; uncommon.
50, 51. SH 87, 97. BO, CC. (2) 1863.

Phlebia radiata Fr. *Wrinkled Crust*
Phlebia merismoides (Fr.) Fr.
On rotten trunks, especially of beech; common.
50, 51. SH 97; SJ 05, 07, 08, 16, 17, 24–27, 34, 35. AV, BO, CE, ER, LH, LO, MF, MO, NF, NM, PP, WW. (24) 1910. (48, 49, 52)

Phlebia rufa (Pers.) M.P. Christ.
On rotten wood, especially oak; common.
50, 51. SH 87; SJ 16, 26. CC, CN, LO. (3) 1880. (48, 49)

Resinicium bicolor (Alb. & Schwein.) Parmasto
Grandinia mucida auct.
On rotten conifer wood; common.
50. SH 85; SJ 34. ER. (2) 1910. (48, 49, 52)

Scopuloides hydnoides (Cooke & Massee) Hjortstam & Ryvarden
On damp, rotten wood; frequent.
50, 51. SH 87; SJ 17, 26. CC, CE, DU. (3) 1880. (48, 49)

Scopuloides rimosa (Cooke) Jülich
On rotten wood of deciduous trees; uncommon.
51. SJ 17, 36. DU, GW. (2) 1985.

Family Phanerochaetaceae

Phanerochaete sanguinea (Fr.) Pouzar
On rotten beech wood; rare.
50. Recorded only from Erddig (SJ 34) in 1910. (48, 49) Last seen in North Wales in 1924.

Phanerochaete tuberculata (P. Karst.) Parmasto
On rotten beech wood; rare.
50, 51. SJ 26, 34. ER, WW. (2) 1910.

Phanerochaete velutina (DC.) Fr.
On rotten wood; common.
50. SH 87; SH 16, 34. CC, ER, MA (3) 1880. (48, 49, 52)

Phlebiopsis gigantea (Fr.) Jülich
Phlebia gigantea (Fr.) Donk
On decaying conifer stumps; common.
50, 51. SH 87; SJ 24–26, 34. CC, ER, NM, WW. (6) 1880. (48, 49, 52) This species is often used to control the spread of *Heterobasidion annosum* in conifer plantations.

Terana caerulea (Lam.) Kuntze *Cobalt Crust*
Pulcherricium caeruleum (Lam.) Parmasto
On rotten wood of deciduous trees; uncommon – a Mediterranean species which is slowly moving northwards in the British Isles.
50. SH 85; SJ 16, 34. ER. (4) 1924. (48, 49, 52)

Family Polyporaceae

Cerrena unicolor (Bull.) Murrill
On stumps and fallen trunks of hardwoods; uncommon.
50, 51. SJ 06, 08, 17, 26, 34. DU, ER, LO. (5) 1977.

Daedaleopsis confragosa (Bolton) J. Schröt. *Blushing Bracket*
On dead wood of deciduous trees and shrubs, especially willows and birch; common.
50, 51. SH 76, 87, 96, 97; SJ 05–08, 13, 15–17, 23–27, 34–36, 43, 54. AV, CC, CF, CN, DU, ER, GW, HC, LH, LO, MA, ML, NF, NM, WW. (63) 1880. (48, 49, 52)

Species Accounts

Datronia mollis (Sommerf.) Donk *Common Mazegill*
On dead branches of deciduous trees; common.
50, 51. SH 85, 96, 97; SJ 05, 07, 08, 16, 17, 23-25, 26, 34, 35. AV, BO, CE, CN, DU, ER, LO, ML, NF, WW. (29) 1910. (48, 49, 52)

Diplomitroporus lindbladii (Berk.) Gilb. & Ryvarden
On decayed conifer wood; uncommon.
50. Recorded only from Marford Quarry (SJ 35) in 2016.

Fomes fomentarius (L.) J. Kickx f. *Tinder Bracket*
Parasitic on birch and beech, rarely on other hardwoods; uncommon, except in Scotland.
50. SJ 34. ER. (2) 1910. (49) The 1910 records were the only records from Wales and were regarded as doubtful, but the species was confirmed from near Conwy in 1983, and is recorded as spreading in parts of England.

Ischnoderma benzoinum (Wahlenb.) P. Karst.
On decayed conifer wood; uncommon.
50, 51. SJ 06, 26, 36. CF, GW, LO. (3) 2016.

Laetiporus sulphureus (Bull.) Bondartsev & Sing. *Chicken of the Woods*
Parasitic on trees, especially oak, yew and cherry, but on many species; common.
50, 51. SH 85; SJ 06, 15-18, 23, 24, 26, 34, 36. AV, CN, ER, GW, MA, MF, MO, PP. (23) 1910. (48, 49, 52)

Lentinus brumalis (Pers) *Winter Polypore*
Polyporus brumalis (Pers.) Fr.
On fallen branches in winter; common.
50, 51. SH 97; SJ 15, 25, 26, 35, BO, LO, ML, NF. (7) 1910. (48, 49, 52)

Panus conchatus (Bull.) Fr. *Lilac Oysterling*
Panus torulosus (Pers.) Fr.
On beech stumps; frequent.
50, 51. SJ 17, 26, 34. DU, ER, LO. (4)

Perenniporia fraxinea (Bull.) Ryvarden
On dying ash trunk; uncommon.
50. Recorded only from Chirk Castle (SJ 23) in 1993. (52)

Perenniporia medulla-panis (Jacq.) Donk
On rotten wood of deciduous trees; rare.
50. Recorded only from Erddig (SJ 34) in 1934. **RDL/CR**

Phaeolus schweinitzii (Fr.) Pat. *Dyer's Mazegill*
Parasitic on roots of conifers; common.
50, 51. SH 76, 77, 97; SJ 05, 07, 13, 26, 34. BO, CF, ER, WW. (8) 1977. (48, 49, 52)

Picipes badius (Pers.) Zimtr & Kovalenko *Bay Polypore*
Polyporus durus (Timmerm.) Kreisel; *P. badius* Fr.; *P. picipes* (Fr.) P. Karst.
On dead wood of deciduous trees; common.
50, 51. SH 87, 95, 97; SJ 05-07, 16-18, 24, 26, 34, 36. AV, CC, CE, CF, CN, ER, GW, LO, MA, ML, MO, PP, WW. (19) 1880. (52)

Picipes melanopus (Pers.) Zmitr. & Kovalenko
Polyporus melanopus (Pers.) Fr.
On soil, attached to tree roots and twigs; frequent.
50, 51. SJ 15, 17, 25, 26, 35. DU, LO, MA. (6) 1985.

Polyporus ciliatus Fr. *Fringed Polypore*
Polyporus lepideus Fr.
On dead beech wood; uncommon.
50, 51. SH 95, 97; SJ 08, 16, 17, 25, 26, 35. AV, CE, DU, LH, LO. (11) 1984. (48)

Polyporus leptocephalus (Jacq.) Fr. *Blackfoot Polypore*
Polyporus varius (Pers.) Fr.; *P. nummularius* (Bull.) Pers.
On dead wood of deciduous trees; common.
50, 51. SH 96, 97; SJ 15-17, 24-26, 34, 35, 43. AV, BO, CE, CN, DU, ER, LH, LO, MA, MF, ML, PP, WW. (32) 1977. (48, 49, 52)

Polyporus squamosus (Huds.) Fr. *Dryad's Saddle*
Parasitic on many deciduous trees, especially sycamore; common.
50, 51. SH 76, 87, 88, 96; SJ 05-08, 15-18, 23-27, 34-36, 44. AV, BO, BW, CC, CE, CF, CN, DU, ER, GW, HC, HM, LO, MO, NF, PP, WW. (63) 1880. (48, 49, 52)

Polyporus tuberaster (Jacq.) Fr. *Tuberous Polypore*
Polyporus lentus Berk.
On fallen branches and stumps, especially of beech; uncommon.
50, 51. SJ 15, 26, 36. GW, WW. (3) 1993. (48, 49, 52)

Polyporus umbellatus (Pers.) Fr. *Umbrella Polypore*
Dendropolyporus umbellatus (Pers.) Jülich
On the ground in deciduous woodland, arising from tree rots; rare.
50. SJ 24, 34. (2) 1910. The records from Wynnstay Park and World's End are the only records from Wales. **RDL/NT but RE in Wales**

Skeletocutis amorpha (Fr.) Kotl. & Pouzar
On rotten conifer wood; common.
50, 51. SH 85; SJ 26, 34-36. ER, GW, LO. (6) 1924. (49, 52)

Skeletocutis nivea (Jungh.) Jean Keller *Hazel Bracket*
Incrustoporia semipileata (Peck) Donk
On fallen branches of hazel; common.
50, 51. SJ 05-07, 15, 26. BO, CE, CF, LO, MA, WW. (9) 1981. (48, 49, 52)

Trametes betulina (L.) Pilát *Birch Mazegill*
Lenzites betulinus (L.) Fr.
On dead wood, especially of birch; frequent.
50, 51. SH 97; SJ 05, 17, 26, 27, 34, 35. ER, LH, LO, MA, ML, MO. (10) 1910. (48, 49)

Trametes gibbosa (Pers.) Fr. *Lumpy Bracket*
Pseudotrametes gibbosa (Pers.) Bondartsev & Sing.
On stumps and trunks of hardwoods, especially sycamore and beech; common.
50, 51. SH 77, 88, 97; SJ 15, 16, 23, 26, 27, 34-36. AV, CE, ER, GW, HC, LO, MA, ML, WW. (19) 1972. (48, 49, 52)

Species Accounts

Trametes hirsuta (Fr.) Pilát *Hairy Bracket*
Coriolus hirsutus (Fr.) Quél.; *Polystictus hirsutus* (Fr.) Cooke
On fallen trunks and stumps of deciduous trees; uncommon.
50, 51. SH 97; SJ 08, 15–17, 23–26, 34, 35, 36. CE, CN, DU, ER, GW, LO, NF, PP, WW. (27) 1910.

Trametes ochracea (Pers.) Gilb. & Ryvarden
Coriolus zonatus (Nees) Quél.
On dead wood of deciduous trees; uncommon.
50, 51. SJ 08, 24, 26, 34. ER, MO. (4) (49)

Trametes versicolor (L.) Pilát *Turkeytail*
Coriolus versicolor (L.) Quél.; *Polystictus versicolor* (L.) Fr.
On dead wood of deciduous trees, including stumps, trunks and fallen branches; common.
50, 51. SH 77, 85, 87, 95–97; SJ 05–08, 15–18, 23–26, 34–36, 43. AV, BO, CC, CE, CN, DU, ER, HC, HM, LH, LO, MA, MF, ML, MO, NF, NM, PP, WW. (107) 1880. (48, 49, 52)

Tyromyces wynnei (Berk. & Broome) Donk
On woody debris on forest floor; rare.
50. Recorded only from Coed Coch (SH 87) in 1858, from where it was described. The species is named for the Wynne family of Coed Coch.

Family Sistotremataceae

Sistotrema brinkmannii (Bres.) J. Erikss.
On rotten wood; common.
51. Recorded only from Coed y Felin Nature Reserve (SJ 16) in 1989. (48, 49, 52)

Family Sparassidaceae

Sparassis crispa (Wulf.) Fr. *Wood Cauliflower*
On conifer stumps; uncommon.
50, 51. SH 76, 77; SJ 05, 17, 26. CF. (8) (48, 49, 52)

Family Stecherrinaceae

Junghuhnia nitida (Pers.) Ryvarden
On rotten wood; uncommon.
51. Recorded only from Bodelwyddan (SH 97) in 1988. (48, 49)

Steccherinum fimbriatum (Pers.) J. Erikss.
On rotten wood; common.
50, 51. SH 97; SJ 08, 16, 26. AV, BO, MA, WW. (5) 1988. (48, 49, 52)

Steccherinum ochraceum (Pers.) Gray
On fallen beech branches; uncommon.
50, 51. SH 87; SJ 07, 08, 16, 24, 26, 34. CC, CN, ER, PP, WW. (7) 1880. (52)

Family Xenasmataceae

Phlebiella sulphurea (Pers.) Ginns & Lefebvre *Yellow Cobweb*
Phlebiella vaga (Fr.) P. Karst.
On very rotten wood of beech; common.
50. SH 85, 87. CC. (2) 1924. (49)

Order RUSSULALES
Family Auriscalpiaceae

Auriscalpium vulgare Gray *Earpick Fungus*
On decayed pine cones; frequent.
50, 51. SH 97; SJ 07, 15, 16, 36. BO, CN, GW. (6) 1974. (48, 49, 52)

Clavicorona taxophila (Thom) Doty *Yew Club*
In litter under yews; rare.
50. Recorded only from Cilygroeslwyd Nature Reserve (SJ 15) in 1994. **RDL/NT**

Lentinellus cochleatus (Pers.) P. Karst. var. **cochleatus** *Aniseed Cockleshell*
On stumps of deciduous trees; frequent.
50, 51. SH 85; SJ 05, 26, 27, 36. CE, CF, GW, MA. (6) 1924. (48, 49, 52)

Lentinellus cochleatus var. **inolens** Konr. & Maubl. *Scentless Cockleshell*
On stumps; rare.
50. Recorded only from near Llangollen (SJ 24) in 2018. New to Wales.

Lentinellus laurocerasi (Berk. & Broome) P.D. Orton
On dead wood of cherry laurel; rare.
50. Recorded only from the type collection, made at Coed Coch (SH 87) in 1879. This species, which has not been found again, may not be a *Lentinellus*.

Family Bondarzewiaceae

Heterobasidion annosum (Fr.) Bref. *Root Rot*
Parasitic, mainly on conifers, and also found on stumps; common.
50, 51. SH 77, 87, 97; SJ 02, 05–07, 13, 15–17, 23–27, 34–36. AV, CC, CE, CF, CN, ER, GW, HC, LH, LO, MF, NF, PP, WW. (42) 1880. (48, 49, 52)

Family Gloeocystidiellaceae

Gloeocystidiellum porosum (Berk. & M.A. Curt.) Donk
On rotten beech wood; common.
50, 51. SH 87; SJ 16, 34. AV, CC, ER. (3) 1880. (48, 49, 52)

Family Hericiaceae

Hericium erinaceus (Bull.) Pers. *Bearded Tooth*
On recently fallen beech trunk; rare.
50. Recorded only from Big Wood, Erddig (SJ 34) since 2002. (49) This and the Caernarfon site, since 1999, are the only records from Wales, but suggest a possible spread of this rare and spectacular fungus. **RDL/VU**

Species Accounts

Family Peniophoraceae

Peniophora cinerea (Pers.) Cooke
On fallen branches, especially of beech and sycamore; common.
50, 51. SH 87, 97; SJ 06, 16, 17, 25, 26, 34–36. BO, CC, CN, DU, ER, GW, LO, NF, WW. (13) 1880. (48, 49)

Peniophora incarnata (Pers.) P. Karst. *Rosy Crust*
On dead branches, often still attached to the tree or shrub, especially of gorse; common.
50, 51. SH 87; SJ 05, 15, 17, 25, 26, 34, 35. CC, CE, ER, LO, NM, WW. (15) 1880. (48, 49, 52)

Peniophora limitata (Chaillet) Cooke
On fallen branches of ash; common.
50, 51. SJ 05, 15, 16, 24, 26, 34–36. ER, GW, LO, MA, WW. (14) 1985. (48, 49, 52)

Peniophora lycii (Pers.) Höhn. & Litsch.
On fallen branches of a wide range of deciduous trees, especially sycamore; common.
50, 51. SH 97; SJ 05, 07, 08, 15–18, 25–27, 35, 36. AV, BW, CE, GW, LO, MF, PA, WW. (31) 1981. (48, 49, 52)

Peniophora nuda (Fr.) Bres.
On dead attached branches of deciduous trees; uncommon.
50. Recorded only from Erddig (SJ 34) in 1910. (48) Last seen in North Wales in 1913.

Peniophora quercina (Pers.) Cooke
On dead attached branches of oak; common.
50, 51. SH 77, 85, 87; SJ 15–17, 23, 25, 26, 34–36. AV, CC, CE, ER, GW, LO, MA, NF, WW. (29) 1880. (48, 49, 52)

Peniophora violaceolivida (Sommerf.) Massee
On the bark of fallen sallows; uncommon.
51. Recorded only from Rhydymwyn Nature Reserve (SJ 26) in 2007. Possibly new to Wales.

Family Russulaceae

Lactarius acerrimus Britz.
Lactarius insulsus auct.
Mycorrhizal with oak; uncommon.
50, 51. SH 97; SJ 25. BO. (2) 1979. (48, 49, 52)

Lactarius aurantiacus (Pers.) Gray *Orange Milkcap*
Lactarius aurantiofulvus J. Blum; *L. mitissimus* (Fr.) Fr.
Mycorrhizal with a variety of trees; common.
50, 51. SH 85, 87, 97; SJ 16, 24–26, 34. AV, BO, CC, CE, ER, LO, ML, MO, NM. (12) 1880. (48, 49, 52)

Lactarius blennius (Fr.) Fr. *Beech Milkcap*
Mycorrhizal with beech; common.
50, 51. SH 76, 77, 85, 97; SJ 05, 15, 16, 17, 24–27, 34, 36. AV, CE, CF, ER, GW, HC, LH, LO, MA, MF, ML, MO, NM, PP, WW. (32) 1910. (48, 49, 52)

Lactarius camphoratus (Bull.) Fr. *Curry Milkcap*
Mycorrhizal with birch and pine; common.
50, 51. SH 95; SJ 05, 16, 26, 36, 43. AV, CE, GW. (6) 1986. (48, 49, 52)

Lactarius chrysorrheus Fr. *Yellowdrop Milkcap*
Mycorrhizal with oak; common.
50, 51. SH 85; SJ 05, 18. (4) 1924. (48, 49, 52)

Lactarius circellatus Fr. *'Hornbeam Milkcap'*
Mycorrhizal with hornbeam; uncommon.
50, 51. SH 87; SJ 26. CC, MO. 1880. (49)

Lactarius citriolens Pouzar
Mycorrhizal with beech; rare.
50. Recorded only from Marford Quarry (SJ 35) in 2016.

Lactarius controversus Pers.
Mycorrhizal with willows; uncommon.
50. Recorded only from Coed Coch (SH 87) in 1880. (48, 49, 52)

Lactarius decipiens Quél.
Mycorrhizal with hazel; uncommon.
50. SJ 34, 35. ER. (2) 1981. (49)

Lactarius deliciosus (L.) Gray *Saffron Milkcap*
Mycorrhizal strictly with pine, usually on base-rich soils; frequent.
50, 51. SH 87, 95, 97; SJ 05, 16, 17, 24, 26, 34, 35. CC, CE, CF, ER, LO, ML. (19) 1880. (48, 49, 52)

Lactarius deterrimus Gröger *False Saffron Milkcap*
Mycorrhizal strictly with spruce; common.
50, 51. SH 95, 97; SJ 05, 07, 16, 17, 26, 34. AV, BO, CE, ER, LO, MA, ML. (16) 1796. (48, 49, 52)

Lactarius evosmus Kühn. & Romagnesi
Mycorrhizal with oak; rare.
50. Recorded only from the Aberduna Nature Reserve, Maeshafn (SJ 26) in 2001. (49, 52)

Lactarius flexuosus (Pers.) Gray
Mycorrhizal with pine; rare.
51. Recorded only from Nercwys Mountain (SJ 25) in 1980. (48, 49)

Lactarius fluens Boud.
Mycorrhizal with beech; uncommon.
51. Recorded only from Celyn Woods (SJ 26) in 1984.

Lactarius fuliginosus (Fr.) Fr. *Sooty Milkcap*
Lactarius azonites auct.
Mycorrhizal with various trees; uncommon.
50, 51. SH 97; SJ 05, 24, 26, 34. BO, CE, ER. (5) 1910. (48, 49, 52)

Lactarius fulvissimus Romagn. *Tawny Milkcap*
Mycorrhizal with beech; uncommon.
50, 51. SH 77; SJ 16, 17, 26, 34. ER, LO, MF. (5) 1978. (49, 52)

Lactarius glyciosmus (Fr.) Fr. *Coconut Milkcap*
Mycorrhizal with birch; common.
50, 51. SH 76, 77, 85, 95, 97; SJ 05, 07, 15-17, 24-26, 35. AV, BO, CE, CF, LH, LO, MF, ML, NF, NM, PP, WW. (26) 1910. (48, 49, 52)

Lactarius helvus (Fr.) Fr. *Fenugreek Milkcap*
Mycorrhizal with pine; uncommon.
50, 51. SJ 16, 17, 25. LH, MF, NM. (3) 1977.

Lactarius hepaticus Plowr. *Liver Milkcap*
Mycorrhizal with pine; uncommon.
50, 51. SJ 05, 07, 15, 17, 25-27, 36. CE, CF, GW, HC, LH, MA, LO, NM. (10) 1977. (48, 49, 52)

Lactarius lilacinus Lasch
Mycorrhizal with alder; rare.
50. Recorded only from Hafod Wood Nature Reserve, Erddig (SJ 34) in 2003. (48, 49, 52)

Lactarius mammosus Fr.
Lactarius fuscus Roll.
Mycorrhizal with birch; rare.
51. Recorded only from Ysceifiog (SJ 17) in 1978.

Lactarius obscuratus (Lasch) Fr. *Alder Milkcap*
Mycorrhizal with alder; uncommon.
50, 51. SJ 16, 17, 34. AV, ER. (3) 1977. (48, 49, 52)

Lactarius omphaliformis Romagn.
Mycorrhizal with alder; uncommon.
50. Recorded only from Mostyn Uchaf, near Llansannan (SH 96) in 1992.

Lactarius pallidus Pers. *Pale Milkcap*
Mycorrhizal with beech; uncommon.
50, 51. SJ 05, 06, 16, 24, 26, 34. CF, CN, ER, LO, MA, PP, WW. (10) 1910. (48, 49, 52)

Lactarius pterosporus Romagn.
Mycorrhizal with beech; uncommon.
50. Recorded only from Bontuchel (SJ 05) in 1986. (48, 49, 52)

Lactarius pubescens (Fr.) Fr. *Bearded Milkcap*
Mycorrhizal with birch; common.
50, 51. SH 96, 97; SJ 07, 15-17, 24-27, 34, 35. AV, BW, CE, ER, LH, WW. (24) 1910. (49, 52)

Lactarius pyrogalus (Bull.) Fr. *Fiery Milkcap*
Mycorrhizal with hazel; common.
50, 51. SH 76, 85, 87, 96; SJ 05-08, 15-18, 25, 26, 34, 36. AV, BP, CC, CE, ER, GW, LO, MA, WW. (28) 1880. (48, 49, 52)

Lactarius quietus (Fr.) Fr. *Oakbug Milkcap*
Mycorrhizal with oak; common.
50, 51. SH 77, 85; SJ 05–08, 15, 17, 18, 23–26, 34–36, 43. AV, BP, CE, ER, GW, LH, LO, MA, NM, PP, WW. (31) 1910. (48, 49, 52)

Lactarius rufus (Scop.) Fr. *Rufous Milkcap*
Mycorrhizal with conifers; common.
50, 51. SH 87, 95, 97; SJ 05, 07, 16, 17, 24–27, 34, 36, 43. AV, CC, CE, CF, ER, GW, HC, HM, LH, LO, MF, MO, NF, PP, WW. (27) 1880. (48, 49, 52)

Lactarius serifluus (DC.) Fr. *Watery Milkcap*
Mycorrhizal with oak; uncommon.
50, 51. SH 85, 87; SJ 34, 36. CC, ER, GW. (5) 1880. (49)

Lactarius subdulcis (Pers.) Gray *Mild Milkcap*
Mycorrhizal with beech; common.
50, 51. SH 77, 85, 87, 97; SJ 06, 07, 15–17, 24–26, 34, 36, 43. AV, BO, CC, CE, CF, CN, ER, GW, LH, LO, MA, MF, MO, NM, PP, WW. (34) 1880. (48, 49, 52)

Lactarius subombonatus Lindgr. *Bedbug Milkcap*
Mycorrhizal with beech; uncommon.
Lactarius cimicarius auctt.
50. Record only from Erddig Park (SJ 34) in 2018.

Lactarius tabidus Fr. *Birch Milkcap*
Mycorrhizal with birch and oak; common.
50, 51. SJ 05, 16, 17, 26, 36. AV, CE, GW, LH, LO, WW. (10) 1977. (48, 49, 52)

Lactarius torminosus (Schaeff.) Pers. *Woolly Milkcap*
Mycorrhizal with birch; common.
50, 51. SH 76, 87, 96, 97; SJ 05, 16, 23, 25–27, 34. AV, BO, CC, CE, ER, LO, MO, NF. (22) 1880. (48, 49, 52)

Lactarius trivialis (Fr.) Fr.
Mycorrhizal with birch and conifers; rare.
50. Reported from Coed Coch (SH 87) in 1880.

Lactarius turpis (Weinm.) Fr. *Ugly Milkcap*
Lactarius necator auct.; *L. plumbeus* auct.
Mycorrhizal with birch; common.
50, 51. SH 85, 95, 96; SJ 05, 07, 15–17, 24–27, 34–36, 43. AV, CE, ER, GW, HC, LH, LO, MF, MO, NM, PP, WW. (36) 1902. (48, 49, 52)

Lactarius uvidus (Fr.) Fr.
Mycorrhizal with birch in damp woodland; uncommon.
50, 51. SH 97; SJ 05, 25, 26. CF, ML, NF. (4) 1986. (49)

Lactarius vietus (Fr.) Fr. *Grey Milkcap*
Mycorrhizal with birch in damp woodland; common.
50, 51. SH 76, 85, 96, 97; SJ 05, 07, 16, 17, 24–26, 34, 36. AV, CE, ER, GW, LH, LO, ML, NF, PP, WW. (20) 1924. (48, 49, 52)

Species Accounts

Lactifluus piperatus (L.) Roussel *Peppery Milkcap*
Lactarius piperatus (L.) Pers.
Mycorrhizal with beech and oak; uncommon.
50, 51. SH 85; SJ 05, 07, 16, 17. LO. (5) 1796. (49, 52)

Lactifluus vellereus (Fr.) Kuntze *Fleecy Milkcap*
Lactarius vellereus (Fr.) Fr.
Mycorrhizal with birch and beech; uncommon.
50, 51. SH 87, 96; SJ 05, 16, 18, 24, 25, 34. AV, CC, ER, NF. (8) 1880. (48, 49, 52)

Lactifluus volemus (Fr.) Kuntze
Lactarius volemus (Fr.) Fr.
Mycorrhizal with birch and oak; uncommon.
50, 51. SH 85; SJ 06, 07. (3) 1924. (49)

Russula acrifolia Romagn.
Russula densifolia auct. non Gillet
Mycorrhizal with beech and oak; uncommon.
50, 51. SJ 05, 26. LO, MA, MO, WW. (6) 1973. (48)

Russula adusta (Pers.) Fr. *Winecork Brittlegill*
Mycorrhizal with a variety of trees; uncommon.
50, 51. SH 85, 87; SJ 05, 23, 26, 34. CC, CF, ER, LO. (8) 1880. (48, 49)

Russula aeruginea Fr. *Green Brittlegill*
Mycorrhizal with birch; common.
50, 51. SH 96; SJ 16, 25, 26, 36. AV, CE, GW, MA, NF, WW. (7) 1981. (48, 49, 52)

Russula amoenolens Romagn.
Mycorrhizal with oak; rare.
51. Recorded only from Hawarden Woods (SJ 36) in 1972. (49, 52)

Russula aquosa Leclair
Russula carminea (Jul. Schaeff.) Romagn.
Mycorrhizal with birch in damp woodland, often with *Sphagnum*; uncommon.
50. Recorded only from Coed Hafod near Betws-y-Coed (SH 85) in 1988. (48)

Russula atropurpurea (Krombh.) Britz. *Purple Brittlegill*
Russula krombholzii Shaffer; *R. undulata* Velen.; *R. rubra* auct.
Mycorrhizal with oak; common.
50, 51. SH 76, 87, 96, 97; SJ 05–08, 15–18, 23–26, 34, 35, 36. AV, CC, CE, ER, GW, LH, LO, MF, NF, PP, WW. (32) (48, 49, 52)

Russula azurea Bres.
Mycorrhizal with conifers; rare.
50. Recorded only from Erddig (SJ 34) in 1910. (49)

Russula betularum Hora *Birch Brittlegill*
Mycorrhizal with birch; common.
50, 51. SH 95, 96; SJ 05, 16, 17, 24–26, 35, 36, 43. AV, CF, LH, LO, MF, NF, NM, WW. (201) 1976. (48, 49, 52)

Russula brunneoviolacea Crawshay
Mycorrhizal with oak; uncommon.
51. Recorded only from Celyn Woods (SJ 26) in 1984. (48, 49)

Russula caerulea (Pers.) Fr. *Humpback Brittlegill*
Mycorrhizal with pine; uncommon.
50, 51. SH 77; SJ 07, 16, 24, 26. AV, LO, MF. (8) 1976. (48, 49, 52)

Russula chloroides (Krombh.) Bres. *Blue Band Brittlegill*
Mycorrhizal with oak; uncommon.
50, 51. SJ 26, 34. ER, LO. (2) 1910. (48, 49, 52)

Russula claroflava Grove *Yellow Swamp Brittlegill*
Russula flava Lindblad
Mycorrhizal with birch in *Sphagnum* areas in damp woodland; common.
50, 51. SH 76, 96, 97; SJ 05, 15, 17, 43. CF, LH, ML. (7) 1977. (48, 49, 52)

Russula cuprea Krombh.
Russula firmula auct.
Mycorrhizal with oak; rare.
50, 51. SH 97; SJ 16. AV, BO, LO. (4) 1977. (52)

Russula curtipes F.H. Møller
Mycorrhizal with beech; uncommon.
50. Recorded only from Erddig (SJ 34) in 1985. (49)

Russula cyanoxantha (Schaeff.) Fr. *Charcoal Burner*
Mycorrhizal with beech and oak; common.
50, 51. SH 76, 77, 85, 87, 95-97; SJ 05, 07, 08, 15-18, 24-27, 35, 36, 43. AV, BO, BW, CC, CE, CN, ER, GW, HC, LH, LO, MA, MF, ML, MO, NF, NM, PP, WW. (51) 1880. (48, 49, 52)

Russula delica Fr. *Milk White Brittlegill*
Mycorrhizal with oak; uncommon.
50, 51. SH 87, 97; SJ 05, 15-17, 24, 26, 34, 36. AV, BO, CC, CE, ER, GW, MO. (14) 1880. (49, 52)

Russula densifolia Gillet *Crowded Brittlegill*
Mycorrhizal with beech and oak; uncommon.
50. SH 76, 97; SJ 34. ER. (4) 1977. (48, 49, 52)

Russula emetica (Schaeff.) Pers. *Sickener*
Mycorrhizal with pine in damp woodland; common.
50, 51. SH 87, 95; SJ 05, 25. CC, CF, NM. (11) 1880. (48, 49, 52)

Russula exalbicans (Pers.) Melzer & Zvára *Bleached Brittlegill*
Russula pulchella I.G. Borshch.
Mycorrhizal with birch; uncommon.
50, 51. SH 97; SJ 17, 25, 26, 34-36. AV, BO, NF. (9) 1910. (49) This species appears early in the season; in 2007 it was mature in the first week of July.

Species Accounts

Russula faginea Adamčik
Mycorrhizal with beech; rare.
51. Recorded only from Hawarden (SJ 36) in 2014.

Russula farinipes Romell
Mycorrhizal with birch; frequent.
51. SJ 18, 27. HC. (2) 1984. (49, 52)

Russula fellea (Fr.) Fr. *Geranium Brittlegill*
Mycorrhizal with beech; common.
50, 51. SH 77, 85, 87; SJ 05, 07, 15–17, 24–26, 34, 36. CC, CE, CF, ER, GW, LH, LO, MA, MF, NM, PP, WW. (22) 1880. (48, 49, 52)

Russula foetens Pers. *Stinking Brittlegill*
Mycorrhizal with beech and oak; uncommon.
50, 51. SJ 05, 06, 15–17, 24, 26, 34, 36. AV, ER, GW, LH, LO, MF, WW. (16) 1972. (48, 49, 52)

Russula fragilis (Pers.) Fr. *Fragile Brittlegill*
Mycorrhizal with birch; common.
50, 51. SH 85, 87; SJ 05, 07, 15–17, 24–26, 34–36. AV, CC, CE, ER, GW, LO, MF, NF, PP. (20) 1880. (48, 49, 52)

Russula gracillima Jul. Schäff.
Mycorrhizal with birch; frequent.
50. Recorded only from Moel Famau (SJ 16) in 2015.

Russula grisea (Pers.) Fr.
Mycorrhizal with a variety of trees; uncommon.
50. Recorded only from Chirk Castle (SJ 23) in 1979. (49, 52)

Russula heterophylla (Fr.) Fr. *Greasy Green Brittlegill*
Mycorrhizal with birch and beech; uncommon.
50, 51. SH 87; SJ 26, 35, 36. CC, GW. (4) 1880. (48, 49, 52)

Russula illota Romagn.
Mycorrhizal with broad-leaved trees; rare.
51. Recorded only from Bodelwyddan (SH 97) in 2001.

Russula ionochlora Romagn. *Oilslick Brittlegill*
Mycorrhizal with beech; frequent.
50, 51. SJ 16, 17, 24–26, 34, 36. AV, CE, GW. (8) 1972. (49)

Russula laurocerasi Melzer *Bitter Almond Brittlegill*
Russula grata Britz.
Mycorrhizal with beech and oak; uncommon.
50, 51. SH 95, 97; SJ 15, 26, 34, 36. BO, ER, GW, LO, MA, ML. (9) (48, 49, 52)

Russula lepida Fr. *Rosy Brittlegill*
Russula rosea Pers.
Mycorrhizal with beech; common.
50, 51. SJ 16, 26, 34, 36. AV, ER, GW, WW. (5) 1910. (48, 49, 52)

Russula luteotacta Rea
Mycorrhizal with beech; uncommon.
50. SJ 05, 06. CF. (2) 1996.

Russula mairei Sing. *Beechwood Sickener*
Russula nobilis Velen.
Mycorrhizal with beech; common.
50, 51. SH 77, 87, 95, 97; SJ 05, 06, 15–18, 24–27, 34, 36. AV, CC, CF, CN, ER, GW, HC, LH, LO, MA, MO, NM, PP, WW. (27) 1972. (48, 49, 52)

Russula melitodes Romagn
Russula integra auct.
Mycorrhizal with broad-leaves; uncommon.
50. Recorded only from Coed Coch (SH 87) in 1880. (49)

Russula nauseosa (Pers.) Fr.
Mycorrhizal with conifers; uncommon.
50. SH 87; SJ 34. CC, ER. (2) 1880.

Russula nigricans (Bull.) Fr. *Blackening Brittlegill*
Mycorrhizal with beech and oak; common.
50, 51. SH 76, 85, 87, 95, 97; SJ 05–08, 15–18, 23–26, 34–36. AV, BP, CC, CE, CF, ER, GW, LH, LO, MA, ML, NF, NM, PP, WW. (40) 1880. (48, 49, 52)

Russula nitida (Pers.) Fr. *Purple Swamp Brittlegill*
Mycorrhizal with birch in damp woodland; common.
50, 51. SH 96, 97; SJ 16, 26. AV, LO, ML. (4) 1972. (48, 49, 52)

Russula ochroleuca Pers. *Ochre Brittlegill*
Mycorrhizal with a wide range of deciduous trees; common.
50, 51. SH 76, 77, 85, 87, 95–97; SJ 05–07, 13, 15–18, 23–26, 34–36, 43, 54. AV, CE, CF, CN, DU, ER, GW, LH, LO, MA, MF, ML, MO, NF, NM, PP, WW. (70) 1910. (48, 49, 52)

Russula olivacea (Schaeff.) Fr. *Olive Brittlegill*
Mycorrhizal with beech; uncommon.
51. SJ 16, 36. AV, GW. (2) 1976. (48, 49)

Russula parazurea Jul. Schaeff. *Powdery Brittlegill*
Mycorrhizal with beech and oak; common.
50, 51. SH 85, 97; SJ 16, 26, 34. AV, BO, BW, CE, ER, MO. (7) 1910. (48, 49, 52)

Russula pectinatoides Peck
Russula pectinata auct.
Mycorrhizal with lime; rare.
50, 51. SJ 07, 35. (2) 2005. (49, 52)

Russula puellaris Fr. *Yellowing Brittlegill*
Mycorrhizal with beech and birch; frequent.
51. SJ 26, 36. CE, LO, MO, GW. (6) 1972. (48, 49, 52)

Species Accounts

Russula queletii Fr. *Fruity Brittlegill*
Mycorrhizal with conifers; uncommon.
50, 51. SH 87, 97; SJ 17, 26. AV, CC, ML. (4) 1880. (48, 49, 52)

Russula raoultii Quél.
Mycorrhizal with birch; rare.
50. Recorded only from Alyn Waters Country Park (SJ 35) in 1998. (52)

Russula risigallina (Batsch) Kuyper & Vuure *Golden Brittlegill*
Russula chamaeleontina Fr.; *R. lutea* auct.
Mycorrhizal with beech and oak; uncommon.
50. Recorded only from Wynnstay Park (SJ 34) in 1910. (48, 49)

Russula sanguinaria (Schumach.) Rauschert *Bloody Brittlegill*
Russula sanguinea Fr.
Mycorrhizal with pine; common.
50, 51. SH 76, 97; SJ 05, 07, 24, 26. AV, CE, CF, ML, PP. (9) 1984. (48, 49, 52)

Russula sardonia Fr. *Primrose Brittlegill*
Russula drimeia Cooke
Mycorrhizal with pine; common.
50, 51. SH 97; SJ 05, 16, 17, 26, 35, 36. CF, GW, LH, LO, MF. (15) 1974. (48, 49, 52)

Russula silvestris (Sing.) Reumaux
Russula emeticella (Sing.) J. Blum
Mycorrhizal with oak; uncommon.
50, 51. SJ 16, 17, 26, 34, 36. AV, ER, GW, LH, LO, WW. (7) 1962. (48, 49)

Russula sororia Fr. *Sepia Brittlegill*
Mycorrhizal with oak; uncommon.
50, 51. SJ 34, 36. ER, GW. (2) 1984. (48, 49, 52)

Russula subfoetens Wm. G. Sm.
Mycorrhizal with beech; uncommon.
50. Recorded only from Erddig (SJ 34) in 1985.

Russula velenovskyi Melzer & Zvára *Coral Brittlegill*
Mycorrhizal with birch and oak; uncommon.
50. SH 76, 96; SJ 26. MA. (3) 1979. (48, 49)

Russula versicolor Jul. Schaeff. *Variable Brittlegill*
Mycorrhizal with birch; uncommon.
50, 51. SH 96, 97; SJ 16. AV, BO. (3) 1979. (49)

Russula vesca Fr. *The Flirt*
Mycorrhizal with oak; common.
50, 51. SH 76, 77, 96; SJ 06, 16, 23, 26, 34, 36. AV, ER, GW, LO, MF, MO, WW. (13) 1972. (48, 49, 52)

Russula virescens (Schaeff.) Fr. *Greencracked Brittlegill*
Mycorrhizal with beech and oak; uncommon.
50, 51. SJ 06, 26, 34. CE. (3) 1910. (49, 52)

Russula xerampelina (Schaeff.) Fr. *Crab Brittlegill*
Mycorrhizal with pine; common.
50, 51. SH 97; SJ 16, 26, 34, 36. AV, BO, CE, ER, GW, LO, MF, WW. (10) 1910. (48, 49, 52)

Family Stereaceae

Amylostereum chailletii (Pers.) Boidin
On rotten conifer trunk; uncommon.
50. Recorded only from Erddig (SJ 34) in 1985. This is the only Welsh record.

Stereum gausapatum (Fr.) Fr. *Bleeding Oak Crust*
On fallen oak branches; common.
50, 51. SH 85, 87, 97; SJ 05–07, 15–17, 23–27, 34, 36. AV, BO, CC, CE, CN, ER, GW, LH, LO, MF, NM, WW. (28) 1880. (48, 49, 52)

Stereum hirsutum (Willd.) Gray *Hairy Curtain Crust*
On living branches and on fallen wood; common.
50, 51. SH 77, 85, 87, 95, 97; SJ 05–08, 15–18, 23–27, 34–36, 43, 54. AV, BO, BP, BW, CC, CE, CF, CN, DU, ER, GW, HC, LH, LO, MA, MF, ML, MO, NF, NM, PA, PP, WW. (115) 1880. (48, 49, 52) This species probably occurs in every piece of woodland in Wales.

Stereum rameale (Pers.) Burt.
Stereum ochraceoflavum auct.; *S. sulphuratum* auct.
On fallen small branches of oak; uncommon.
50, 51. SJ 05, 15, 16, 25, 26, 36. AV, CE, CN, GW, WW. (8) 1987. (48, 49, 52)

Stereum rugosum (Pers.) Fr. *Bleeding Broadleaf Crust*
On stumps and trunks, especially of beech; common.
50, 51. SH 77, 85, 87; SJ 05–07, 15, 16, 25–27, 34, 36. AV, CC, CE, ER, GW, LO, NF. (23) 1880. (48, 49, 52)

Stereum sanguinolentum (Alb. & Schwein.) Fr. *Bleeding Conifer Crust*
On rotten conifer wood; common.
50, 51. SH 87; SJ 16, 17, 23, 24, 26, 34, 36. AV, CC, CE, ER, GW, LH, LO, MF, PP, WW. (12) 1910. (48, 49, 52)

Stereum subtomentosum Pouzar *Yellowing Curtain Crust*
On fallen wood of deciduous trees, especially beech and sycamore; once rare, now frequent.
50, 51. SH 97; SJ 07, 26, 34, 35. BO, CE, ER. (5) 1992. (49, 52)

Order SEBACINALES
Family Sebacinaceae

Sebacina epigaea (Berk. & Br.) Neuhoff
On woodland soil; uncommon.
51. Recorded only from Hawarden (SJ 36) in 2019.

Sebacina incrustans (Pers.) Tul. *Enveloping Crust*
On soil and encrusting the stems of living herbs and seedling trees; common.
50, 51. SJ 05, 15, 16, 25, 26, 34. CE, ER, LO, NM. (9) 1910. (48, 49, 52)

Species Accounts

Order THELEPHORALES
Family Bankeraceae

Hydnellum spongiosipes (Peck) Pouzar *Velvet Tooth*
In litter under oak; uncommon.
50. Recorded only from Loggerheads Country Park (SJ 16) in 2001. (48) **RDL/DD**

Phellodon melaleucus (Schwartz) P. Karst. *Grey Tooth*
In woodland litter; uncommon.
51. Recorded only from Coed Felin-Blwm, Ffynnongroew (SJ 18) in 1996. (48, 49, 52) **RDL/LC**

Family Thelephoraceae

Pseudotomentella tristis (P. Karst.) M.J. Larsen
On rotten wood; rare.
50. Recorded only from Coed Hafod near Betws-y-Coed (SH 85) in 1924. (48, 49)

Thelephora caryophyllea (Schaeff.) Pers.
On woodland soil; rare.
50. Recorded only from Coed Coch (SH 87) in 1880.

Thelephora palmata (Scop.) Fr. *Stinking Earthfan*
On acid soil in pine plantation; uncommon.
50. Recorded only from Coed Clwyd on Moel Famau (SJ 16) in 1981. (52)

Thelephora penicillata Fr.
Thelephora spiculosa Fr.
In woodland soil; uncommon.
50. SH 97; SJ 05. (2) 2001.

Thelephora terrestris Ehrh. *Earthfan*
On soil in acid woodlands, mycorrhizal with many species of tree seedlings; common.
50, 51. SH 87, 95; SJ 05, 15–18, 24–26, 34, 36. CC, CE, GW, MF, NM, PP. (14) 1880. (48, 49, 52)

Tomentella crinalis (Fr.) M.J. Larsen
Caldesiella ferruginosa (Pers.) Sacc.
On rotten beech wood; rare.
50. Recorded only from Erddig (SJ 34) in 1910.

Tomentella ellisii (Sacc.) Jülich & Stalpers
Tomentella luteomarginata M.P. Christ.
On rotten wood; common.
50, 51. SJ 26, 34. ER, LO. (2) 1985. (48)

Tomentella ferruginea (Pers.) Pat.
On rotten wood of deciduous trees; uncommon.
50. SH 85, 87. CC. (2) 1880. (48, 49)

Tomentella sublilacina (Ellis & Holw.) Wakef.
Tomentella fusca auct.
On soil and very rotten wood in damp areas; common.
50, 51. SJ 16, 26, 34. AV, ER, LO. (3) 1977. (48, 49, 52)

Order TRECHISPORALES
Family Trechisporaceae

Trechispora confinis (Bourdot & Galzin) Liberta
On very decayed wood in old forests; rare.
50. Recorded only from Hafod Wood (SH 85) and Fairy Glen (SH 85) in 1924. (49)

Trechispora farinacea (Pers.) Liberta
Cristella farinacea (Pers.) Donk
On rotten wood; common.
50, 51. SH 85; SJ 17, 26, 34, 35. DU, ER, LO. (6) 1910. (48, 49)

Trechispora fastidiosa (Pers.) Liberta
On sawdust; uncommon.
50. Reported only from Coed Coch (SH 87) in 1879.

Trechispora mollusca (Pers.) Liberta
On rotten wood; common.
50. Recorded only from Erddig (SJ 34) in 1910. (48, 49)

Trechispora nivea (Pers.) K.H. Larss.
On rotten beech wood; rare.
50. Recorded only from Erbistock (SJ 34) in 1910. (48) Last reported from North Wales in 1913.

Class DACRYMYCETES
Order DACRYMYCETALES
Family Dacrymycetaceae

Calocera cornea (Batsch) Fr. *Small Stagshorn*
On fallen trunks and branches, mostly on hardwoods, especially beech; common.
50, 51. SH 77, 85, 95, 97; SJ 05–07, 15–18, 24–27, 34, 36. AV, BO, CE, CF, CN, GW, LH, LO, MA, ML, MO, NF, NM, PP, WW. (39) 1924. (48, 49, 52)

Calocera furcata (Fr.) Fr.
On rotten conifer wood; rare.
51. Recorded only from Hawarden (SJ 36) in 2019.

Calocera glossoides (Pers.) Fr.
On rotten wood, especially of oak; uncommon.
50, 51. SJ 18, 23, 25, 34, 36. ER, GW, NF. (6) 1972. (48)

Species Accounts

Calocera pallidospathulata D.A. Reid *Pale Stagshorn*
On rotten wood, especially of conifers but also on a wide range of hardwoods; common.
50, 51. SH 76, 77, 86, 87, 95, 97; SJ 05–08, 13, 15–18, 24, 26, 34–36, 43. BO, CE, CF, CN, ER, GW, LH, LO, MF, PP. (35) 1981. (48, 49, 52) This species was first discovered in north-east Yorkshire in 1969 and has spread rapidly since then. It was described as new to science in 1974 but its origin was a mystery. The present author discovered it in natural forest in the central highlands of Mexico in August 2005. It had always been suspected that it was an introduction from an under-worked region of North America so this find was especially exciting. In Great Britain it has spread at about 20 km per annum and is out-competing both the common species, *C. cornea* and *C. viscosa*, as it produces spores all year round, not just in the autumn. It has reached most of Wales and has spread as far north as the Scottish Highlands and is common throughout England, except for the South West. When it colonises a new area it is first found on conifers, then it moves to birch and other hardwoods.

Calocera viscosa (Pers.) Fr. *Yellow Stagshorn*
On rotten conifer stumps, trunks and branches; common.
50, 51. SH 85, 87, 95–97; SJ 05, 07, 13, 15–17, 23–27, 34–36, 43. AV, BW, CC, CE, CF, ER, GW, HC, LH, LO, MF, NF, PP, WW. (44) 1880. (48, 49, 52)

Dacrymyces chrysocoma (Bull.) Tul.
On rotten pine wood; rare.
51. Recorded only from Loggerheads Country Park (SJ 26) in 1978.

Dacrymyces punctiformis Neuhoff
On rotten pine wood; rare.
50. Recorded only from Coed Clwyd on Moel Famau (SJ 16) since 1976. (49)

Dacrymyces stillatus Nees *Common Jellyspot*
Dacrymyces deliquescens auct.
On damp, rotten wood of all kinds; common.
50, 51. SH 85, 87, 95, 97; SJ 02, 05–08, 13, 15–18, 24–27, 34–36. AV, BO, BW, CC, CE, CF, CN, BU, ER, GW, HC, LH, LO, MF, ML, MO, NF, NM, PP, WW. (89) 1880. (48, 49, 52)

Ditiola peziziformis (Lév.) D.A. Reid
Femsjonia luteoalba Fr.
On fallen oak branches; uncommon.
50. Recorded only from Foxhall, Denbighshire (SJ 06) in 1996.

Class TREMELLOMYCETES
Order CYSTOFILOBASIDIALES
Family Mrakiaceae

Mrakia aquatica (E.B.G. Jones & Sloof) Xin Zhan, F.Y. Bai, M. Groenew. & Boekhart
Cryptococcus aquaticus (E.B.G. Jones & Sloof) Rodr. Mir. & Neijman
Inside living and moribund stems of *Equisetum fluviatile*, and in river foam; frequent.
51. Recorded only from Llyn Helyg (SJ 17) in 1977. Found as the yeast-like anamorph *Vanrija aquatica* (E.B.G. Jones & Sloof) R.T. Moore (*Candida aquatica* E.B.G. Jones & Sloof).

Order TREMELLALES
Family Tremellaceae

Phaeotremula foliacea *Leafy Brain*
Tremella foliacea Pers.
Parasitic on the mycelium of *Stereum* spp. on conifers and broad-leaved trees; common.
50, 51. SH 85, 97; SJ 06, 15-17, 24-26, 35, 36. AV, CE, CN, GW, LH, LO, NF, NM, PP, WW. (23) 1902. (48, 49, 52)

Tremella aurantia Schwein.
Parasitic on fruit bodies of *Stereum hirsutum*; uncommon.
50, 51. SJ 34, 36. ER, GW. (2) 2014.

Tremella encephala Pers.
Parasitic on fruit bodies of *Stereum sanguinolentum*; uncommon.
50, 51. SH 87; SJ 16. AV, CC. (5) 1880. (48, 49, 52)

Tremella exigua Desm.
Parasitic on *Diaporthe* spp. on dead gorse stem; uncommon.
50. Recorded only from Erddig (SJ 34) in 1977.

Tremella indecorata Sommerf.
Parasitic on pyrenomycetes on twigs of trees and shrubs; rare.
50. Recorded only from Erddig (SJ 34) in 1910.

Tremella mesenterica Retz. *Yellow Brain*
Tremella lutescens Pers.
Parasitic on the mycelium of *Peniophora* spp., especially on gorse and hazel; common.
50, 51. SH 77, 85, 87, 95, 97; SJ 06, 15-17, 24-27, 34-36. AV, BO, CC, CE, CN, DU, ER, GW, HC, LH, LO, MA, MF, ML, NF, NM, PP. (51) 1880. (48, 49, 52)

Sub-phylum PUCCINIOMYCOTINA
Class MICROBOTRYOMYCETES
Order MICROBOTRYALES
Family Microbotryaceae

Microbotryum coronarium (Liro) Denchev & T. Denchev
Ustilago violacea (Pers.) Roussel in part
Anther smut in false anthers of female flowers of *Silene flos-cuculi*; frequent.
50, 51. SJ 25, 34. NF. (3) 1910. (49) **RDL/VU**

Microbotryum lychnidis-dioicae (Liro) G. Demi & Oberw. *Campion Anther Smut*
Ustilago violacea (Pers.) Roussel in part
Anther smut in false anthers of female flowers *Silene dioica*, *S. x hampeana* and *S. latifolia*; common.
50, 51. SH 97; SJ 06-08, 15-17, 23, 25, 26, 34, 36. AV, BO, DU, ER, GW, LO, MO, WW. (24) 1972. (49, 52) **RDL/LC**

Microbotryum silenes-inflatae (Liro) G. Demi & Oberw.
Ustilago violacea (Pers.) Roussel in part
51. Recorded by Woods et al. (2018) without details. **RDL/LC**

Species Accounts

Microbotryum stellariae (Sow.) G. Demi & Oberw.
Anther smut in *Stellaria graminea*; frequent.
50. Recorded by Woods et al. (2018) without details. **RDL/LC**

Microbotryum stygium (Liro) Vánky
Bauhinus kuhneanus (R. Wolff.) Denchev; *Ustilago kuhneana* R. Wolff misident.
Ovary smut on *Rumex acetosa*; rare.
51. Recorded only from Northop (SJ 26) in 1931. **RDL/CR**

Microbotryum succisae (Magnus) R. Bauer & Oberw.
Bauhinus succisae (Magnus) Dencher; *Ustilago succisae* Magnus
Anther smut of *Succisa pratensis*; frequent.
50, 51. SH 88; SJ 17, 26, 34. ER, LO, MO. (5) 1972. **RDL/VU**

Microbotryum tragopogonis-pratensis (Pers.) R. Bauer & Overw.
Bauhinus tragopogonis-pratensis (Pers.) R.T. Moore; *Ustilago tragopogonis-pratensis* (Pers.) Roussel
Ovary smut on *Tragopogon pratensis*; uncommon.
51. SJ 08, 18, 26. LO, MO, PA. (4) 1974. (52) **RDL/VU**

Sphacelotheca hydropiperidis (Schum.) de Bary
Ovary smut of *Persicaria hydropiper*; common.
50. Recorded only from Erbistock (SJ 34) in 1910. (49, 52) **RDL/LC**

Class PUCCINIOMYCETES
Order HELICOBASIDIALES
Family Tuberculinaceae

Tuberculina sbrozzii Cavara & Sacc.
Parasitic on the sori of *Puccinia vincae*; uncommon.
50, 51. SH 87; SJ 08. 1990.

Order PUCCINIALES
Family Chaconiaceae

Ochropsora ariae (Fuckel) Ramsb.
Rust on leaves of *Anemone nemorosa*; rare.
50, 51. SJ 16, 17, 25, 34. LO, NF. (4) 1813. (48, 49, 52) **RDL/NT**

Family Coleosporiaceae

Chrysomyxa empetri Cummins
Rust on leaves of *Empetrum nigrum*; uncommon.
50. Recorded only from Eglwyseg Rocks (SJ 24) in 2012. (48, 49) **RDL/NT**

Coleosporium tussilaginis (Pers.) Berk.
Rust on leaves and stems of *Campanula glomerate, C. persicifolia, C. rotundifolia. Calendula officinalis, Petasites hybridus, Tussilago farfara, Senecio inaequidens, S. vulgaris, Euphrasia nemorosa* and on needles of *Pinus nigra*; common.
50, 51. SH 85, 87, 97; SJ 05–08, 14–17, 23–27, 34–36. AV, CC, CF, DU, ER, LO, MF, MO, NF, NM. (41) 1880. (48, 49, 52) **RDL/LC**

Family Melampsoraceae

Melampsora allii-populina Kleb.
Rust on leaves of *Populus trichocarpa*; rare.
50. Recorded only from Marford Quarry (SJ 35) in 2016. **RDL/VU**

Melampsora capraearum Thuem.
Rust on leaves of *Salix caprea* and *S. cinerea* spp. *oleifolia*; common.
50, 51. SH 77, 94, 97; SJ 06–08, 14–18, 24–27, 34–36, 43. AV, BO, DU, ER, GW, LO, MA, MF, MO, NF, PA, WW. (36) 1902. (48, 49, 52) **RDL/LC**

Melampsora epitea Thuem.
Rust on leaves of *Salix repens* and *S. x smithiana*; frequent.
51. SJ 08, 17, 18, 26. DU, MO, PA. (4) 1972. (48, 49, 52) **RDL/LC**

Melampsora euphorbiae Cast.
Rust on leaves of *Euphorbia characias, E. cyparissias, E. dulcis, E. griffithii, E. helioscopia, E. lathyris* and *E. peplus*; common.
50, 51. SH 77, 87; SJ 06–08, 18, 26, 35, 36. MO. (19) 1974. (48, 49, 52) **RDL/LC**

Melampsora hypericorum Wint.
Rust on leaves of *Hypericum androsaemum, H. calycinum, H. hirsutum, H. x inodorum* and *H. perforatum*; common.
50, 51. SH 77, 87, 88, 96, 97; SJ 08, 16, 18, 26, 35. (20) 1902. (48, 49, 52) **RDL/LC**

Melampsora larici-populina Kleb.
Rust on leaves of *Populus x canadensis*; uncommon.
50, 51. SJ 05, 07, 08, 15, 16, 26, 27, 36. CE, MO. (9) 1989. (48, 49, 52) **RDL/LC**

Melampsora lini (Ehrenb.) Desm. var. **lini**
Rust on leaves of *Linum catharticum*; common.
50, 51. SJ 18, 24, 26. LO, PA. (3) 1977. (48, 49, 52) **RDL/LC**

Melampsora magnusiana G.H. Wagner
Rust on *Chelidonium majus*; rare.
50. Recorded only from Maeshafn (SJ 16) in 2019. This appears to be the first British record.

Melampsora populnea (Pers.) Karst.
Incl. *Melampsora laricis-tremulae* Kleb.
Rust on leaves of *Mercurialis perennis, Populus alba* and *P. tremula*; common.
50, 51. SJ 07, 08, 15–18, 24–26, 34–36, 43. BW, DU, ER, LO, NF. (23) 1910. (48, 49, 52) **RDL/LC**

Melampsora ribesii-viminalis Kleb.
Rust on leaves of *Salix viminalis*; uncommon.
51. Recorded only from the Hawarden Woods (SJ 36) in 1972. **RDL/DD**

Melampsora salicis-albae Kleb.
Rust on leaves of *Salix alba*; uncommon.
50. Recorded only from the Alyn Waters Country Park, Gwersyllt (SJ 35) in 1998. **RDL/NT**

Species Accounts

Family Phragmidiaceae

Frommeëlla tormentillae (Fuckel) Cummins & Y. Hiratsuka
Rust on leaves of *Potentilla erecta*; rare.
50. Reported from Llansannan (SH 96) by Green (1902). (49, 52) **RDL/LC**

Kuehneola uredinis (Link) Arth. *Pale Bramble Rust*
Rust on stems and leaves of *Rubus* species; common.
50, 51. SJ 15-18, 24-26, 34, 36. AV, ER, NF, WW. (14) 1976. (48, 49, 52) **RDL/LC**

Phragmidium bulbosum (Str.) Schlecht.
Rust on leaves of *Rubus* species; frequent.
50, 51. SH 85, 87, 97; SJ 07, 08, 15-18, 24-26, 34, 35. AV, BP, BO, CC, ER, LO, MA, MO, NF, PA, WW. (27) 1880. (48, 49, 52) **RDL/LC**

Phragmidium fragariae (DC.) Rabenh.
Rust on leaves of *Potentilla sterilis*; frequent.
50, 51. SH 96; SJ 15-17, 25, 26. DU, MO, NF. (7) 1902. (48, 49, 52) **RDL/LC**

Phragmidium mucronatum (Pers.) Schlecht.
Rust on leaves of *Rosa arvensis, R. canina* and cultivated species; common.
50, 51. SH 77, 96, 97; SJ 07, 08, 13, 15-18, 24, 26, 36, 45. AV, BO, GW, LO, MO. (26) 1902. (48, 49, 52) **RDL/LC**

Phragmidium rosae-pimpinellifoliae Diet.
Rust on leaves of *Rosa pimpinellifolia*; uncommon.
50, 51. SH 88; SJ 07, 15, 17, 26. AV, LO. (7) 1984. (48, 49, 52) **RDL/LC**

Phragmidium rubi-idaei (DC.) Karst.
Rust on leaves of *Rubus idaeus*; uncommon.
50, 51. SJ 07, 08, 17, 23, 26, 35. (6) 1977. (48, 49, 52) **RDL/LC**

Phragmidium sanguisorbae (DC.) Schröt.
Rust on leaves of *Sanguisorba minor*; uncommon.
50, 51. SH 88, 97; SJ 07, 08, 15-17, 26, 35. AV, BP, LO. (12) 1972. (49, 52) **RDL/LC**

Phragmidium tuberculatum J. Müller
Rust on leaves of *Rosa rugosa*; uncommon.
50, 51. SH 77, 97; SJ 18, 25, 26. MO, PA. (5) 1985. (48, 49, 52) **RDL/LC**

Phragmidium violaceum (C.F. Schultz.) Wint. *Violet Bramble Rust*
Rust on leaves of *Rubus* species; very common.
50, 51. SH 77, 85, 97; SJ 05-08, 15-18, 23, 25-27, 34-36. AV, BO, CF, CN, ER, GW, LO, MO, NF, WW. (53) 1902. (48, 49, 52) **RDL/LC**

Family Pucciniaceae

Cumminsiella mirabilisssima (Peck) Nannf.
Rust on leaves of *Mahonia aquifolium*; frequent.
50, 51. SH 77, 85, 97; SJ 06, 26, 34, 35. BO, MO. (8) 1926. (49, 52) **RDL/LC**

Gymnosporangium sabinae (Dicks.) G. Wint.
Rust on leaves of *Juniperus sabina* and *Pyrus communis*; rare.
51. Recorded only from a garden in Leeswood (SJ 26) in 2001; this is the first record for Wales. The common *Gymnosporangium cornutum*, which alternates between *Sorbus aucuparia* and *Juniperus communis*, has not been recorded locally – the juniper is very rare in North East Wales. (49, 52) **RDL/NE**

Miyagia pseudosphaeria (Monmt.) Jorst.
Rust on leaves of *Sonchus arvensis, S. asper* and *S. oleraceus*; frequent.
50, 51. SH 88, 97, 98; SJ 07, 08, 15–18, 23, 26. AV, BO, LO, PA. (19) 1954. (48, 49, 52) **RDL/LC**

Puccinia acetosae Kornicke
Rust on leaves of *Rumex acetosa*; common.
50, 51. SH 97; SJ 06, 08, 17, 26, 35. LO. (7) 1976. (48, 49, 52) **RDL/LC**

Puccinia adoxae DC.
Rust on leaves of *Adoxa moschatellina*; frequent.
50, 51. SJ 07, 08, 15–17, 26, 34–36. AV, DU, ER, GW, LO, MO. (17) 1902. (48, 49, 52) **RDL/LC**

Puccinia aegopodii Röhl
Rust on leaves of *Aegopodium podagraria*; uncommon.
50, 51. SH 77; SJ 16, 17, 24, 26, 34–36. AV, DU, ER, GW, LO. (10) 1972. (48, 49, 52) **RDL/LC**

Puccinia albescens Plowr.
Rust on leaves of *Adoxa moshatellina*; uncommon.
51. Recorded only from Hawarden (SJ 36) in 2014. (48, 49, 52) **RDL/LC**

Puccinia allii Rud.
Rust on leaves of *Allium vineale*; uncommon.
50, 51. SH 88; SJ 08, 34. ER. (3) 1989. (48, 49, 52) **RDL/LC**

Puccinia angelicae (Schum.) Fuckel
Rust on leaves of *Angelica sylvestris*; uncommon.
50, 51. SJ 26, 34. AV, ER. (2) 2005. (48, 49, 52) **RDL/EN**

Puccinia annularis (Str.) Röhl
Rust on leaves of *Teucrium scorodonia*; frequent.
50, 51. SH 77, 85, 87; SJ 15, 36. GW. (5) 1972. (48, 49, 52) **RDL/LC**

Puccinia antirrhini Diet. & Holw.
Rust on leaves of *Antirrhinum majus*; common.
50, 51. SH 77, 88, 97; SJ 06–08, 16, 18, 26, 36. BO, MO. (17) 1972. (48, 49, 52) **RDL/LC**

Puccinia arenariae (Schum.) Wint.
Rust on leaves of *Moehringia trinervia, Silene dioica* and *S. latifolia*; common.
50, 51. SH 77, 97; SJ 05, 07, 08, 16–18, 23, 24–26, 34, 36. AV, BO, DU, ER, GW, LO, MO, WW. (30) 1910. (48, 49, 52) **RDL/LC**

Puccinia behenis Otth
Rust on leaves of *Silene dioica* and *S. latifolia*; uncommon, but increasing.
50, 51. SH 77, 86; SJ 07, 23, 34. (7) 2011. (49, 52) **RDL/LC**

Species Accounts

Puccinia betonicae DC.
Rust on leaves of *Stachys officinalis*; uncommon.
51. SJ 17, 26. LO. (2) 1986. **RDL/LC**

Puccinia brachypodii Otth var. **brachypodii**
Rust on leaves of *Brachypodium sylvaticum*; frequent.
50, 51. SH 97; SJ 16, 26, 34, 36. BW, ER, GW, LO, WW. (7) 1972. (48, 49, 52) **RDL/LC**

Puccinia brachypodii var. **arrhenatheri** (Kleb.) Cummins & Greene
Rust on leaves of *Arrhenatherum elatius* and *Deschampsia caespitosa*; common.
50, 51. SJ 16, 17. DU, MF. (2) 1978. (49, 52) **RDL/LC**

Puccinia buxi DC.
Rust on leaves of *Buxus sempervirens*; uncommon.
50, 51. SH 97; SJ 07, 14. BO. (3) 1996. (48, 49, 52) **RDL/LC**

Puccinia calcitrapae DC.
Rust on leaves of *Centaurea montana*, *C. nigra* and *Cirsium palustre*; common.
50, 51. SJ 16, 17, 23, 26, 36, 45. AV, DO, GW, LO, MO. (8) 1902. (48, 49, 52) **RDL/LC**

Puccinia caricina DC. var. **pringsheimiana** (Kleb.) D.M. Henderson
Rust on leaves of *Ribes uva-crispa*; rare.
50. SH 77; SJ 16. (2) 1988. (48, 49, 52) **RDL/DD**

Puccinia caricina var. **ribesii-pendulae** (Hasler) D.M. Henderson
Rust on leaves of *Carex pendula*; frequent.
50, 51. SH 77; SJ 07, 16, 18, 26, 34, 35. ER, WW. (8) 1988. (49, 52) **RDL/LC**

Puccinia chaerophylli Purton
Rust on leaves of *Anthriscus sylvestris* and *Myrrhis odorata*; frequent.
50, 51. SH 96; SJ 16, 17, 26, 34, 35. CN, ER, LO, MA. (8) 1902. (49, 52) **RDL/DD**

Puccinia chrysosplenii Grev.
Rust on leaves of *Chrysosplenium* spp.; rare.
50, 51. SH 97; SJ 16, 17, 26. DU, LO, MO. (3) 1985. The colonies at Loggerheads and Mold have since been destroyed, one by river flooding and the other by road building. (48, 49, 52) **RDL/LC**

Puccinia circaeae Pers.
Rust on leaves of *Circaea lutetiana*; common.
50, 51. SH 77, 85, 97; SJ 05, 08, 16, 17, 26, 34–36. BW, ER, GW, LO. (16) 1910. (48, 49, 52) **RDL/LC**

Puccinia cnici Mart.
Rust on leaves of *Cirsium vulgare*; common.
50, 51. SJ 16, 17, 23, 25. AV, NF. (4) 1972. (48, 49, 52) **RDL/LC**

Puccinia cnici-oleracei Desm.
Rust on leaves of *Achillea millefolium* and *Cirsium palustre*; frequent.
50, 51. SH 85; SJ 23, 36. GW. (3) 1972. (48, 49, 52) **RDL/LC**

Puccinia coronata Corda
Rust on leaves of *Rhamnus catharticus, Arrhenatherum elatius, Festuca gigantea, Holcus lanatus, H. mollis, Lolium perenne* and *Phalaris arundinacea*; common on the grasses, uncommon on the buckthorn.
50, 51. SH 94, 97, 98; SJ 07, 08, 15–18, 25–27, 34–36. AV, BO, DU, ER, GW, LO, MO, NF, PA, WW. (30) 1972. (48, 49, 52) **RDL/LC**

Puccinia crepidicola Syd.
Rust on leaves of *Crepis vesicaria* ssp. *taraxacifolia*; frequent.
51. Recorded only from Mold (SJ 26) in 1974. (48, 49, 52) **RDL/NT**

Puccinia dioicae Magn. var. **extensicola** (Plowr.) D.M. Henderson
Rust on leaves of *Tripolium pannonicum* and *Carex extensa*; uncommon.
51. Recorded only from Point of Ayr (SJ 18) in 1985. (49, 52) **RDL/LC**

Puccinia distincta McAlpine
Rust on leaves of *Bellis perennis*; rapidly spreading since its introduction from Australia in 1997, causing much damage to both wild and cultivated forms of the species.
50, 51. SH 76, 87, 88, 97; SJ 04, 16, 24, 26, 34–36, 43. BO, ER, LO, MA, MO. (17) 1998. (49, 52) **RDL/DD**

Puccinia elymi West.
Rust on leaves of *Leymus arenarius*; frequent.
51. Recorded only from Point of Ayr (SJ 18) in 1978. (49, 52) **RDL/VU**

Puccinia epilobii DC.
Rust on leaves of *Epilobium hirsutum, E. obscurum* and *E. palustre*; uncommon.
50, 51. SH 77; SJ 08, 26, 34, 36. ER, GW, MO. (5) 1972. (49, 52) **RDL/LC**

Puccinia festucae Plowr.
incl. *Uredo festica* DC.
Rust on leaves of *Festuca filiformis, F. ovina* and *F. rubra*; frequent.
51. SJ 07, 16, 17. DU, MF. (4) 1976. (48, 49, 52) **RDL/LC**

Puccinia galii-verni Ces.
Rust on leaves of *Cruciata laevipes* and *Galium saxatile*; frequent.
50, 51. SH 85; SJ 16, 26. LO, MF. (4) 1972. (48, 49, 52) **RDL/LC**

Puccinia glechomatis DC.
Rust on leaves of *Glechoma hederacea*; common.
50, 51. SH 87; SJ 06, 07, 15, 17, 18, 23, 27, 34, 35. DU, ER, PA. (13) 1980. (48, 49, 52) **RDL/LC**

Puccinia glomerata Grev.
Rust on leaves of *Jacobaea vulgaris*; rare.
50, 51. SJ 15, 17, 26. MO. (4) 1976. (49, 52) **RDL/NT**

Puccinia graminis Pers.
Rust on leaves of many species of grasses; frequent.
50. Reported only from the Fairy Glen (SH 85) in 1924. (48, 49, 52) Surprisingly uncommon in the area. **RDL/DD**

Species Accounts

Puccinia hieracii Mart. var. **hieracii**
Rust on leaves of *Hieracium, Leontodon* and *Taraxacum* spp.; common.
50, 51. SH 77, 85, 87, 97; SJ 08, 16, 17, 24-26, 34, 36 BP, DU, ER, GW, LO, MF, MO, NF. (20) 1924. (48, 49, 52) **RDL/LC**

Puccinia hieracii var. **hypochaeridis** (Oud.) Jorst.
Rust on leaves of *Hypochaeris radicata*; frequent.
51. SJ 17, 26. DU, NF. (2) 1978. (48, 52) **RDL/LC**

Puccinia hieracii var. **pilosellidarum** (Probsr.) Jorst.
Rust on leaves of *Pilosella officinarum*; uncommon.
51. Recorded only from Lixwm (SJ 17) in 1985. (49, 52) **RDL/LC**

Puccinia hordei Otth
Rust on leaves of *Hordeum murinum* and *Trisetum flavescens*; uncommon.
50, 51. SJ 07, 23, 27, 35. (5) 1978. (48, 49, 52) **RDL/LC**

Puccinia iridis Wallr.
Rust on leaves of *Iris foetidissima* and *I. germanica*; uncommon.
50, 51. SH 87, 88, 97; SJ 26, 35. MO. (7) 1980. (49, 52) **RDL/LC**

Puccinia lagenophorae Cooke
Rust on leaves of *Senecio cambrensis, S. squalidus* and *S. vulgaris*; it has become very common in all parts of these islands since its arrival from Australia in 1961.
50, 51. SH 87, 88, 96-98; SJ 05-08, 15-18, 25-27, 34-36. AV, BO, CN, DU, ER, LO, MO, PA. (45) 1972. (48, 49, 52) **RDL/LC**

Puccinia lapsanae Fuckel
Rust on leaves of *Lapsana communis*; common.
50, 51. SH 87; SJ 07, 08, 16-18, 26, 34, 36. AV, ER, LO, MO. (12) 1974. (48, 49, 52) **RDL/LC**

Puccinia luzulae Lib.
Rust on leaves of *Luzula pilosa*; frequent.
50. Recorded only from Nant Mill Country Park (SJ 24) in 1991. (48, 49) **RDL/VU**

Puccinia maculosa (Str.) Röhl
Rust on leaves of *Mycelis muralis*; uncommon.
50, 51. SJ 16, 26. LO. (2) 1972. (48, 49) **RDL/EN**

Puccinia magnusiana Kornicke
Rust on leaves of *Ranunculus acris* and *Phragmites australis*; frequent.
51. Recorded only from Mold (SJ 26) in 1972. (48, 49, 52) **RDL/LC**

Puccinia malvacearum Mont.
Rust on leaves of *Alcea rosea, Lavatera arborea, Malva neglecta* and *M. sylvestris*; common. The hollyhock rust was introduced from South America in the late nineteenth century and has become a serious obstacle to the cultivation of this popular plant.
50, 51. SH 88, 97, 98; SJ 06-08, 17, 18, 24-27, 34-36, 45. BP, DU, MO, NF. (45) 1910. (49, 52) **RDL/LC**

Puccinia menthae Pers.
Rust on leaves of *Clinopodium vulgare, Mentha aquatica, M. arvensis* and *M. x villosa*; common.
50, 51. SJ 07, 08, 16, 17, 26, 34–36. AV, ER, GW, LH, LO, MO. (20) 1972. (48, 49, 52) **RDL/LC**

Puccinia obscura Schröt.
Rust on leaves of *Luzula sylvatica*; common.
50, 51. SH 77; SJ 05, 07, 16, 18, 24, 26, 35. PP, WW. (8) 1982. (48, 49, 52) **RDL/LC**

Puccinia oxalidis (Lév.) Diet. & Ellis
Rust on leaves of *Oxalis articulata* and *O. adenophylla*; becoming common and moving northwards after its introduction in the 1960s from Mexico – it is confined to Mexican and South African species of *Oxalis* in Europe and does not infect native European species.
51. SJ 08, 18, 26. (5) 1989. (49, 52) **RDL/LC**

Puccinia pelargonii-zonalis Doidge
Rust on leaves of *Pelargonium x hybridum*; becoming common since its arrival in 1965.
51. Recorded only from Mynydd Isa (SJ 26) in 1981. (49, 52) **RDL/NE**

Puccinia phragmitis (Schumach.) Körn.
Rust on leaves of *Rumex obtusifolius* and *Phragmites australis*; frequent.
51. Recorded only from Point of Ayr (SJ 18) in 2012. (48, 49, 52) **RDL/LC**

Puccinia pimpinellae (Str.) Röhl.
Rust on leaves of *Pimpinella saxifraga*; frequent.
50. Reported from Llansannan (SH 96) by Green, 1902. (49) **RDL/CR**

Puccinia poarum Niels.
Rust on leaves of *Tussilago farfara, Poa nemoralis* and *P. pratensis*; common.
50, 51. SH 87, 96, 97; SJ 05, 07, 08, 14–18, 24–27, 34–36. AV, BO, CC, CF, DU, ER, GW, LH, LO, MA, MF, MO, NF, NM, PA, WW. (43) 1880. (48, 49, 52) **RDL/LC**

Puccinia porri G. Wint.
Rust on leaves of *Allium porrum*; common.
51. Recorded only from Mold (SJ 26) where it has been common on shop-bought leeks since 1981, and is likely to be common throughout the region. (49, 52) **RDL/NE**

Puccinia primulae Duby
Rust on leaves of *Primula vulgaris*; uncommon.
51. Recorded only from Ddol Uchaf (SJ 17) in 2017. (48, 49, 52) **RDL/NT**

Puccinia pulverulenta Grev.
Rust on leaves of *Epilobium hirsutum*; frequent.
50, 51. SJ 08, 16, 17, 24, 26, 35. AV, LO, MO. (10) 1972. (48, 49, 52) **RDL/LC**

Puccinia punctata Link
Rust on leaves of *Cruciata laevipes, Galium saxatile* and *G. verum*; uncommon.
50, 51. SH 84; SJ 08, 17, 18. DU, PA. (5) 1985. (48, 49, 52) **RDL/LC**

Puccinia punctiformis (Str.) Röhl
Rust on leaves of *Cirsium arvense*; very common.
50, 51. SH 96, 98; SJ 07, 08, 15–18, 24–27, 34–36. AV, CE, DU, ER, GW, HC, LO, MO, PA, WW. (37) 1971. (48, 49, 52) **RDL/LC**

Puccinia pygmaea Eriks. var. **ammophilina** (Mains) Cummins
Rust on leaves of *Ammophila arenaria*; uncommon.
51. Recorded only from Point of Ayr (SJ 18) since 1978. (48, 49, 52) **RDL/LC**

Puccinia recondita Rob. & Desm.
Rust on leaves of *Anisantha sterilis, Bromopsis ramosus* and *Elytrigia repens*; uncommon.
51. SJ 16, 26. AV, LO, MO. (3) 1971. (48, 49, 52) **RDL/LC**

Puccinia saniculae Grev.
Rust on leaves of *Sanicula europaea*; uncommon.
50, 51. SJ 16, 17, 26, 36. GW, LO. (5) 1902. (48, 49, 52) **RDL/LC**

Puccinia saxifragae Schlecht.
Rust on leaves of cultivated *Saxifraga spathularis*; rare.
51. SJ 17, 26. (2) 1991. (48, 49) **RDL/EN**

Puccinia sessilis Schröt.
Rust on leaves of *Allium ursinum, Arum maculatum* and *Phalaris arundinacea*; frequent.
50, 51. SJ 15–18, 24, 26, 34, 36. AV, CN, DU, GW, LO, WW. (16) 1976. (48, 49, 52) **RDL/LC**

Puccinia smyrnii Biv.-Bernh.
Rust on leaves of *Smyrnium olusatrum*; frequent.
50, 51. SH 87, 88, 97; SJ 06–08, 17, 18. PA. (17) 1989. (48, 49, 52) **RDL/LC**

Puccinia striiformis Westend.
Rust on leaves of *Elytrigia repens* and *Hordeum marinum*; frequent.
51. SJ 08, 18. PA. (2) 1978. (48, 49, 52) **RDL/LC**

Puccinia tanaceti DC.
Rust on leaves of *Artemisia absinthium*; uncommon.
50, 51. SJ 27. 1971. (49, 52) Reported from 50 by Woods et al. (2015) without detail. **RDL/LC**

Puccinia tumida Grev.
Rust on leaves of *Conopodium majus*; frequent.
50, 51. SJ 14–17, 25, 26, 34. AV, ER, HM, MO. (11) 1976. (48, 49) **RDL/LC**

Puccinia umbilici Duby
Rust on leaves of *Umbilicus rupestris*; frequent in the west.
50. Recorded only from Fairy Glen (SH 85) in 1924. (48, 49, 52) This is the furthest east that the species has been found in North Wales. **RDL/LC**

Puccinia urticata Kern *sensu lato Nettle Clustercup Rust*
Puccinia caricina DC. in part
Rust on leaves of *Urtica dioica* and *Carex* species; common.
50, 51. SJ 05, 07, 08, 15–18, 26, 34. CN, DU, ER, LO, MO, PA. (19) 1910. (48, 49, 52) **RDL/LC**

Puccinia urticata var. **urticae-acutiformis** (Kleb.) Zwetko
Rust on leaves of *Urtica dioica* and *Carex acutiformis*; frequent.
51. Recorded only from Mold (SJ 26) in 1972. (48, 49, 52) **RDL/LC**

Puccinia urticata var. **urticae-flaccae** (Hasler) Zwetko
On leaves of *Urtica dioica* and *Carex flacca*; frequent.
50. Recorded by Woods et al. (2015) without locality. (49, 52) **RDL/LC**

Puccinia urticata var. **urticae-hirtae** (Kleb.) Zwetko
Rust on leaves of *Urtica dioica* and *Carex hirta*; frequent.
51. Recorded only from Mold (SJ 26) in 1977. (49, 52) **RDL/LC**

Puccinia variabilis Grev.
Rust on leaves of *Taraxacum* species; common.
50, 51. SH 97, 98; SJ 06–08, 16–18, 25–27, 34–37, 45. BO, ER, LO, PA. (19) 1974. (48, 49, 52) **RDL/LC**

Puccinia veronicae Schröt.
Rust on leaves of *Veronica montana*; frequent.
50, 51. SH 77, 97; SJ 05, 07, 08, 16, 17, 23, 26, 34–36. AV, DU, ER, GW, LO, MO, WW. (19) 1950. (48, 49, 52) **RDL/LC**

Puccinia veronicae-longifoliae Savile
Rust on leaves of *Veronica spicata*; rare.
50. Recorded only from Cefn Rocks (SJ 07) in 1950. **RDL/RE**

Puccinia vincae Berk.
Rust on leaves of *Vinca major*; uncommon but apparently increasing, especially since 2001.
50, 51. SH 87, 97; SJ 08, 14, 17, 26, 34. ER, MO. (6) 1998. (49, 52) **RDL/LC**

Puccinia violae DC.
Rust on leaves of *Viola odorata* and *V. riviniana*; common.
50, 51. SH 77, 87; SJ 16–18, 26, 34, 36. AV, BW, ER, LO, MO, PA. (12) 1910. (48, 49, 52) **RDL/LC**

Uromyces acetosae Schröt.
Rust on leaves of *Rumex acetosa*; frequent.
51. Recorded only from Sychdyn (SJ 26) in 1977. (48, 49, 52) **RDL/LC**

Uromyces airae-flexuosae Ferd. & Winge.
Rust on leaves of *Deschampsia flexuosa*; frequent.
51. Recorded only from Nant-y-Ffrith (SJ 25) in 1976. (48, 49, 52) **RDL/LC**

Uromyces anthyllidis Schröt.
Rust on leaves of *Anthyllis vulneraria*; uncommon.
50. Recorded by Woods et al. (2015) without locality. (48, 49, 52) **RDL/LC**

Uromyces appendiculatus (Pers.) Unger
Rust on leaves of *Phaseolus vulgaris*; uncommon.
51. Recorded only from Rhyl (SJ 08) in 1979. **RDL/NE**

Species Accounts

Uromyces armeriae Kickx.
Rust on leaves of *Armeria formosa*; uncommon.
51. Recorded only from a garden in Gronant (SJ 08) in 1990. (49, 52) **RDL/LC**

Uromyces beticola (Bellynck) Boeremea, Loer. & Hamers
Uromyces betae (Pers.) Lév.
Rust on leaves of *Beta vulgaris* sspp. *vulgaris* and *maritima*; frequent.
50, 51. SH 88; SJ 06, 08, 18. PA. (4) 1978. (49, 52) **RDL/LC**

Uromyces chenopodii (Duby) J. Schröt.
Rust on leaves of *Suaeda maritima*; rare.
51. Recorded only from Point of Ayr (SJ 18) in 2011. (48, 49, 52) **RDL/LC**

Uromyces dactylidis Otth *Celandine Clustercup Rust*
Rust on leaves of *Ranunculus bulbosus, R. ficaria, R. repens, Dactylis glomerata* and *Festuca rubra*; common.
50, 51. SH 97; SJ 07, 08, 14–18, 23–26, 34–36. BO, DU, ER, LO, MO, NF, PA, WW. (41) 1902. (48, 49, 52) **RDL/LC**

Uromyces dianthi (Pers.) Niessl.
Rust on leaves of *Dianthus barbatus* and *D. deltoides*; frequent.
51. Recorded only from gardens in Mold (SJ 26) since 1973. (49) **RDL/NE**

Uromyces fallens (Arth.) Barth.
Rust on leaves of *Trifolium pratense*; frequent.
51. Recorded only from Ddol Uchaf Nature Reserve (SJ 17) in 1985. (49, 52) **RDL/LC**

Uromyces ficariae Tul. *Bitter Chocolate Rust*
Rust on leaves of *Ranunculus ficaria*; common.
50, 51. SH 97; SJ 07, 08, 15–17, 24, 26, 34–36. AV, CN, DU, ER, GW, LO, MO. (26) 1902. (48, 49, 52) **RDL/LC**

Uromyces geranii (DC.) Fr.
Rust on leaves of *Geranium pratense* and *G. pyrenaicum*; common.
50, 51. SH 77; SJ 07, 08, 16, 26, 35. LO, MO. (10) 1972. (48, 49, 52) **RDL/LC**

Uromyces inaequialtus Lasch
Rust on leaves of *Silene nutans*; rare.
51. Recorded from Prestatyn (SJ 08) in 1921, but presumed to be extinct in the region. **RDL/RE**

Uromyces lineolatus (Desm.) Schroet.
Rust on leaves of *Bulboschoenus maritimus*; rare.
51. Recorded only from Morfa Rhuddlan (SJ 07) in 1935. (49, 52) **RDL/LC**

Uromyces muscari (Duby) Graves
Rust on leaves of *Hyacinthoides hispanicus* and *H. non-scriptus*; common.
50, 51. SJ 07, 08, 15–17, 23, 26, 34–36. AV, CN, DU, ER, GW, LO, MO, WW. (24) 1902. (48, 49, 52) **RDL/LC**

Uromyces pisi-sativi (Pers.) Liro
Rust on leaves of *Lathyrus pratensis, Medicago lupulina* and *M. sativa*; common.
51. SJ 17, 18, 26. MO, PA. (3) 1978. (48, 49, 52) **RDL/LC**

Uromyces polygoni-avicularis (Pers.) Karst.
Rust on leaves of *Polygonum aviculare*; frequent.
50, 51. SJ 08. (2) 1989. Recorded from 50 by Woods et al. (2015) without locality. (49, 52) **RDL/LC**

Uromyces rumicis (Schum.) Wint.
Rust on leaves of *Rumex crispus* and *R. obtusifolius*; frequent.
50, 51. SH 96, 98; SJ 08, 18, 25, 27. (10) 1902. (48, 49, 52) **RDL/LC**

Uromyces salicorniae de Bary
Rust on leaves and stems of *Salicornia europaea*; rare.
51. Recorded only from Morfa Rhuddlan (SJ 07) in 1935. (49, 52) **RDL/LC**

Uromyces trifolii (Hedw.) Fuckel
Rust on leaves of *Trifolium repens*; uncommon.
50, 51. SJ 36. GW. (1) 2014. Recorded by Woods et al. (2015) without locality. (49, 52) **RDL/LC**

Uromyces trifolii-repentis Liro
Uromyces trifolii (DC.) Fuckel in part
Rust on leaves of *Trifolium repens*; uncommon.
50, 51. SJ 26, 45. MO. (2) 1980. (49) **RDL/LC**

Uromyces viciae-fabae (Pers.) Schröt. var. **viciae-fabae**
Rust on leaves of *Lathyrus pratensis* and *Vicia sepium*; frequent.
50, 51. SJ 15, 18, 34. ER, PA. (3) 1985. (48, 49, 52) **RDL/LC**

Uromyces viciae-fabae var. **orobi** (Schum.) Jorst.
Rust on leaves of *Lathyrus montanus*; uncommon.
51. Recorded only from Maes-y-Groes (SJ 17) in 1978. (48) **RDL/VU**

Family Pucciniastraceae

Hyalopsora polypodii (Diet.) Magn.
Rust on fronds of *Cystopteris fragilis*; rare.
50, 51. SH 85; SJ 16, 25, 26. LO. (4) 1924. (48) **RDL/NT**

Melampsorella caryophyllacearum Schroet.
Rust on leaves of *Cerastium*; uncommon.
50. Recorded only from Maes Mynan (SJ 17) in 1994. (48, 49, 52) **RDL/NT**

Melampsorella symphyti Bubak
Rust on leaves of *Symphytum officinale*; frequent.
50, 51. SJ 16, 26, 34. LO, MO. (3) 1978. (49, 52) **RDL/LC**

Melampsoridium betulinum (Fr.) Kleb.
Rust on leaves of *Betula* spp.; common.
50, 51. SH 85; SJ 14, 16, 25, 26, 35, 36. AV, GW, LO, NF, NM, WW. (16) 1972. (48, 49, 52) **RDL/LC**

Species Accounts

Melampsoridium hiratsukanum Ito
Rust on leaves of *Alnus glutinosa*; uncommon but spreading, especially in Wales.
50, 51. SH 95; SJ 25, 35. (3) 2001. (48, 49, 52) **RDL/LC**

Milesina blechni Syd.
Rust on fronds of *Blechnum spicant*; frequent.
50, 51. SJ 16, 25, 26. LO, NM, WW. (3) 1989. (48, 49, 52) **RDL/LC**

Milesina dieteliana (Syd.) Magn.
Rust on fronds of *Polypodium vulgare*; frequent.
50, 51. SJ 12, 16. AV. (2) 1977. (48, 49, 52) **RDL/LC**

Milesina kriegeriana (Magn.) Magn
Rust on fronds of *Dryopteris dilatata*; common.
50, 51. SH 87; SJ 26. LO, WW. (3) 1989. (48, 49, 52) **RDL/LC**

Milesina murariae P. & H. Sydow
Rust on fronds of *Asplenium ruta-muraria*; uncommon.
50. SH 88; SJ 15, 16. LO. (3) 1989. (48, 49, 52) **RDL/LC**

Milesina scolopendrii (Faull) D.M. Henderson
Rust on fronds of *Asplenium scolopendrium*; frequent.
50, 51. SH 77, 87; SJ 16, 17, 26, 36. AV, CN, DU, GW, LO, WW. (10) 1978. (48, 49, 52) **RDL/LC**

Milesina whitei (Faull) Hirats.
Rust on fronds of *Polystichum setiferum*; uncommon.
50. Recorded by Woods et al. (2015) without locality. (49, 53) **RDL/LC**

Naohidemyces vacciniorum (J. Schröt.) Spooner
Pucciniastrum vaccinii (Wint.) Jorst.
Rust on leaves of *Vaccinium myrtillus*; common.
50, 51. SH 85; SJ 16, 25, 43. HM, MF, NF, NM. (7) 1924. (48, 49, 52) **RDL/LC**

Pucciniastrum agrimoniae (Dietel) Tranzschel
Rust on leaves of *Agrimonia eupatoria*: uncommon.
50, 51. SJ 07. (2) 2002. (48, 49) Woods et al. (2015) recorded the species from v.c. 50, without locality. **RDL/NT**

Pucciniastrum circaeae (Wint.) De Toni
Rust on leaves of *Circaea lutetiana*; common.
50, 51. SH 97; SJ 16, 23, 26, 34, 36. BW, CN, ER, GW, LO. (9) 1972. (48, 49, 52) **RDL/LC**

Pucciniastrum epilobii Otth
Rust on leaves of *Chamerion angustifolium*; common.
50, 51. SH 77, 97; SJ 16, 17, 26, 34–36. CN, DU, ER, GW, HC, LO, MA, MO. (15) 1972. (48, 49, 52) **RDL/LC**

Family Sphaerophragmiaceae

Triphragmium ulmariae (DC.) Link
Rust on leaves of *Filipendula ulmaria*; common.
50, 51. SH 85; SJ 16, 17, 25, 26, 34. AV, DU, ER, NF. (7) 1910. (48, 49, 52) **RDL/LC**

Family Uropyxidaceae

Tranzschelia anemones (Pers.) Nannf.
Rust on leaves of *Anemone nemorosa*; frequent.
50, 51. SJ 07, 14–16, 24, 34, 35. ER, LO. (8) 1902. (48, 49, 52) **RDL/LC**

Tranzschelia discolor (Fuckel) Tranz. & Litv.
Rust on leaves of *Prunus spinosa*; uncommon.
50. SH 97; SJ 34. ER. (2) 1999. (49) **RDL/LC**

Sub-phylum USTILAGINOMYCOTINA
Class USTILAGINOMYCETES
Sub-class Exobasidiomycetidae
Order EXOBASIDIALES
Family Exobasidiaceae

Exobasidium arescens Nannf.
Leaf spot on *Vaccinium myrtillus*; rare.
50, 51. SJ 16, 25. HM, MF, NM. (3) 1978. (48) **RDL/EN**

Exobasidium japonicum Shirai
Leaf gall on evergreen azaleas; common.
50, 51. SH 77, 97; SJ 25, 26, 35, 36. MO. (6) 1992. (49, 52) **RDL/NE**

Exobasidium juelianum Nannf.
Shoot gall on *Vaccinium vitis-idaea*; uncommon.
50, 51. SJ 24, 25. (2) 1991. **RDL/DD**

Exobasidium karstenii Sacc. & Trotter
Shoot gall on *Andromeda polifolia*; frequent with its rare host.
50. Recorded only from Fenn's Moss (SJ 43) in 1992. **RDL/LC**

Exobasidium myrtilli Siegm.
Shoot gall on *Vaccinium myrtillus*; frequent.
50, 51. SH 85, 95; SJ 16, 17, 24–27. HM, MF, NM. (14) 1897. (48, 49) **RDL/LC**

Exobasidium oxycocci Rostr. ex Shear
Shoot gall on *Vaccinium oxycoccus*; uncommon.
50. Recorded by Woods et al. without details. (48, 49) **RDL/NT**

Exobasidium rostrupii Nannf.
Leaf spot on *Vaccinium oxycoccus*; frequent.
50, 51. SH 96; SJ 25, 35, 53. (4) 1991. (48, 49) **RDL/LC**

Exobasidium vaccinii (Fuckel) Woron. *Cowberry Redleaf*
Leaf gall on *Vaccinium vitis-idaea*; frequent.
50, 51. SH 87; SJ 13, 24, 25. CC. (4) 1880. (48) **RDL/NT**

Order MICROSTROMATALES
Family Microstromataceae

Microstroma album (Desm.) Sacc.
On leaves of *Quercus x rosacea*; rare.
51. Recorded only from Celyn Wood (SJ 26) in 2002. (48) **RDL/LC**

Pseudomicrostroma juglandis (Bereng.) Kijparn. & Ains
Microstroma juglandis (Bereng.) Sacc.
On young leaves of *Juglans regia*; rare.
51. Recorded only from the author's garden in Mold (SJ 26) in 2001. (48, 49) The Caernarvonshire record is from 2009, adding to the impression that this species is spreading westwards across Europe.

Sub-class Ustilaginomycetidae
Order ENTYLOMATALES
Family Entylomataceae

Entyloma chrysosplenii (Berk. & Broome) Schröt.
Leaf smut of *Chrysosplenium oppositifolium*; rare.
51. Recorded only from Ddol Uchaf (SJ 17) in 2017. **RDL/CR**

Entyloma ficariae (Cornu & Rose) Fischer v. Wald
Leaf smut of *Ranunculus ficaria*; common.
50, 51. SH 97; SJ 14-17, 23, 24, 26, 34-36. AV, DU, ER, GW, LO, MO, WW. (19) 1972. (48, 49, 52) **RDL/LC**

Entyloma microsporum (Unger) J. Schröt.
Leaf smut of *Ranunculus repens*, where it is common and, rarely, on *R. auricomus*.
50, 51. SH 85; SJ 16, 26, 34, 45. ER, MO. (5) 1924. (48, 49, 52) **RDL/LC**

Order TILLETIALES
Family Tilletiaceae

Tilletia holci (Westend.) Schröt.
Ovary smut of *Holcus mollis*; rare.
51. Recorded only from Penycloddiau (SJ 16) in 1992. Listed as regionally extinct in Woods et al. (2018), but they were unaware of the Flintshire record, so it should be **RDL/CR**

Tilletia sphaerococca (Rabemh.) A. Fischer v. Wald
Ovary smut of *Agrostis tenuis*; uncommon.
51. Recorded only from Penycloddiau (SJ 16) in 1992. **RDL/LC**

Order UROCYSTIDIALES
Family Floromycetaceae

Antherospora hortensis M. Piętek & M. Lutz
Ustilago vaillantii Tul. & C. Tul., in part
Anther smut of *Muscari* cultivar; uncommon.
51. Recorded only from Mold (SJ 26) in 2011. **RDL/NE**

Family Melanotaeniaceae

Melanotaenium cingens (Berk.) Magn.
Stem smut of *Linaria vulgaris*; rare.
50, 51. SJ 04, 08, 14. (3) 1902. (48) **RDL/EN**

Family Urocystidiaceae

Urocystis anemones (Pers.) G. Wint.
Leaf smut of *Anemone nemorosa*; frequent.
50, 51. SJ 14, 17, 24, 26, 34. AV, ER. (5) 1910. (49) **RDL/LC**

Urocystis ranunculi (Lib.) Moesz.
Leaf smut of *Ranunculus repens*; common.
50, 51. SH 85; SJ 16, 17, 26, 34. DU, ER, LO, MO. (7) 1972. (48, 49, 52) **RDL/LC**

Urocystis trollii Nannf.
Leaf smut on cultivated *Trollius europaeus*; rare.
51. Recorded only from the author's garden in Mold (SJ 26) in 1980. **RDL/CR** in the wild.

Urocystis violae (Sow.) A.A. Disch. Waldh.
Leaf spot of *Viola riviniana*; uncommon.
50. Recorded only from Legacy (SJ 24) in 1994. **RDL/LC**

Order USTILAGINALES
Family Ustilaginaceae

Anthracoidea arenariae (H. Syd.) Nannf.
Ovary smut on *Carex arenaria*; uncommon.
51. Recorded only from Point of Ayr (SJ 18) in 1997. (48, 52) **RDL/NT**

Anthracoidea paniceae Kukkonen
Ovary smut on *Carex panicea*; uncommon.
50. Recorded only from Minera Quarry (SJ 25) in 2016. **RDL/EN**

Tranzscheliella hypodytes (Schlecht.) Vánky & McKenzie
Ustilago hypodytes (Schlecht.) Fr.
Stem smut of *Leymus arenarius*; uncommon.
51. Recorded only from Point of Ayr (SJ 18) in 1978. (52) **RDL/CR**

Species Accounts

Ustilago avenae (Pers.) Rostr.
Ustilago segetum (Bull.) Roussel var. *avenae* (Pers.) Brun
Ovary smut of *Arrhenatherum elatius*; common.
50, 51. SH 87, 97; SJ 08, 17, 18, 26, 27, 36. DU, MO, PA. (11) 1902. (48, 49, 52) **RDL/LC**

Ustilago filiformis (Schrank) Rostr.
Ustilago longissima (Sow.) Meyen
Leaf stripe smut of *Glyceria* and *G. fluitans*; common.
50, 51. SJ 25, 26, 34, 35. ER, MO, NM. (5) 1972. (52) **RDL/LC**

Ustilago hordei (Pers.) Lagerh.
Flower smut of *Hordeum vulgare*; once common, now rare.
50. Recorded only from Fairy Glen (SH 85) in 1924. (49)

Ustilago serpens (P. Karst.) B. Lindeb.
Stripe smut on leaves of *Elytrigia repens*; rare.
51. SJ 18, 26. MO, PA. (3) 1974. Last seen in 1985. **RDL/EN**

Ustilago striiformis (Westend.) Niessl.
Stripe smut of the leaves of grasses, including *Agrostis stolonifera, Holcus lanatus* and *H. mollis*; common.
50, 51. SJ 16, 26. LO, MF, MO. (4) 1974. (49) **RDL/LC**

Ustilago tritici (Pers.) Rostr.
Ustilago segetum (Bull.) Roussel var. *tritici* (Pers.) Brun; *Ustilago nuda* auct.
Loose smut of the ovaries of wheat and barley; frequent in the past but now rare.
51. SJ 16, 26. (2) 1976. (49, 52)

ILLUSTRATIONS

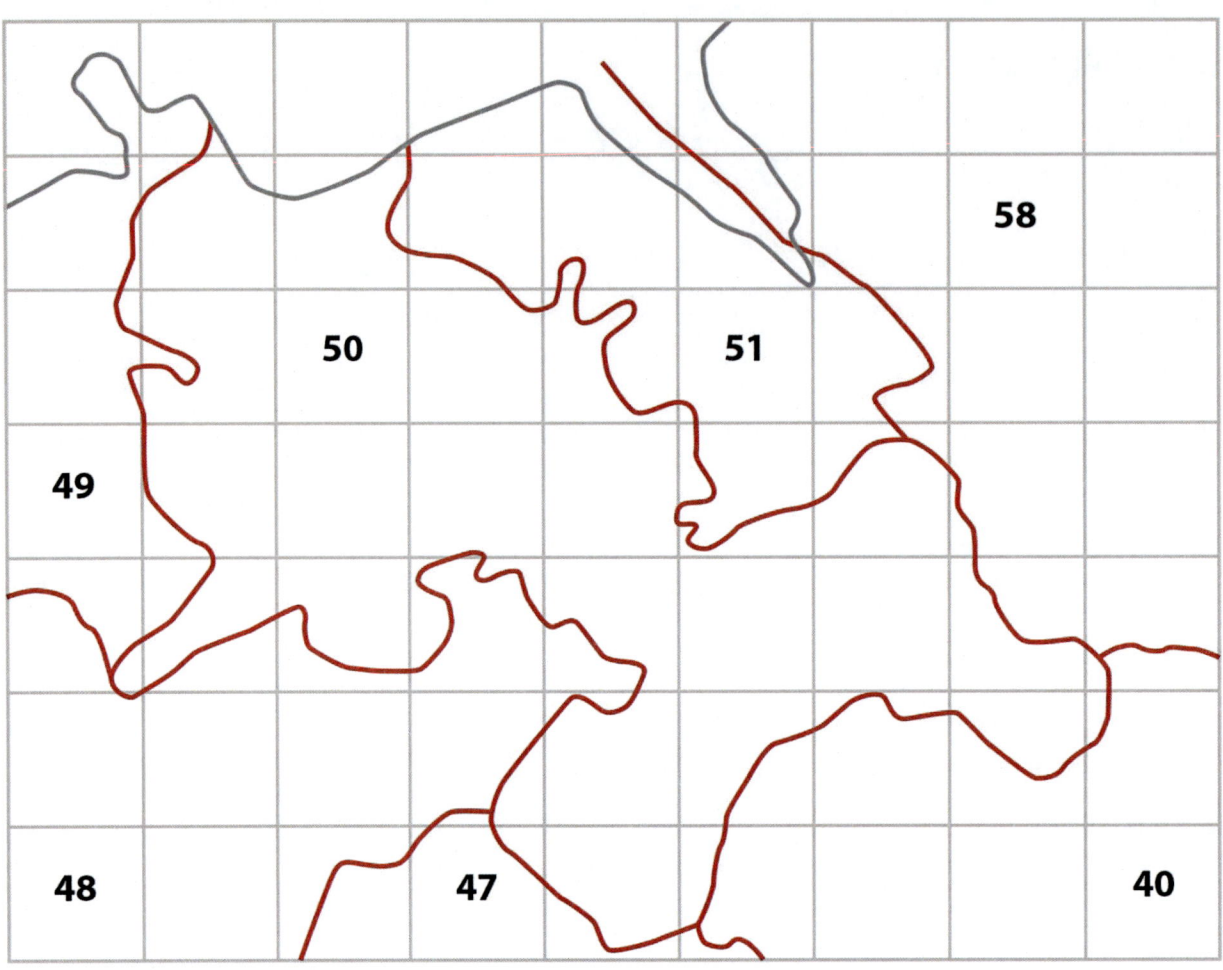

Figure 1: The boundaries of vice-counties 50 and 51

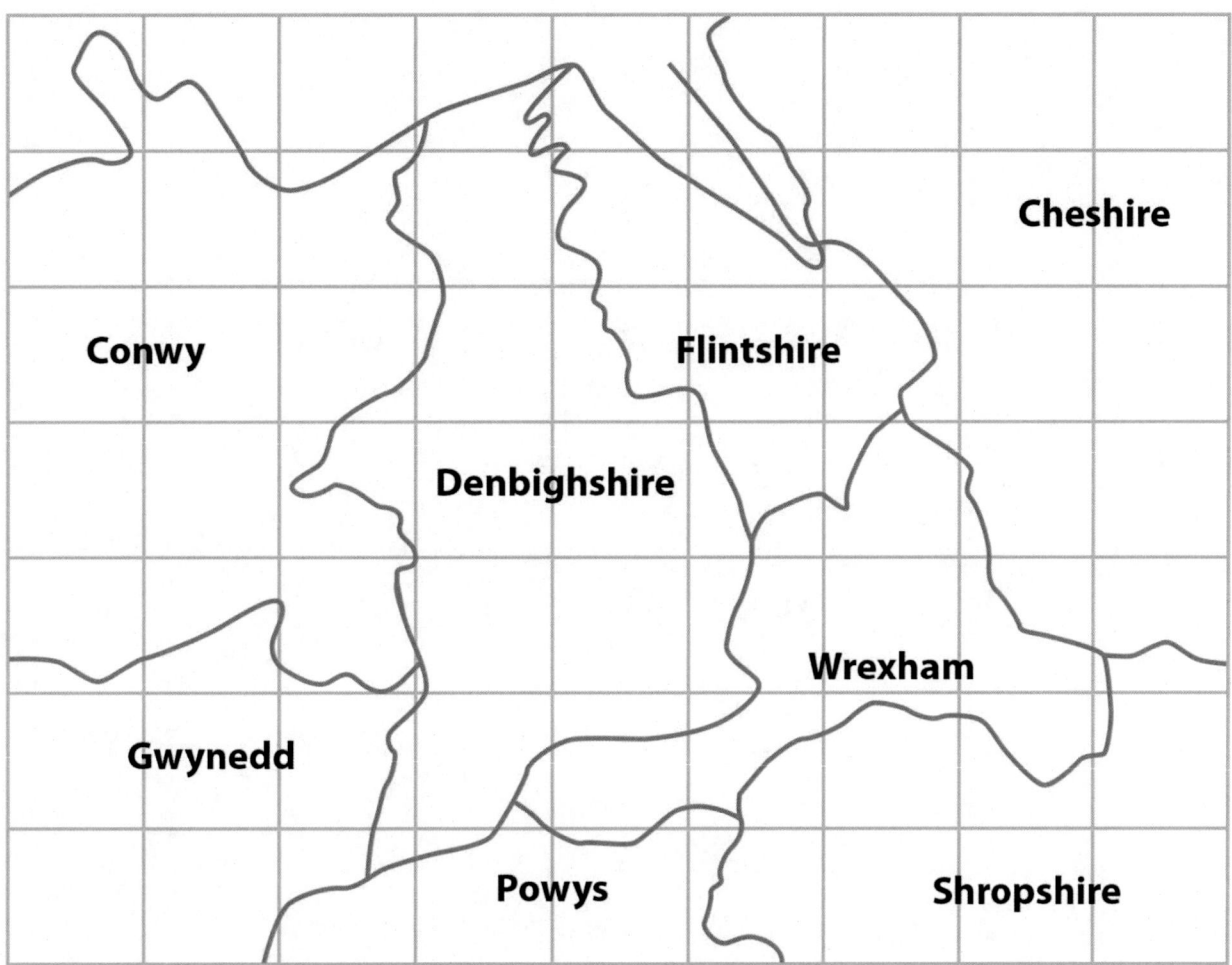

Figure 2: The local authority areas

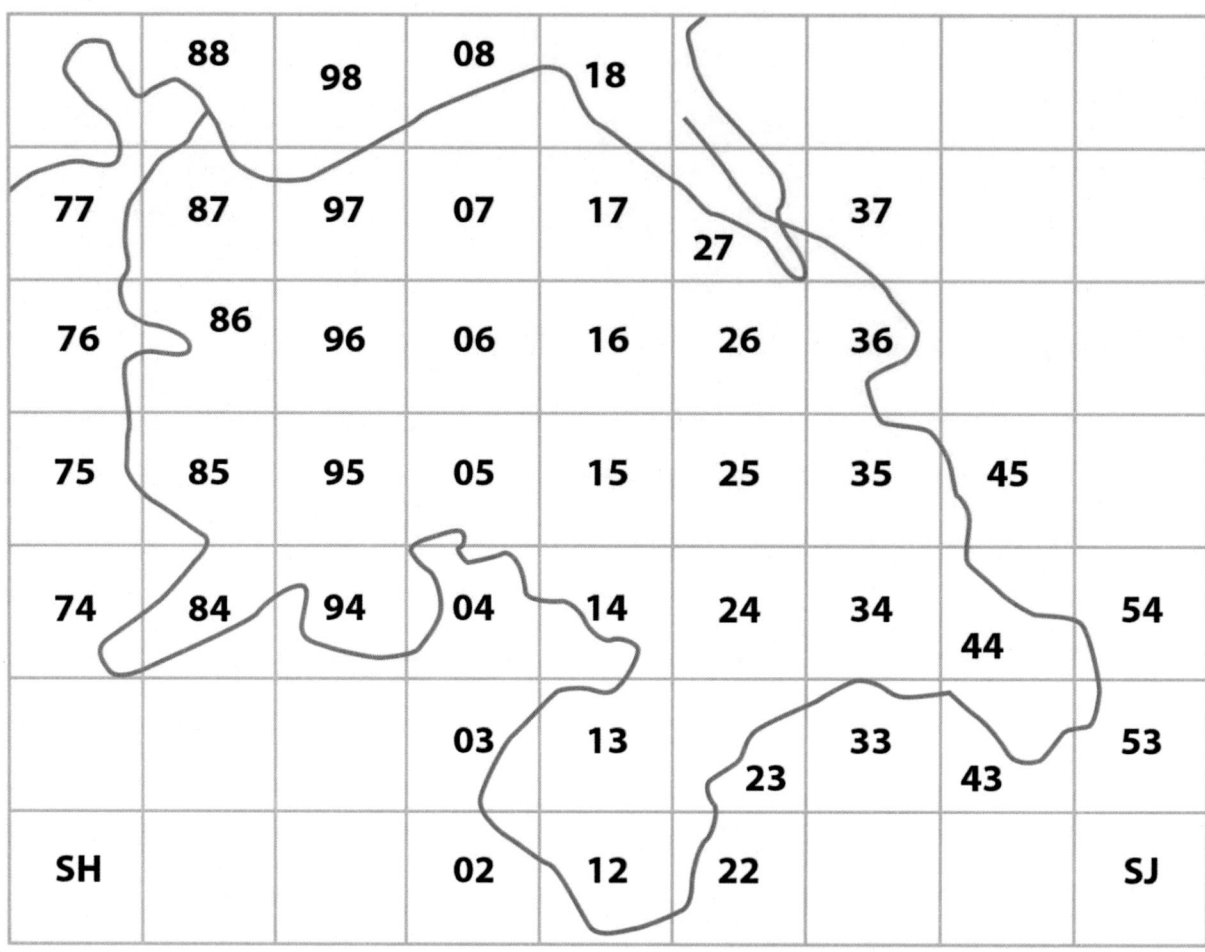

Figure 3: The 10 km squares of the National Grid

Figure 4: *Erysiphe guarinonii*, a rare powdery mildew on laburnum, from Mold

Figure 5: *Sarcoscypha austriaca* and its rare variety *lutea*, from Abergele (Roger Brown)

Figure 6: *Mitrophora semilibera*, an uncommon morel, on coal waste, Ewloe

Figure 7: *Morchella crassipes*, an uncommon morel, from Cilygroeslwyd Wood

Illustrations

Figure 8: *Taphrina pruni*, causing pocket plums on blackthorn, from Mold

Figure 9: *Amanita phalloides*, the Death Cap, from Erddig

Figure 10: *Cyathus stercoreus*, a rare bird's-nest fungus, from Point of Ayr

Figure 11: *Coprinus picaceus*, the rare Magpie Ink Cap, from Alyn Valley

Illustrations

Figure 12: *Hericium erinaceum*, the rare Monkey-head, from Erddig

Figure 13: *Psathyrella ammophila*, the Dune Brittlestem, from Point of Ayr

Figure 14: *Hyalopsora polypodii*, the rare rust on Brittle Bladder fern, from Loggerheads

Figure 15: *Puccinia smyrnii*, the rust on Alexanders, from Point of Ayr

Figure 16: *Exobasidium karstenii* galling Bog Rosemary, from Fenn's Moss